Paraconsistency: Logic and Applications

LOGIC, EPISTEMOLOGY, AND THE UNITY OF SCIENCE

VOLUME 26

Logic, Epistemology, and the Unity of Science aims to reconsider the question of the unity of science in light of recent developments in logic. At present, no single logical, semantical or methodological framework dominates the philosophy of science. However, the editors of this series believe that formal techniques like, for example, independence friendly logic, dialogical logics, multimodal logics, game theoretic semantics and linear logics, have the potential to cast new light on basic issues in the discussion of the unity of science.

This series provides a venue where philosophers and logicians can apply specific technical insights to fundamental philosophical problems. While the series is open to a wide variety of perspectives, including the study and analysis of argumentation and the critical discussion of the relationship between logic and the philosophy of science, the aim is to provide an integrated picture of the scientific enterprise in all its diversity.

For further volumes:
http://www.springer.com/series/6936

Koji Tanaka • Francesco Berto
Edwin Mares • Francesco Paoli
Editors

Paraconsistency: Logic and Applications

Springer

Editors
Koji Tanaka
Department of Philosophy
University of Auckland
Arts 2, 18 Symonds Street
Auckland
New Zealand

Francesco Berto
Department of Philosophy
University of Aberdeen
King's College, High Street
Aberdeen
United Kingdom

Edwin Mares
Department of Philosophy
Victoria University of Wellington
Murphy 518, Kelburn Parade
Wellington
New Zealand

Francesco Paoli
Department of Education
University of Cagliari
Via Is Mirrionis 1
Cagliari
Italy

ISBN 978-94-007-4437-0 ISBN 978-94-007-4438-7 (eBook)
DOI 10.1007/978-94-007-4438-7
Springer Dordrecht Heidelberg London New York

Library of Congress Control Number: 2012943587

Printed on acid-free paper

Springer is part of Springer Science+Business Media (www.springer.com)

Acknowledgements

In 2008, the Fourth World Congress on Paraconsistency was held at the University of Melbourne in Australia. The occasion brought together scholars from around the globe to showcase their latest developments. New systems of paraconsistent logic and applications were presented and discussed. This book collects some of the papers presented at the Congress. Papers that are technical in nature were published in the Special Issue on Paraconsistent Logic of *Logic and Logical Philosophy* 19(1–2): 2010.

We would like to thank Graham Priest, Greg Restall and David Sweeney who organised the Congress. We would also like to thank Andrew Withy for his editorial assistance, which was indispensable to the publication of this book. The generous financial support from the Faculty of Arts and the Department of Philosophy at the University of Auckland allowed Koji Tanaka to have Andrew as an editorial assistant.

We are grateful to the editors of Logic, Epistemology, and the Unity of Science series who helped with the publication of this volume, especially Shahid Rahman, and to Ties Nijssen and to the team at Springer, particularly Hendrikje Tuerlings. We are also indebted to Heinrich Wansing who was instrumental in finding an appropriate publisher.

Koji Tanaka
Francesco Berto
Edwin Mares
Francesco Paoli

Contents

Chapter 1
Paraconsistency: Introduction

Koji Tanaka, Francesco Berto, Edwin Mares, and Francesco Paoli

1.1 Logic

It is a natural view that our intellectual activities should not result in positing contradictory theories or claims: we ought to keep our theories and claims as consistent as possible. The rationale for this comes from the venerable *Law of Non-Contradiction*, to be found already in Aristotle's *Metaphysics*, and which can be formulated by stating: for any truth-bearer A, it is impossible for both A and $\neg A$ to be true. *Dialetheism*, the view that some true truth-bearers have true negations, challenges this orthodoxy.[1] If some contradictions can be true, as dialetheists have argued, then it may well be rational to accept and assert them. For example, one may think that the naïve account of truth, based on the unrestricted T-schema: $\langle A \rangle$ is true

[1]Dialetheism itself has a venerable tradition in the history of Western philosophy: Heraclitus and other pre-Socratic philosophers were arguably dialetheists, for instance; and so were Hegel and Marx, who placed the obtaining and overcoming (*Aufhebung*) of contradictions at the core of their 'dialectical method'. For an introduction to dialetheism, see Berto and Priest (2008). A notable collection of essays on the Law of Non-Contradiction is Priest et al. (2004).

K. Tanaka (✉)
University of Auckland, Auckland, New Zealand
e-mail: k.tanaka@auckland.ac.nz

F. Berto
University of Aberdeen, Aberdeen, UK
e-mail: f.berto@abdn.ac.uk

E. Mares
Victoria University of Wellington, Wellington, New Zealand
e-mail: Edwin.Mares@vuw.ac.nz

F. Paoli
University of Cagliari, Cagliari, Italy
e-mail: paoli@unica.it

K. Tanaka et al. (eds.), *Paraconsistency: Logic and Applications*, Logic, Epistemology, and the Unity of Science 26, DOI 10.1007/978-94-007-4438-7_1,

if and only if A, should be accepted on a rational ground because of its virtues of adequacy to the data, simplicity, and explanatory power. However, the account is inconsistent, due to its delivering semantic paradoxes, such as the Liar.[2]

A dialetheist had better not be a classical logician. Classical logical consequence supports the principle often called *ex contradictione quodlibet* (ECQ): $\{A, \neg A\} \models B$ for any A and B. We are licensed by classical logic to infer anything whatsoever when we end up with a contradiction. To use a lively expression, classical logic is *explosive*: the truth of everything—a view often called *trivialism*—is classically entailed by the obtaining of a single contradiction; and trivialism is rationally unacceptable if anything is.[3]

A necessary condition for a logic to be *paraconsistent* is that its logical consequence relation, $\models$, is not explosive, invalidating ECQ. Although there is no general consensus on a definition of paraconsistent logic among researchers in the area, more often than not this necessary condition is taken to be a sufficient one too. Some logicians,[4] on the other hand, have argued that this 'negative' constraint should be supplemented by appropriate additional 'positive' properties. Be it as it may, since paraconsistent logics do not allow us to infer anything arbitrarily from a contradiction, their treatment of inconsistencies appears more sensible than the one in classical logic. But whereas a dialetheist should go paraconsistent, one does not need to accept that there are true contradictions to adopt a paraconsistent logic.[5] Dialetheism is a controversial view and many people find it counterintuitive. But, regardless of whether there are some true contradictions, it may be that in most cases when we find that we hold inconsistent beliefs or make inconsistent claims, we should revise them to be consistent.[6]

Whether or not there are *no* true contradictions, inconsistency is pervasive in our rational life. We often find that we have inconsistent beliefs or make inconsistent claims, and we are often subject to inconsistent information. Any philosopher who thinks that we may use a logic to make inferences from, determine commitments of, or otherwise logically examine the contents of people's beliefs, theories, or stories, should therefore think twice before being committed to explosion. For example, telling someone who has contradictory beliefs that they are committed to believing every proposition would be a very unproductive move in most debates, and do little more than merely pointing out that the person has inconsistent beliefs. Such considerations provide independent motivations for the development of paraconsistent logics: we need subtle, non-classical logical techniques to analyse the features of inconsistent theories and beliefs.

---

[2]See Priest (2005), Chap. 7.

[3]Trivialism finds, however, a recent, brilliant defence in Kabay (2010).

[4]See Béziau (2000).

[5]See Berto (2007) Chap. 5 and Priest and Tanaka (2009).

[6]Even dialetheists accept this. See, for example, Priest (2005) Chap. 8. For paraconsistent belief revision, see Mares (2002) and Tanaka (2005).

The history of paraconsistent logic has taught us that just taking classical logic and barring ECQ is not sufficient to produce an interesting non-explosive logic. In fact, a number of distinct logical techniques to invalidate ECQ have been proposed. As the interest in paraconsistent logic has grown, different people at different times and places have developed different non-explosive perspectives independently of each other. As a result, the development of paraconsistent logics has somewhat a regional flavour. This book is not a technical survey of the variety of paraconsistent logics[7]: it aims at illustrating their philosophical motivations, applications, and spin-offs. Since these logics are little known to non-specialists, though, in what follows we briefly summarise the most prominent logical strategies to achieve paraconsistency which feature in, or are presupposed by, the essays in this volume.

1.1.1 Discursive Logic

The first formal paraconsistent logic was developed in 1948 by the Polish logician Jaśkowski, in the form of *discussive* (or *discursive*) logic.[8] Jaśkowski's approach addressed situations involving distinct cognitive agents each putting forth her own beliefs, opinions, or reports on some event or other. Each participant's opinions may be self-consistent. However, the resultant discourse or set of data as a whole, taken as the sum of the assertions put forward by the participants, may be inconsistent.

Jaśkowski formalised this idea by modelling the inconsistent dialogical situation in a modal logic. For simplicity, Jaśkowski chose S5. We think of each participant's belief set (or set of opinions, assertions, etc.) as the set of sentences true at a world in a S5 model $\mathscr{M}$. Thus, a sentence A asserted by a participant in a discourse is interpreted as "It is possible that A" ($\Diamond A$). That is, a sentence A of discussive logic can be translated into a sentence $\Diamond A$ of S5. Then A holds in a discourse iff A is true at some world in $\mathscr{M}$. Since A may hold in one world but not in another, both A and $\neg A$ may hold in a discourse. In this volume, however, Marek Nasieniewski and Andrzej Pietruszczak show how Jaśkowksi's discussive logic can also be expressed via normal and regular modal logics weaker than S5 in their essay *On Modal Logics Defining Jaśkowski's D2-Consequence*.

1.1.2 Preservationism

In a discursive logic, a consequence relation can be thought of as defined over maximally consistent subsets of the premises. Given a set of premises, we can measure its degree of (in)consistency in terms of the number of its maximally consistent subsets.

---

[7]For surveys, besides Priest and Tanaka (2009), see Priest (2002) and Brown (2002).

[8]See Jaśkowski (1948).

For example, the level of $\{p, q\}$ is 1 since the maximally consistent subset is the set itself. The level of $\{p, \neg p\}$ is 2 since there are two maximally consistent subsets. If we define a consequence relation over some maximally consistent subset, then the relation can be thought of as preserving the level of consistent fragments. This is the approach which has come to be called *preservationism*. It was first developed by the Canadian logicians Ray Jennings and Peter Schotch.[9] In this volume, Bryson Brown's essay *Consequence as Preservation: Some Refinements* moves within this tradition, but proposes a more general view of the features a logical consequence relation can be seen as preserving.

1.1.3 Adaptive Logics

One may think that we should treat a sentence or a theory as consistently as possible. However, once we encounter a contradiction in reasoning, we should adapt to the situation. *Adaptive logics*, developed by Diderik Batens and his collaborators in Belgium, are logics that 'adapt' themselves to the (in)consistency of a set of premises available at the time of application of inference rules. As new information becomes available expanding the premise set, consequences inferred previously may have to be withdrawn. However, as our reasoning proceeds from a premise set, we may encounter a situation where we infer a consequence provided that no abnormality, in particular no contradiction, obtains at some stage of the reasoning process. If we are forced to infer a contradiction at a later stage, our reasoning has to adapt itself so that an application of the previously used inference rules is withdrawn. Adaptive logics model the dynamics of our reasoning as it may encounter contradictions in its temporal development.[10] In this volume, Diderik Batens' essay *New Arguments for Adaptive Logics* presents four new arguments vindicating the utility of the adaptive approach.

1.1.4 Logics of Formal Inconsistency

The approaches to paraconsistency we have referred to so far retain as much classical machinery as possible (many paraconsistent logicians believe that the full inferential power of classical logic ought to be retained as much as possible, insofar as we find ourselves in consistent contexts). One way to make this aim explicit is to extend the expressive power of our logic by encoding the metatheoretical notions of consistency and inconsistency in the object language. The *Logics of Formal Inconsistency* (*LFIs*) are a family of paraconsistent logics which constitute consistent fragments of classical logic, yet reject explosion where a contradiction

[9]See for instance Schotch and Jennings (1980).

[10]For a general overview of adaptive logics, see Batens (2001).

is present. The investigation of this family of logics was initiated by the Brazilian logician Newton da Costa. An effect of encoding consistency and inconsistency as object language operators on sentences is that we can explicitly separate inconsistency from triviality. With a language rich enough to express consistency and inconsistency, we can study inconsistent theories without assuming that they are necessarily trivial, but at the same time admitting that *some* inconsistencies are so bad that they can trivialize a theory, whereas others are not. This makes it explicit that the presence of a contradiction is a separate issue from the non-trivial nature of paraconsistent inferences.

Prominent among the LFIs are the so-called *positive-plus systems*, which bear this name because they are paraconsistent logics whose negation-free fragment is just positive intuitionistic logic. The paraconsistent features of these systems are obtained by placing on top of the orthodox positive logic a profoundly modified treatment of negation, which turns out to be non-truth-functional: at least one of A and $\neg A$ has to be true, but given that A is true, $\neg A$ may be true or may be false. As a consequence, whereas Excluded Middle, $A \vee \neg A$, is logically valid, the Law of Non-Contradiction in the form of $\neg(A \wedge \neg A)$ is not. The negation of positive-plus systems displays some notable dualities with respect to intuitionistic negation.[11] In this volume, Walter Carnielli and Marcelo Coniglio provide a defense of the LFI approach and its epistemic viability in their essay *On Discourses Addressed by Infidel Logicians*.

1.1.5 Many-Valued Logics

In the standard semantics for classical logic there are exactly two truth values, namely true, 1 and false, 0. Many-valued logics allow more than two truth values. Not all many-valued logics are paraconsistent. Perhaps the most famous—Kleene's and Łukasiewicz's three-valued logics—are explosive. These logics admit, besides truth and falsity, a third value, say $\frac{1}{2}$, which can be thought of as *indeterminate*, or *neither true nor false*.

A many-valued paraconsistent logic typically allows inconsistent values to be designated, i.e., preserved in valid inferences (many-valued approaches to paraconsistency were first proposed by the Argentinian logician Florencio Asenjo[12]). The simplest strategy is to use three values. Suppose we start with the classical set of truth values, $\{1, 0\}$, and consider its power set, i.e., the set of all its subsets, minus the empty set, $\varnothing$: $\mathscr{P}\{1, 0\} - \varnothing = \{\{1\}, \{0\}, \{1, 0\}\}$. The three remaining items can be read as $\{1\}$ = true (only), $\{0\}$ = false (only), which can function as in classical logic, and $\{1, 0\}$ = *both true and false*, which, naturally enough, is a fixed point for negation: if A is both true and false, $\neg A$ is as well. Both $\{1\}$ and $\{1, 0\}$ are

[11] A classic paper in this tradition is Da Costa (1974).

[12] See Asenjo (1966).

designated, the idea being that a designated value must have *some* truth, 1, in it. ECQ is invalidated by having a propositional parameter p which is both true and false; then $\neg p$ is both true and false as well, and the inference to a q which is false (only) does not preserve the designated values. This is the approach of the paraconsistent logic LP (the *Logic of Paradox*) developed by Graham Priest.[13]

If one lets $\varnothing$ play the role of a fourth (and non-designated) value, to be read as *neither true nor false*, which behaves in an appropriate way, one obtains *Belnap's four valued logic* and, in particular, its linguistic fragment FDE (*First Degree Entailment*), a basic *relevant* logic.[14] In this volume, innovative informational models for FDE are proposed by R.E. Jennings and Yue Chen's essay *Articular Models for First Degree Entailment.*

1.1.6 Relevant Logics

Relevant (or *relevance*) *logics* are perhaps the most developed and discussed among paraconsistent logics. The approaches to paraconsistency we have mentioned above target ECQ on the basis of the pervasive presence of inconsistencies in our inferential practices. One may think, though, that ECQ is just one of a set of inferences that are problematic for a more general reason, having to do with the lack of *relevance* between the premises and the conclusion. $(A \wedge \neg A) \rightarrow B$, an 'object-language' counterpart of ECQ, is called, not accidentally, a 'paradox' of the (material or strict) conditional even within classical logic. The problem with such entailments as 'If it is both raining and not raining, then the moon is made of green cheese' is that rain (even inconsistent rain!) seems to have little to do with the material constitution of the moon. Other paradoxes of the conditional, such as $A \rightarrow (B \vee \neg B)$ ('If the moon is made of green cheese, then either it is raining or not'), and $A \rightarrow (B \rightarrow B)$ ('If all instances of the Law of Identity fail, then (if it is raining, then it is raining)') are also taken in this approach as 'fallacies of relevance', due to the lack of a connection between antecedents and consequents.

Relevant logics were pioneered by the American logicians Anderson and Belnap, in order to provide accounts of conditionality free from such fallacies.[15] Anderson and Belnap motivated the development of relevant logics using natural deduction systems; yet they developed a family of relevant logics in axiomatic systems. As research on relevance proceeded and was carried out also in Australia, more focus was given to semantics and model theory. The mainstream approach consists in developing worlds semantics including, besides ordinary possible worlds, also so-called *non-normal* or *impossible* worlds, to be thought of, roughly, as worlds

[13]See Priest (1979).

[14] For Belnap's logic, see Belnap (1977). The interpretation of the truth values of FDE in terms of sets of classical truth values has been suggested by Dunn (1976).

[15]See Anderson and Belnap (1975) and Anderson et al. (1992).

where the truth conditions of logical operators are non-classical. The main semantic tool to obtain a relevant conditional consists in specifying its truth conditions in terms of a three-place accessibility relation on worlds, due to the logicians Richard Routley and Robert Meyer. By accessing worlds which are locally inconsistent or incomplete, one can also invalidate $(A \wedge \neg A) \rightarrow B$ and $A \rightarrow (B \vee \neg B)$.[16]

The core of the philosophical debate on these models is what intuitive sense one is to give them. In this volume, Koji Tanaka's essay *Making Sense of Paraconsistency* addresses the issue in a general setting, turning tables around and challenging the classical logician to make intuitive sense of ECQ, while Ed Mares' *Information, Negation, and Paraconsistency* proposes an informational interpretation that, in a sense, dispenses with possible and impossible worlds altogether, in favour of situations interpreted *à la* Barwise and Perry. In his *Assertion, Denial and Non-Classical Theories*, a notable exponent of the relevantist tradition like Greg Restall provides innovative insights to paraconsistency by considering what he calls 'bitheories'—formal theories based on assertion and denial operators. The expressive powers of bitheories allow them to abstract away from much logical vocabulary whose meaning is controversial in the debate between classical and non-classical logicians.

Relevant logics belong to the family of *substructural* logics, which, besides rules of inference for the logical operators, have structural rules allowing one to operate on the structure of the premises and conclusions.[17] In this volume, the topic is addressed by Francesco Paoli's *A Paraconsistent and Substructural Conditional Logic* via a formal system providing an innovative approach to *ceteris paribus* conditionals. Patrick Allo's work, *Noisy vs. Merely Equivocal Logics*, connects substructural logics to ambiguities of logical connectives that are overlooked within classical logic, in order to shed new light on the issue of rivalry between logics.

1.2 Applications

We claimed that the main motivation for paraconsistency, apart from dialetheism, is the need to model, and account for, non-trivial inferences from inconsistent theories, data bases, and belief sets. It is therefore no surprise that paraconsistency has many applications, given how pervasive these phenomena can be. They can manifest themselves in ordinary life reasoning (a paraconsistent approach to commonsensical inference is proposed in this volume by Michael Anderson, Walid Gomaa, John Grant and Bon Perlis, in their essay *An Approach to Human-Level Commonsense Reasoning*). But they also show up in more theoretical contexts. Working scientists can and have worked productively with inconsistent theories

---

[16]For a general introduction to relevant logics, see Mares (2006) and, for a philosophical interpretation, Mares (2004). On non-normal or impossible worlds, see Berto (2009).

[17]On substructural logics, see Restall (2000) and Paoli (2002).

(which they could not do if they merely inferred that, then, everything is true according to such theories).[18] Readers of fiction understand and appreciate stories that are inconsistent, and at times not accidentally (because of authorial inaccuracy), but essentially so.[19] Similarly, we may have real moral dilemmas, in which we have inconsistent obligations; and we do have inconsistent legal codes. Other examples of inconsistent but intuitively non-trivial information and theories traditionally suggested are: quantum mechanical phenomena on the micro-scale; predicates with over-determined criteria of application; the intuitive metaphysics of change and becoming.[20] The relation between quantum mechanics and paraconsistency is addressed in this volume by Ross Brady and Andrea Meinander's essay, *Distribution in the Logic of Meaning Containment and in Quantum Mechanics*.

We have singled out two paradigmatic (sets of) cases for closer, albeit still rapid, inspection: the role of paraconsistency in the philosophy of mathematics, and its application to the modeling of vagueness in natural language. Many of the papers in the second part of this volume can be located within these two areas.

1.2.1 Philosophy of Mathematics

Historically speaking, paraconsistency comes into the philosophy of mathematics via the celebrated paradoxes of naïve set theory, such as Russell's (the set of non-self-membered sets does and does not belong to itself) and Cantor's (the set of all sets is, via Cantor's Theorem, and of course is not, larger than itself). There are various axiomatised set theories, such as ZF-ZFC or VNB, that are free from these paradoxes; it is well-known, though, that they all introduce more or less *ad hoc* limitations to the unrestricted Comprehension Principle for sets, stating that any well-formed condition, $A[x]$, delivers a set of all and only the items satisfying $A[x]$. Also given Gödel's Incompleteness Theorem, a consistent theory capable of representing basic arithmetical truths cannot represent its own consistency proof. And since theories of sets like ZFC can represent such truths, they cannot therefore represent their own consistency proofs. In fact, the situation is worse: ZFC can formalize *all* of standard mathematics; therefore, a consistency proof for ZFC, not being representable in ZFC by Gödel's result, would be, in some sense, beyond

---

[18]For instance, Bohr's atomic theory assumed that energy comes in discrete quanta, and also assumed Maxwell electromagnetic equations to make predictions on atomic behaviour. The two assumptions are inconsistent, but the theory was quite successful—and, more importantly, nobody would find intuitively acceptable that the theory entails that everything is true. On this story, see Brown (1993).

[19]For instance, Priest (1997a) is a story centred on an inconsistent box which is both empty and not empty; the contradiction is only true in the fiction, of course, but if we bracketed the inconsistency we would miss the whole point of the narration. And intuitively, not everything happens in the story.

[20]For an overview of applications of paraconsistency, see Priest and Routley (1989). Specifically on the metaphysics of change, see Priest (1987), Chaps. 11, 12 and 15.

standard mathematics (e.g., by including so-called large cardinal axioms whose epistemic status may be more problematic than that of the consistency of ZFC itself).

This landscape has motivated the development of paraconsistent theories of sets which retain the full Comprehension Principle of naïve set theory. This delivers inconsistent sets like Cantor's and Russell's, but the underlying non-explosive logic prevents the inconsistencies from trivializing the theory. Whereas consistency proofs are not at issue for such formal theories, there exist non-triviality proofs for paraconsistent set theories, and they are representable within the theories themselves.[21] Interesting new results in this tradition are provided in this volume by Zach Weber's essay, *Notes on Inconsistent Set Theory*.

Paraconsistent arithmetics have also been developed. The first such theory, the system of *relevant arithmetic* R#, had an underlying relevant logic and was proposed in the 1970s by Robert Meyer. Its most interesting feature is that it can be proved absolutely consistent (i.e. nontrivial) by finitary means. However, Friedman and Meyer somewhat downplayed the significance of this result by showing that there are (purely mathematical) theorems of classical Peano arithmetic that cannot be proved in R#. Classes of inconsistent arithmetical theories were later explored by Meyer and Chris Mortensen, and they proved capable of representing also algebraic structures like rings and fields. Their inconsistency and 'finitary' features allow them to escape from Church's undecidability result: they are, that is, provably decidable.[22] The topic of paraconsistent arithmetic is addressed in this volume by Chris Mortensen's essay, *Arithmetic Starred*, while Francesco Berto's *Wittgenstein on Incompleteness Makes Paraconsistent Sense* attempts to make sense of Wittgenstein's (in)famous remarks on Gödel's First Incompleteness Theorem by advocating a paraconsistent reading of Wittgenstein's deeply finitistic philosophy of mathematics.

Just as the issue of logical pluralism is turned on by the development of paraconsistent logic, the one of pluralism in the philosophy of mathematics is triggered by the development of paraconsistent and radically non-classical formal mathematical theories. In this volume, Michelle Friend's *Pluralism and 'Bad' Mathematical Theories* defends such a form of pluralism, in the light of paraconsistency as well as in that of Stewart Shapiro's structuralism.

1.2.2 Philosophy of Language: Vagueness

Natural language abounds in vague predicates, that is, predicates whose criteria of application admit of borderline cases. What must your age be in order for you to

[21]See Brady (1989) for a proof of the non-triviality of paraconsistent set theory, and Brady (2006) for a general account.

[22]See Meyer (1976), Friedman and Meyer (1992), Meyer and Mortensen (1984) and, for a general characterization, Priest (1997b) and Priest (2000).

be *old*? How much money must you make in a year to be *rich*? How many hairs must you lose to become *bald*? And so on. Vagueness causes notorious problems to classical logic, for the latter licenses paradoxical inferences, like the *Heap* (a form of the Sorites paradox—from the Greek *soros*, which means precisely 'heap'): one million grains of sand form a heap; if n grains of sand form a heap, then also $n-1$ grains form a heap (what difference can one grain make?); apply the latter repeatedly, until you get that one single grain of sand forms a heap, which will not do.

In fact, with the exception of the so-called epistemic solutions, all the main approaches to vagueness, such as the ones based on many-valued logics, or supervaluations, already require some departure from classical logic, in the form of under-determinacy of reference, and/or the rejection of Bivalence: if a middle-aged man, m, is a borderline case with respect to the predicate 'is old', $O(x)$, then $O(m)$ may turn out to have an intermediate truth value between truth and falsity, or no truth value at all. But it may be conjectured that a borderline object like m, instead of satisfying neither a vague predicate nor its negation, satisfies them *both*: a middle-aged man, in some sense, can be correctly characterized both as being and as not being old. Similarly, in a borderline rainy day we may safely answer to the question whether it is raining with a 'Yes and no', and get away with it. If these phenomena have, as is usually claimed in this context, a *de re* reading, then actually inconsistent objects may be admitted, together with vague objects. To the satisfaction of the dialetheist, this would spread inconsistency all over the empirical world: if borderline cases can be inconsistent, inconsistent objects are everywhere, given how pervasive the phenomenon of vagueness notoriously is: teen-agers, borderline bald people, middle-age men, etc. Again, however, it is an open option for the paraconsistent logician to assume that the inconsistencies due to vague predicates and borderline objects are only *de dicto*: they may be due to merely semantic under- and/or over- determination of ordinary language predicates.

Whatever one's attitude on this issue is, given the obvious dualities between Excluded Middle, $A \vee \neg A$, and the Law of Bivalence, $T\langle A\rangle \vee T\langle\neg A\rangle$ (with T the relevant truth predicate), on the one side, and the Law of Non-Contradiction in 'syntactic' ($\neg(A \wedge \neg A)$) and 'semantic' ($\neg(T\langle A\rangle \wedge T\langle\neg A\rangle)$) formulations on the other, it has not been too difficult for authors in the paraconsistent tradition to envisage a 'sub-valuational' paraconsistent semantic approach, dual to the supervaluational strategy.[23] However, it is not uncontroversial that super- and sub-valuational approaches are the right paraconsistent way to address the phenomena at issue. In this volume, David Ripley's essay, *Sorting out the Sorites*, proposes an alternative paraconsistent strategy, based on Priest's logic LP.

In fact, also the connections between the paradoxes of self-reference (taken by dialetheists, as we have claimed, as a decisive motivation for their view) and the paradoxes of vagueness may be quite tighter than expected. In this volume, Graham Priest's essay *Vague Inclosures* shows how the Sorites can fit into Priest's general

[23]Sub-valuational semantics have been proposed by Hyde (1997) and Varzi (1997).

'Inclosure Schema' for the paradoxes of self-reference. Dominic Hyde's *Are the Sorites and Liar Paradox of a Kind?* also addresses the issue of the structural similarities and differences between the two kinds of paradox, finding their common source in the under-determinacy of the relevant predicates in a paraconsistent setting.

References

Anderson, A.R., and N.D. Belnap. 1975. *Entailment: The logic of relevance and Necessity*, vol. 1. Princeton: Princeton University Press.

Anderson, A.R., N.D. Belnap, and J.M. Dunn. 1992. *Entailment: The logic of relevance and necessity*, vol. 2. Princeton: Princeton University Press.

Asenjo, F.G. 1966. A calculus of antinomies. *Notre Dame Journal of Formal Logic* 7: 103–105.

Batens, D. 2001. A General characterization of adaptive logics. *Logique et Analyse* 173–175: 45–68.

Belnap, N.D. 1977. How a computer should think. In *Contemporary aspects of philosophy*, ed. G. Ryle, 30–55. Boston: Oriel Press.

Berto, F. 2007. *How to sell a contradiction: The logic and metaphysics of inconsistency*. London: College Publications.

Berto, F. 2009. Impossible worlds. In *The stanford encyclopedia of philosophy*, Fall 2009th ed, ed. E.N. Zalta. Stanford: Stanford University.

Berto, F., and G. Priest. 2008. Dialetheism. In *The stanford ncyclopedia of philosophy*, Summer 2010th ed, ed. E.N. Zalta. Stanford: Stanford University.

Béziau, J.-Y. 2000. What is paraconsistent logic? In *Frontiers in paraconsistent logic*, ed. D. Batens, 95–112. London: Wiley.

Brady, R. 1989. The nontriviality of dialectical set theory. In *Paraconsistent logic: Essays on the inconsistent*, ed. G. Priest, R. Routley, and J. Norman, 437–471. München: Philosophia Verlag.

Brady, R. 2006. *Universal logic*. Stanford: CSLI Publications.

Brown, B. 1993. Old quantum theory: A paraconsistent approach. *Proceedings of the Philosophy of Science Association* 2: 397–441.

Brown, B. 2002. On paraconsistency. In *A companion to philosophical logic*, ed. D. Jacquette, 628–650. Oxford: Blackwell.

Da Costa, N.C.A. 1974. On the theory of inconsistent formal systems. *Notre Dame Journal of Formal Logic* 15: 497–510.

Dunn, J.M. 1976. Intuitive semantics for first-degree entailments and coupled trees. *Philosophical Studies* 29: 149–168.

Friedman, H., and R.K. Meyer. 1992. Whither relevant arithmetic? *Journal of Symbolic Logic* 57: 824–831.

Jaśkowski, S. 1948. Rachunek zdań dla systemów dedukcyjnych sprzecznych. *Studia Societatis Scientiarun Torunesis (Sectio A)* 1(5): 55–77. (trans) Propositional calculus for contradictory deductive systems, Studia Logica, 24(1969): 143–157.

Hyde, D. 1997. From heaps and gaps to heaps of gluts. *Mind* 106: 641–660.

Kabay, P. 2010. *On the plenitude of truth: A defense of trivialism*. Saarbrücken: Lambert Academic Publishing.

Mares, E.D. 2002. A paraconsistent theory of belief revision. *Erkenntnis* 56: 229–224.

Mares, E.D. 2004. *Relevant logic: A philosophical interpretation*. Cambridge: Cambridge University Press.

Mares, E.D. 2006. Relevance logic. In *The stanford encyclopedia of philosophy*, Spring 2009th ed, ed. E.N. Zalta. Stanford: Stanford University.

Meyer, R.K. 1976. Relevant arithmetic. *Bulletin of the Section of Logic of the Polish Academy of Sciences* 5: 133–137.
Meyer, R.K., and C. Mortensen. 1984. Inconsistent models for relevant arithmetics. *The Journal of Symbolic Logic* 49: 917–929.
Paoli, F. 2002. *Substructural logics: A primer*. Dordrecht: Kluwer.
Priest, G. 1979. Logic of paradox. *Journal of Philosophical Logic* 8: 219–241.
Priest, G. (1987). In *Contradiction. A study of the transconsistent*. Dordrecht: Martinus Nijhoff. 2nd and expanded edition, Oxford: Oxford University Press, 2006.
Priest, G. 1997a. Sylvan's box: A short story and ten morals. *Notre Dame Journal of Formal Logic* 38: 573–582.
Priest, G. 1997b. Inconsistent models for arithmetic: I, finite models. *The Journal of Philosophical Logic* 26: 223–235.
Priest, G. 2000. Inconsistent models for arithmetic: II, the general case. *The Journal of Symbolic Logic* 65: 1519–1529.
Priest, G. 2002. Paraconsistent logic. In *Handbook of philosophical logic*, vol. 6, ed. D. Gabbay and F. Guenthner. Dordrecht: Kluwer.
Priest, G. 2005. *Doubt truth to be a liar*. Oxford: Oxford University Press.
Priest, G., J. Beall, and B. Armour-Garb (eds.). 2004. *The law of non contradiction*. Oxford: Oxford University Press.
Priest, G., and R. Routley. 1989. Applications of paraconsistent logic. In *Paraconsistent logic: Essays on the inconsistent*, ed. G. Priest, R. Routley, and J. Norman, 367–393. München: Philosophia Verlag.
Priest, G., and K. Tanaka. 2009. Paraconsistent logic. In *The stanford encyclopedia of philosophy*, Summer, 2009th ed, ed. E.N. Zalta. Stanford: Stanford University.
Restall, G. 2000. *An Introduction to substructural logics*. London: Routledge.
Schotch, P.K., and R.E. Jennings. 1980. Inference and necessity. *Journal of Philosophical Logic* 9: 327–340.
Tanaka, K. 2005. The AGM theory and inconsistent belief change. *Logique et Analyse* 48: 113–150.
Varzi, A. 1997. Inconsistency without contradiction. *Notre Dame Journal of Formal Logic* 38: 621–639.

Part I
Logic

Chapter 2
Making Sense of Paraconsistent Logic: The Nature of Logic, Classical Logic and Paraconsistent Logic

Koji Tanaka

2.1 Paraconsistent Logic Doesn't Make Sense

Over the last few decades, there have been great advances in the development of paraconsistent logic.[1] It now has well-developed proof-theories and semantics. While it has still not found a good basis in *classical* mathematics (although even this situation is changing as a result of the development of paraconsistent set theory satisfying theorems of classical mathematics, see Weber 2010a,b), paraconsistent logic has arguably been successfully applied to many areas both within philosophy as well as such rapidly developing areas as computer science (see for example Priest and Tanaka 2009). Nonetheless, paraconsistent logicians are often faced with the remark that, despite all of this development, paraconsistent logic still does not really 'make sense'. Even someone as friendly towards non-classical logics as Putnam is able to write:

> I am aware that some people think such a logic—paraconsistent logic—has already *been* put in the field. But the lack of any convincing application of that logic makes it, at least at present, a *mere* formal system, in my view (Putnam 1994, p. 262, footnote 12.)

Putnam here expresses his concern about the status of paraconsistent logic in terms of 'application'. I take it, however, that finding an application involves finding how a formal system can be used in a particular context, which, for Putnam, is knowing the 'sense' of the system (see Putnam 1994, pp. 256–257).[2] That is, to give a sense to a

[1] In this paper, I often use the term 'paraconsistent logic' as a mass noun.

[2] Putnam is concerned with the sense of a statement or a question rather than of a formal system. My statement is an application of his claim rather than his claim itself.

K. Tanaka (✉)
Department of Philosophy, University of Auckland, Auckland, New Zealand
e-mail: k.tanaka@auckland.ac.nz

K. Tanaka et al. (eds.), *Paraconsistency: Logic and Applications*, Logic, Epistemology, and the Unity of Science 26, DOI 10.1007/978-94-007-4438-7_2,

formal system is to make it intelligible by, for example, providing an 'interpretation' outside of the formal machinery that shows how the system can be applied. It is one thing to construct an account of logical consequences by providing truth conditions for logical connectives; it is another to give it a sense (that is, to specify its use). The difference here is not just the difference between 'pure' and 'applied' logics, at least in the way that pure/applied distinction is often understood. 'Application', according to this distinction, is understood to refer to the giving of meaning often in non-formal terms (see, for example, Haack 1978, 30ff.). This might suggest that providing truth conditions for logical connectives is enough for specifying an application, since truth conditions specify the meaning of logical connectives in non-formal terms (at least this is how they are usually understood). As we will see, however, Putnam's concern seems to be about the *conception* of logic. On this understanding of Putnam, his charge against paraconsistent logic seems to be that there is no conception of logic that can accommodate paraconsistent logic; that is, no sense has been given to paraconsistent logic.[3]

Now, this charge is reasonable if we are in the business of presenting paraconsistent logic as more than just a mere mathematical tool (even though the development of such a tool itself is an achievement worthy of praise). In this paper, I directly address the charge that paraconsistent logic does not 'make sense'. First, I shall argue that paraconsistent logic does 'make sense' as much as we can 'make sense' of non-normal modal logics as Cresswell (1967) demonstrates. I then turn the tables on classical logicians and ask what sense we can make of explosive reasoning. It is not clear that even classical logicians can answer that question.

2.2 The Tale of Non-normal Worlds

Once upon a time, non-normal worlds were considered as mere technical devices to model non-normal systems of modal logic developed by C.I. Lewis. When Kripke (1963, 1965) provided appropriate semantics in the form of possible world semantics for Lewis systems (some of them at least), he introduced non-normal worlds to model *S2* and *S3*, the systems often called non-normal systems (as well as *E2* and *E3* of Lemmon 1957). One of the main characteristics of a Lewis non-normal system is the failure of the rule of necessitation, i.e., it is not the case that if $\models A$ then $\models \Box A$. In order to develop a semantics for the logic which violates the rule, Kripke introduced non-normal worlds at which any formula of the form $\Box A$ fails to be true. For, then, $\not\models \Box\Box(A \vee \neg A)$, even if $A \vee \neg A$ is true at every world and so $\models \Box(A \vee \neg A)$. (By the interchangeability of $\Box\neg$ and $\neg\Diamond$, every formula of

[3]A similar charge against paraconsistent logic was put to me by my former colleague, Max Cresswell (though his charge was not to do with the conception of logic but with a coherent interpretation of logical connectives). This paper is, in part, a response to him.

the form $\Diamond A$ takes truth as its truth value.) Non-normal worlds were introduced as technical devices to achieve this effect.

The idea that non-normal worlds were mere technical devices was overturned by an ingenious interpretation of non-normal worlds by Cresswell (1967). Cresswell's quest was partly directed by the fact that Lewis thought of *S2* as the true system of modal logic (though Cresswell's aim was not to make Lewis' thought plausible). Cresswell notes that the laws of logic are sometimes regarded as 'laws of thought'. But that is the case only in a world in which there are 'thinking beings'. So a suggestion is that a non-normal world is a world where there are no thinking beings (and that it is not a logical truth that there are thinking beings).[4]

This seems to mean that non-normal worlds are worlds at which there are no laws of thought and hence no laws of logic. But this characterisation fails to capture the non-normal worlds introduced by Kripke. At a non-normal world, A is true if $A \wedge B$ is true. Hence, if it is a law of logic that A follows from $A \wedge B$, then it is not the case that there are no laws of logic at a non-normal world. Thus, having no laws of logic (because there are no laws of thought) is too strong a characterisation of non-normal worlds.

However, depending on how one understands 'laws of logic', the idea that there are no laws of logic seems to intelligibly cash out the notion of non-normal worlds. Consider a non-normal model (for a propositional modal language) $\mathcal{M} = \langle W, N, R, v\rangle$ where W is the set of possible worlds, N is the set of normal worlds, R is the accessibility relation on W and v is an evaluation function such that, for a propositional variable p and a world $w \in W$, $v(p, w) = 1$ or $v(p, w) = 0$, and that $v(\Box p, w) = 0$ if $w \in W - N$. For the sake of simplicity, we assume that R is the accessibility relation of *S5*, i.e., R is universal. Now, if we define validity in terms of all *normal* worlds, i.e., every $w \in N$, then we have a Lewis non-normal system. However, we can define validity in terms of *all* worlds, i.e., every $w \in W$. This gives rise to a Lemmon system of modal logic (Hughes and Cresswell 1996, pp. 205–206). So there are two ways of defining validity in a modal logic. Let's call validity defined in terms of all normal worlds *weak validity*, represented as $\models_w$, and that defined in terms of all worlds *strong validity*, represented as $\models_s$.

A logical truth in a weak sense is defined in terms of normal worlds. So, a sentence, A, is a weak logical truth, i.e., $\models_w A$, if in every model, A is true at every normal world. A logical truth in a strong sense, on the other hand, is defined in terms of all worlds. So $\models_s A$ if A is true at every world in every model.

Now, *prima facie* at least, $\models_s A$ expresses that the necessity of A is *general*. Later in this paper, we will examine how Kant understood and Frege and Wittgenstein appropriated this feature of logic. For now, however, let's take it to mean that logic is

[4]See particularly Cresswell (1967), pp. 202–203. Note that Cresswell doesn't hold the view that the laws of logic are laws of thought. The purpose of his article is to show that different modal logics reflect different interpretations of the necessity operator, just as I am trying to show, in the first half of this paper, that paraconsistent logic (some relevant logics at least) reflects a different interpretation of the conditional operator. Thanks go to Max Cresswell for clarifying his position in personal communication.

general in the sense that a logical truth expresses a truth no matter what the situation turns out to be. So to say that the necessity of A is *general* is to say that $\Box A$ is a logical truth. In order to formalise the idea suggested here, let's represent the generality of logical truth, B, by $\models B$. Then the idea is that $\models \Box A \Leftrightarrow \models_s A$.

If a law of logic is understood to be expressed by a logical truth,[5] then asserting that $\models \Box A$ is asserting that $\Box A$ is a law of logic. Thus, there is a sense in which laws of logic are expressed by logical truths of the form $\Box A$. If laws of logic are delivered by formulas of the form $\Box A$, then the failure of $\Box A$ implies the failure of the laws of logic. So, the fact that any formula of the form $\Box A$ fails to be true invokes the idea that there are no laws of logic. Thus, even though the truth of $A \wedge B$ entails the truth of A, a non-normal world can be characterised as a world where laws of logic fail.

Here then is the tale of non-normal worlds. What was once thought of only as technical devices (i.e., defined only in virtue of formal conditions) have now been given a conception of logic that can tell us how to think of them intelligibly. Non-normal worlds are thus given a sense. At the same time, non-normal systems of modal logic have gained proper recognition.

2.3 ...and Relevant Logic

A similar tale can be told for paraconsistent logic or at least some relevant logics which form a sub-class of paraconsistent logic. When Anderson and Belnap developed relevant logics, they introduced them in an axiomatic form.[6] We had to wait for Routley and Routley (1972) and Urquhart (1972) to provide the appropriate semantics for some of the Anderson and Belnap systems.[7] When the semantics for relevant logics, in particular for *R*, were introduced in the form of the Kripke possible world semantics based on the semantics of Routley and Routley for *First Degree Entailment (FDE)*, Routley and Meyer (1972, 1973) held that the *real world*, denoted by 0, played a distinguished role. The reason is that in order to invalidate the irrelevant formula $A \rightarrow (B \rightarrow B)$, logical truths, e.g., $B \rightarrow B$, have to fail at some world. However, 0, the real world, verifies all logical truths. After all, 0 is the real world: 0 is given a "privileged status" (Routley and Meyer 1973, p. 205).[8]

---

[5]I don't take this statement to be uncontroversial. However, I am concerned with attaching a 'sense' to non-normal worlds and to non-normal modal logics and I take it that there is a sense in which a law of logic is expressed by a logical truth as was held by, for example, Frege, Russell and Hilbert.

[6]These logics are recorded in Anderson and Belnap (1975) and Anderson et al. (1992).

[7]Exactly who came up with the first semantics for the Anderson and Belnap systems is a matter of dispute, just like who came up with the first semantics for Lewis modal logics is, I believe, a matter of dispute. I let historians settle the issue.

[8]Note that there can be more than one world which has this privileged status. However, completeness doesn't force one to assume that there is more than one.

The worlds which do not have this privileged status have come to be known as non-normal worlds among (some) relevant logicians.

It is not just a historical coincidence that some worlds have come to be called non-normal worlds both in the modal logic community and in the relevant logic community. For instance, Priest (1992) claims that the Routleys and Meyer generalised Kripke's notion of non-normal worlds in formulating their semantics for some relevant logics. Whether or not Priest is right, the Routleys and Meyer themselves didn't explicitly refer to Kripke's non-normal worlds. Nor does Priest provide the general characteristics of Kripke's non-normal worlds that can also be attributed to relevant non-normal worlds. If the insight of the Routleys and Meyer was indeed derived from Kripke's semantics for Lewis' weaker modal systems, that insight has not been made widely available.

Whether or not it was an insight of the Routleys and Meyer, the characterisation of non-normal worlds offered in the previous section can be generalised to include the relevant non-normal worlds.[9] Consider a formula of the form $A \rightarrow B$. Assuming that A and B do not contain $\rightarrow$, if the formula is a logical truth, then it expresses truth-preservation between A and B in every situation. This can be shown in a semantics for *FDE* which was developed to study the relationship between antecedents and consequents of implicational sentences in terms of truth-preservation between them in every situation (see for example Routley and Routley 1972; Dunn 1976). Even if A and B contain $\rightarrow$, a formula of the form $A \rightarrow B$, if it is a logical truth, can be seen to express truth-preservation in every situation, since a semantics for a (full) relevant logic may be an extension of that for *FDE* (see for example Tanaka 2000).

So, to say that $A \rightarrow B$ is a logical truth is to say that the (truth-preserving) connection between A and B is *general*. Let $\models A$ be that A is a logical truth. Then, given that the study of relevant logics is the study of the (logical) relationship between A and B, asserting that $\models A \rightarrow B$ is asserting that $A \rightarrow B$ is a law of logic. Thus, by following the line of reasoning in the modal case, there is a sense in which formulas of the form $A \rightarrow B$ express laws of logic in relevant logics.

One can show in a semantics for *FDE* that the truth-preserving connection between B and B (for any formula B) is general. This shows that $B \rightarrow B$ is a logical truth.[10] Hence, a non-normal world where $B \rightarrow B$ fails is a world where a logical truth fails. A non-normal world in relevant logics is, thus, a world where laws of logic fail.

In the case of relevant logics, however, there is no uniform evaluation of $\rightarrow$-formulas across all non-normal worlds. The evaluation is based on the constraints

[9]Priest (1992) provides a different analysis of the relevant non-normal worlds. His analysis does not seem to be a generalisation of Kriple's non-normal worlds, despite his claim that the Routleys and Meyer generalised Kripke's non-normal worlds.

[10]Strictly speaking, there are no logical truths in *FDE*. Even though *FDE* may not be the best paraconsistent logic to be used for the current purpose, it is the easiest to understand the nature of paraconsistency with and, hence, I have used *FDE* in my discussion. One can replace *FDE* with *LP* of Priest (1979) which imposes the exhaustion principle: for all p, either $\langle p, 1\rangle \in \mu$ or $\langle p, 0\rangle \in \mu$.

imposed on the ternary relation between worlds at a relevant non-normal world. Hence a formula of the form $B \rightarrow B$ may turn out to be true at a relevant non-normal world. Nonetheless, $B \rightarrow B$ may fail to be true even if $\models B \rightarrow B$. Hence laws of logic may fail at a relevant non-normal world. And it is because of this that the irrelevant formula of the form $A \rightarrow (B \rightarrow B)$ fails to be a logical truth. Thus, it is the (possible) failure of laws of logic that characterises relevant non-normal worlds too.

By making non-normal worlds in the relevant semantics intelligible, we have attached a sense to (some) relevant logics. Therefore, if we can claim to have made sense of non-normal systems of modal logic, we can also claim to have made sense of relevant logics. Relevant logics 'make sense' as much as non-normal modal logics do.

2.4 Making Sense of Paraconsistent Logic

In the previous sections, I provided an interpretation of non-normal worlds in the possible world semantics for non-normal modal logics of C.I. Lewis which was used to make sense of non-normal worlds in the semantics for some relevant logics. Since not all paraconsistent logics are relevant logics, I now need to provide an intelligible interpretation of paraconsistent logic in general in order to meet the charge that no sense has been given to paraconsistent logic. My strategy will be to show that the conception of logic that Putnam (1994) seems to endorse can reasonably be extended to incorporate paraconsistent logic (at least some paraconsistent logics).

Putnam tries to grasp the early Wittgenstein's thought that 'logical truths do not really say anything, that they are empty of sense (which is not the same thing as being nonsense), *sinnlos* if not *unsinning*' (Putnam 1994, p. 246). He does this by tracing the conception of logic that he thinks runs through the thoughts of Kant and Frege and finally Wittgenstein.

First, Putnam grapples with Kant's view of logic as a maximally general science. For Kant, logic is not *descriptive* of the world, whether actual or possible. That is, Kant does not present a *metaphysical* conception of logic. For Kant, according to Putnam, logic is 'a doctrine of *the form of coherent thought*' (p. 247). It is normative in the sense that 'my thought [in the normative sense of *judgement which is capable of truth*] would not be a thought at all unless it conforms to logic' (p. 247). Putnam overstates Kant's view when he writes that 'illogical thought is not, properly speaking, thought at all [for Kant]' (p. 246). For Kant's point is rather about the normative status of logic. As MacFarlane puts it nicely: for Kant, 'no activity that is not held accountable to [logical] rules can *count* as thought, and not that there cannot be thought that does not conform to these rules' (MacFarlane 2000, p. 87).

Kant's view that logic is a maximally general science may imply, just as Kant himself inferred, the *formal* nature of logic. Since this is exactly the inference Frege rejected (see MacFarlane 2002), it is not clear that Frege would accept Kant's view (as characterised by Putnam) that 'to say that thought, in the normative sense of *judgement which is capable of truth*, necessarily conforms to logic is not to say something which a metaphysics has to *explain*' (Putnam 1994, p. 247).[11] Nonetheless, it seems true that '[l]aws of logic are without content, in the Kant-and-possibly-Frege view, insofar as they do not *describe* the way things are or even the way they (metaphysically) *could* be' (p. 248). Even though Frege rejects Kant's *formal* view of logic that logic is abstracted from objects, Frege agrees with Kant in rejecting the view that logic is only a description of the way things are or could be. In the preface to the *Grundgesetze*, for example, Frege mobilises the distinction between descriptive and prescriptive laws and draws an important implication from this distinction concerning the nature of logical laws by emphasising the importance of prescriptive laws.

Putnam thinks that Wittgenstein expanded on this view of Kant and Frege. Putnam tries to put himself in Wittgenstein's shoes (as he puts it) by elaborating on the difference in status between logical and empirical laws. He invites us to consider the following three sentences:

1. It is not the case that the Eiffel Tower vanished mysteriously last night and in its place there has appeared a log cabin.
2. It is not the case that the entire interior of the moon consists of Roquefort cheese.
3. For all statements p, $\ulcorner\neg(p \wedge \neg p)\urcorner$ is true. (p. 250)

He argues that there is a difference of methodological significance between these three sentences. (1) and (2) are empirical hypotheses. As such, we know how to show them as false or we can adopt a 'conceptual scheme' that falsifies them (though (2) is apparently harder than (1)). However, Putnam argues that we don't know how to even begin showing that (3) is false.

It is here that Putnam appeals to the Kant-Frege-Wittgenstein view of logic. He argues that, for them, logic is not a matter of what the world is or could be like. Instead, logic is prior to all rational activities. It is logic that sets the standard for rationality. Without the logical laws, no activity can be said to be rational. Thus, any thought which violates (3) cannot be counted as rational since such a thought cannot be accounted for by the standard for rationality.

Now, one can question the legitimacy of attributing the above view to Kant, Frege and Wittgenstein. One can even question the coherence of such a view. However, it should be noted that if this is what it is to make sense of logic, then we can easily make sense of paraconsistent logic. For in many (though not all) paraconsistent logics, it is a logical truth that $\neg(p \wedge \neg p)$ for any p and so we can respect the difference in status between (3), on one hand, and (1) and (2) on the

[11]Putnam in fact declines to attribute this view to Frege (p. 247). See also Goldfarb (2001) who presents Frege as holding a different conception of logic.

other.[12] Thus, the Kant-Frege-Wittgenstein view of logic can be used to give sense to paraconsistent logic.[13]

One may argue, nonetheless, that in paraconsistent logic, for some p, $\ulcorner\neg(p \wedge \neg p)\urcorner$ may be true and hence $\ulcorner\neg(p \wedge \neg p)\urcorner$ is false. Since we know that there are false instances of logical truths, we thus know how to 'falsify' (3). The difference between (1) and (2), on one hand, and (3), on the other, is supposed to be that one can know how to falsity (1) and (2) but not (3) and it is supposed to be this difference that separates empirical statements such as (1) and (2) and logical laws such as (3). Hence, so the critique goes, we cannot give a significance to (3) different from (1) and (2) by appealing to paraconsistent logic.

Regardless of how plausible the above critique may sound to classical logicians, it misrepresents the nature of paraconsistent logic. Even if there may be some p such that $\ulcorner\neg(p \wedge \neg p)\urcorner$ is false, (3) does not cease to be a logical truth in paraconsistent logic. In the case of paraconsistent logic, there can be a difference between a statement being false and not true. Consider, for example, *FDE*. Let μ be a relation between a propositional variable and a truth value. An *FDE*-evaluation is thus a *relation* rather than a *function*. Then we can express an evaluation of p to be true as $\langle p, 1\rangle \in \mu$ and false as $\langle p, 0\rangle \in \mu$. An evaluation of p as not true is expressed as $\langle p, 1\rangle \notin \mu$ which is not equivalent to $\langle p, 0\rangle \in \mu$ as μ is a relation. Showing that p is false is no longer the same as showing that p is not true.[14]

The important point of Putnam's thought is that (3) cannot be falsified in the sense that there could not be any rational thought that does not conform to logical laws. The fact that there may be some p such that $\ulcorner\neg(p \wedge \neg p)\urcorner$ is false does not change that. For it is still the case that for all p, $\ulcorner\neg(p \wedge \neg p)\urcorner$ is true in paraconsistent logic.

If we are to focus on the normative nature of logical laws, therefore, we can give a sense to paraconsistent logic which is the same as the sense that Putnam would give to classical laws. If the Kant-Frege-Wittgenstein 'sense' can be attached to classical logic, we can also attach the same sense to paraconsistent logic.

2.5 Classical Logic Doesn't Make Sense!

Having shown that paraconsistent logic has an intelligible interpretation outside of its formal context, I now turn the tables and ask whether classical logic makes sense in the way that Putnam would see it. A logic is paraconsistent if it

---

[12]Notable paraconsistent logics in which $\neg(p \wedge \neg p)$ is not a logical truth are the *Logics of Formal Inconsistency* (*LFIs*). See, for example, Carnielli et al. (2007). My defence of paraconsistent logic doesn't extend to *LFIs*. I let the advocates of *LFIs* provide their own defence.

[13]I note that $\neg(p \wedge \neg p)$, in fact, fails to be a logical truth in *FDE*. But this is because of truth-value gap rather than truth-gap glut that *FDE* allows: p may be assigned no truth value in which case $\neg(p \wedge \neg p)$ lacks truth value too. I take it that the concern of Putnam is with contradiction rather than indeterminacy of truth value.

[14]This is also the case in *LP*.

invalidates *ex contradictione quodlibet* (ECQ): $\{A, \neg A\} \models B$ for any A and B.[15] Paraconsistency is thus a property that an inference may possess. Hence it must be distinguished from *dialetheism* which is a *metaphysical* view that there are true contradictions, since obtainability of a true contradiction in itself is a separate issue from the nature of an inference. With this in mind, we now examine whether or not classical logic actually makes sense.

Putnam's Kant-Frege-Wittgenstein view of logic focuses on the *formal* nature of logic. Frege (and possibly Wittgenstein) did not hold logic to be *formal* in Kant's sense (in fact Frege explicitly rejected Kant's formal view of logic). Nonetheless, Frege's view of logic, just like Kant's, does not involve a metaphysical presupposition. That is, logical laws are not derived from what there is or could be. Rather, they are conceived to be the standard for our rational thought: any activity that can count as rational can be held accountable to logical laws.

If this is the way to make sense of logical laws, what sense can we give to a thought that conforms to the classically valid inference $\{A \wedge \neg A\} \models B$ for any A and B? Can we claim it to be a *rational* thought? Consider a random thought as a result of assuming a contradiction. Could that random thought count as a rational thought?

It may be claimed that what ECQ encapsulates is not really the idea that a random thought based on assuming a contradiction is rational but the idea that a contradiction cannot be rationally obtained: ECQ encodes a 'warning' against having a contradiction in our thought or theory.[16] The issue of true contradiction, i.e., dialetheism, however, is a separate issue, as we saw above. It is primarily a metaphysical issue and not an issue about the nature of inferences as such. Moreover, according to Putnam's Kant-Frege-Wittgenstein view of logic, logic assumes no metaphysical presupposition. Thus, the issue of ECQ cannot be collapsed to that of dialetheism. But if we remove any metaphysical presuppositions from logical consideration just as Putnam claims Kant, Frege and Wittgenstein did, then it is hardly clear that the inference $\{A \wedge \neg A\} \models B$ for any A and B sets the standard for rationality. In fact, there are reasons to think that it is the thought that violates ECQ that should count as rational. For example, consider the paradox of the preface. A rational person, after thorough research, writes a book in which they claim $A_1, \ldots, A_n$, but is aware that no book of any complexity contains only truths and so believes $\neg(A_1 \wedge \ldots \wedge A_n)$ too. It may be rational to revise these inconsistent beliefs. Nonetheless, no random thought should count as rational based on this inconsistent set of beliefs.

If ECQ does not set a standard for rationality and it is an integral part of classical logic, classical logic cannot be made intelligible under Putnam's Kant-Frege-Wittgenstein conception of logic. There may be another sense that we can

[15]Some paraconsistent logicians define this principle for *some* A. It does not make any difference for the purpose of this paper whether it is defined for *some* or *any* A.

[16]As far as I know, no one has explicitly formulated this claim on paper. However, the claim is often put to me by my colleagues, for example.

attach to classical logic. However, the onus is now on classical logicians to provide an intelligible conception of ECQ outside of the formal machinery. Given that ECQ is often claimed to be purely a '*formal*' principle separate from the truth of a claim (or soundness of an argument), as for example Lemmon claims (1965, pp. 60–61), it is not clear that classical logicians can provide such a conception. Thus, it is classical logic that has not been made sense of.

2.6 Conclusion

The quest in this paper has been to present paraconsistent logic as more than a mere mathematical tool. If it were to be accepted as a genuine logic, a paraconsistent logic must be made sense of. I have shown that there are senses that can be attached to paraconsistent logic. Along the way, I have shown that it is, in fact, explosive reasoning that has yet to be made sense of. It has been classical logicians who question the significance of paraconsistent logic. It is now time for paraconsistent logicians to question the legitimacy of classical logic.

References

Anderson, A.R., and N.D. Belnap. 1975. *Entailment: The logic of relevance and necessity*, vol. I. Princeton: Princeton University Press.

Anderson, A.R., N.D. Belnap, and J.M. Dunn. 1992. *Entailment: The logic of relevance and necessity*, vol. II. Princeton: Princeton University Press.

Carnielli, W.A., M.E. Coniglio, and J. Marcos. 2007. Logics of formal inconsistency. In *Handbook of philosophical logic*, vol. 14, 2nd ed., ed. D. Gabbay, and F. Guenthner, 15–107. Berlin: Springer.

Cresswell, M.J. 1967. The interpretation of some lewis systems of modal logic. *Australasian Journal of Philosophy* 45: 198–206.

Dunn, J.M. 1976. Intuitive semantics for first-degree entailment and 'Coupled Trees'. *Philosophical Studies* 29: 149–168. Republished in Anderson et al. (1992).

Goldfarb, W.D. 2001. Frege's conception of logic. In *Future pasts: The analytic tradition in twentieth-century philosophy*, ed. J. Floyd and S. Shieh, 25–41. Oxford: Oxford University Press.

Haack, S. 1978. *Philosophy of logic*. Cambridge: Cambridge University Press.

Hughes, G.E., and M.J. Cresswell. 1996. *A new introduction to modal logic*. London: Routledge.

Kripke, S.A. 1963. Semantical analysis of modal logic I. Normal modal propositional calculi. *Zeitschrift fur mathematische Logik und Grundlagen der Mathematik* 9: 67–96.

Kripke, S.A. 1965. Semantical analysis of modal logic II. Non-normal modal propositional calculi. In *The theory of models: Proceedings of the 1963 international symposium at Berkeley*, ed. J.W. Addison, L. Henkin, and A. Tarski, 206–220. Amsterdam: North-Holland.

Lemmon, E.J. 1957. New foundations for lewis modal systems. *The Journal of Symbolic Logic* 22(2): 176–186.

Lemmon, E.J. 1965. *Beginning logic*. London: Thomas Nelson and Sons.

MacFarlane, J. 2000. *What does it mean to say that Logic is Formal?* Ph.D. Dissertation, University of Pittsburgh.

MacFarlane, J. 2002. Frege, Kant, and the logic in logicism. *The Philosophical Review* 111: 25–65.
Priest, G. 1979. Logic of paradox. *Journal of Philosophical Logic*, 8: 219–241.
Priest, G. 1992. What is a non-normal World? *Logique et Analyse* 35: 219–302.
Priest, G., and K. Tanaka. 2009. Paraconsistent logic. In *Stanford encyclopaedia of philosophy*, Summer 2009 edn., ed. E.N. Zalta. Stanford: Stanford University.
Putnam, H. 1994. Rethinking mathematical necessity. In *Words and Life*, ed. J. Conant, 245–263. Cambridge: Harvard University Press.
Routley, R. and Meyer, R.K. 1972. The semantics of entailment – II. *Journal of Philosophical Logic* 1: 53–73.
Routley, R. and Meyer, R.K. 1973. The semantics of entailment – I. In *Truth, syntax and modality*, ed. H. Leblanc, 199–243. Amsterdam: North-Holland.
Routley, R. and Routley, V. 1972. The semantics of first degree entailment. *Noûs* 6: 335–359.
Tanaka, K. 2000. *The labyrinth of trees and the sound of silence: Topics in dialetheism*. Ph.D. thesis, University of Queensland.
Urquhart, A. 1972. Semantics for relevant logics. *The Journal of Symbolic Logic* 37: 159–169.
Weber, Z. 2010a. Extensionality and restriction in naive set theory. *Studia Logica* 94: 109–126.
Weber, Z. (2010b). Transfinite numbers in paraconsistent set theory. *Review of Symbolic Logic* 3: 1–22.

Chapter 3
On Discourses Addressed by Infidel Logicians

Walter Carnielli and Marcelo E. Coniglio

3.1 On Repugnancies and Contradictions

In his well-known invective, Bishop Berkeley tried to reveal contradictions in the infinitesimal calculus, perplexed as he was by the "evanescent increments" that are neither finite nor infinitely small quantities (and "nor yet nothing", but merely "ghosts of departed quantities"). In Section L, 'Occasion of this Address. Conclusion. Queries.' of *The Analyst; or, a Discourse Addressed to an Infidel Mathematician* (cf. Berkeley 1734), Berkeley asks:

> Whether the Object of Geometry be not the Proportions of assignable Extensions? And whether, there be any need of considering Quantities either infinitely great or infinitely small?

and

> Whether [mathematicians] do not submit to Authority, take things upon Trust, and believe Points inconceivable? Whether they have not their Mysteries, and what is more, their Repugnancies and Contradictions?

In an indirect sense, Berkeley's *The Analyst* was very influential in the development of mathematics. The piece was a direct attack on the foundations and principles of calculus and, in particular, Newton and Leibniz's notion of infinitesimal change (or *fluxions*).[1] It is not to be denied that the resulting controversy gave some impetus to the later foundations of calculus and led them to be reworked, in a much more formal and rigorous form, through the use of limits.

---

[1]It is believed that the "infidel mathematician" in question was either Edmond Halley or Isaac Newton himself.

W. Carnielli (✉) • M.E. Coniglio
Department of Philosophy and Centre for Logic, Epistemology and the History of Science, State University of Campinas, Campinas, Brazil
e-mail: walter.carnielli@cle.unicamp.br; coniglio@cle.unicamp.br

K. Tanaka et al. (eds.), *Paraconsistency: Logic and Applications*, Logic, Epistemology, and the Unity of Science 26, DOI 10.1007/978-94-007-4438-7_3,

However, Berkeley's criticisms did not have an effect on everyone. Leonard Euler, for instance, paid little attention to the invectives against the use of infinite series, and found an astonishing new proof of the fact, originally proved by Euclid, that there are infinitely many prime numbers. Departing from Euclid's original purely combinatorial proof, Euler realized (cf. Sandifer 2006) that a distinction between divergent and convergent series could explain an infinitude. Indeed, by comparing infinite sums and products:

$$\frac{2 \cdot 3 \cdot 5 \cdot 7 \cdot 11 \ldots}{1 \cdot 2 \cdot 4 \cdot 6 \cdot 10 \ldots} = 1 + \frac{1}{2} + \frac{1}{3} + \frac{1}{4} + \frac{1}{5} \cdots$$

or, in contemporary notation:

$$\prod_p \frac{p}{p-1} = \sum_n \frac{1}{n} \quad \text{for } p \text{ primes}, n \geq 1$$

it is easy to see that the right-hand harmonic series is divergent; hence there must be infinitely many primes, otherwise there would be an equality between a divergent and a convergent series. Had Euler been afraid of the label "mathematician" (with its connotations of magician and astrologist—as opposed to "geometer"), he would never have dared to mix the continuous with the discrete. Euler's idea turned out to be extremely fruitful: bringing the "forbidden" analysis into the investigation of prime numbers allows for a much more powerful technique than mere combinatorial counting, as further results by Dirichlet and others have revealed. What was at stake was not so much how many numbers there are, but how they are distributed. Analytic number theory was thus born, owing its existence to free-thinkers like Euler and ultimately paving the way for the Riemann Hypothesis (see Derbyshire 2003).

If we accept that repetitive patterns are to be found in science, it may not be a coincidence that paraconsistent negations, that is, negations such that a contradiction does not imply everything, raise perplexities of an analogous sort. This is certainly the case, in particular, with the kind of negation supported by the *Logics of Formal Inconsistency* (*LFI*s) introduced in Carnielli and Marcos (2002) and further developed in Carnielli et al. (2007). Some criticisms of negations in *LFI*s are based on the idea that paraconsistent negations are not negations (in the same way that infinitesimal numbers would considered not to be numbers). Other criticisms, specifically those concerning the core of *LFI*s, fail to see a central point: how is it possible that A and $\neg A$ can be simultaneously held as true (and be not explosive), and the "consistency" of A also be held as true? How can both something and its negation be true and consistent? Is it not inherent in the nature of consistency to require that anything and its negation necessarily have different truth status?

In fact, we wish to argue that this is not only plainly possible, but quite usual: logicians, infidel or not, perform this type of reasoning very often, and it is precisely the fact that A and $\neg A$ are true, while it is at the same time not the case that A is consistent (see Sect. 3.3) which blocks the deductive explosion. On the other hand, with respect to the claim that paraconsistent negations are not negations, a possible answer is that paraconsistent negations can be seen as generalized negations in the same way that infinitesimal numbers are generalized (non-standard) real numbers.

But let us first examine an example of an argument that greatly baffled Berkeley. The example is a contemporary rephrasing of a more geometric one, given by Berkeley himself and found in Lemma 2 of Sect. 1 of the *Philosophiae Naturalis Principia Mathematica* (Newton 1726). It is a typical argument using infinitesimal quantities: in order to find the derivative $f'(x)$ of the function $f(x) = x^2$, let dx be an infinitesimal; then

$$\begin{aligned} f'(x) &= \frac{f(x+dx) - f(x)}{dx} \\ &= \frac{(x+dx)^2 - x^2}{dx} \\ &= \frac{x^2 + 2x.dx + dx^2 - x^2}{dx} \\ &= \frac{2x.dx + dx^2}{dx} \\ &= 2x + dx \\ &= 2x \end{aligned}$$

since dx is infinitely small.

The fundamental problem pointed out by Berkeley is that dx is first treated as *non-zero* (because we divide by it), but later discarded as if it were zero:

> These Expressions [dx, ddx, etc] *indeed are clear and distinct*, and the Mind finds no difficulty in conceiving them to be continued beyond any assignable Bounds. But if we remove the Veil and look underneath, if laying aside the Expressions we set ourselves attentively to consider the things themselves, which are supposed to be expressed or marked thereby, we shall discover much Emptiness, Darkness, and Confusion; nay, if I mistake not, direct Impossibilities and Contradictions. Whether this be the case or no, every thinking Reader is entreated to examine and judge for himself.

In a sense, Berkeley was right: the idea of *infinitesimal* was at that time a naive concept, namely, "a number whose absolute value is less than any non-zero positive number". Thus, using the order properties of real numbers, it can be easily proved that *there are no non-zero real infinitesimals*! From Berkeley's perspective, infinitesimals are not numbers but aberrant intruders, and so should be removed from mathematical reality. But how can Berkeley be completely right, and infinitesimal calculus be what it is today?

Karl Weierstrass and others, by using the rigorous notion of limit, were able to give a formal mathematical foundation for calculus in the second half of the nineteenth century: the epsilon-delta interpretation. This mathematical formulation, without using "abnormalities" such as infinitesimals, removed all concerns about the illegitimacy of calculus. This was, however, a sort of foundational compromise, with concessions to both sides: the notion of infinitesimal had to be replaced by a process (the limits), and in this Berkeley's criticisms of the eighteenth century mathematicians were in principle correct; but the task was not impossible, and in this he was wrong. Two centuries later, in any case, the fluxions vindicated their reputation at the hands of logicians.

3.2 Infidelities and Perplexities: Are Paraconsistent Negations Genuine Negations?

What Bishop Berkeley couldn't see is that the feasibility of infinitesimals depends on the breadth of the mathematical context: they are not mathematically "impossible" or "wrong", they just find a place in a wider mathematical scenario. As is well known, the original appealing idea of Leibniz and Newton[2] of describing differential calculus by using infinitesimal quantities can be recovered in rigorous mathematical terms by means of the non-standard models of contemporary model theory. What this achievement shows is that infinitesimals are numbers of a new kind, introduced conservatively by extending the previous field of numbers.

Such new species of numbers are nothing else than infinitesimals made possible: Abraham Robinson's *nonstandard analysis* of the 1960s (cf. Robinson 1966) makes use of the set of real numbers extended by the hyperreals and thus containing numbers less (in absolute value) than any positive real number. In this formulation an infinitesimal is a *non-standard* number whose absolute value is less than any non-zero positive *standard* number. Other paradigms were introduced to deal with infinitesimals, such as John Conway's *surreal numbers* (which are algebraically equivalent to hyperreals; cf. Conway 1976 and also Knuth 1974), Edward Nelson's *internal set theory* (cf. Nelson 1977), and *synthetic differential geometry* (also known as *smooth infinitesimal analysis*; cf. Lawvere 1998), which is based on category (topos) theory and provides a new approach to Robinson's nonstandard analysis.

The latter makes use of *nilsquare* or nilpotent infinitesimals, that is, numbers x such that $x^2 = 0$ holds but $x = 0$ is not necessarily true. This allows for rigorous algebraic proofs using infinitesimals, like the one given above which so irritated Bishop Berkeley.

Asides from infinitesimals, there are several other examples of generalized mathematical structures losing "classical" features. Non-Euclidean geometry is one of them. Concerning numbers and their operations, the following are obvious examples (assuming a "classical" perspective in each case):

---

[2]There is some historical evidence that the elements of the infinitesimal calculus, developed between the fourteenth and sixteenth centuries in Kerala, India, may have been transmitted to Europe by Jesuit missionaries (cf. Almeida and Joseph 2007).

- Integer numbers are not numbers from the point of view of natural numbers: there cannot be any x such that $x + 2 = 1$!
- Rational numbers are not numbers from the point of view of integer arithmetic: there cannot be any x such that $2x = 1$!
- Real numbers are not numbers from the point of view of rational numbers: there cannot be any x such that $x^2 = 2$!
- Complex numbers are not numbers from the point of view of real numbers: there cannot be any x such that $x^2 = -1$!

There are certain subtle similarities between Berkeley's criticism of infinitesimals and some criticisms of paraconsistent negations found in the literature. For instance, Slater's well-known criticism in Slater (1995) is based on the contention that negations of paraconsistent logics are not proper negation operators given that they are not a 'contradictory-forming functor', but just a 'subcontrary-forming one'.

But arguing along these lines requires strong presuppositions about what a negation should be. As, supposedly, natural language is our basic source of inspiration for understanding negation, it is advisable to pay close attention to it before sermonizing. Is there really only one negation, and does it necessarily have the role of suppressing whatever comes after it? R. Giora argues against the view that negation is unique in natural language, and against the purported functional asymmetry of affirmation and negation—a view that supports the "suppression hypothesis" which assumes that negation necessarily suppresses what is inside its scope:

> Indeed, many discourse functions assumed to uniquely distinguish negatives from affirmatives, such as denying, rejecting, disagreeing, repairing (both linguistically and metalinguistically), eliminating from memory, communicating the opposite, attenuating or reducing the accessibility of concepts and replacing them with alternative opposites, are equally enabled by affirmatives. Similarly, discourse roles assumed to uniquely distinguish affirmatives from negatives, such as representing events, conveying agreement, confirmation, or affective support, highlighting and intensifying information, introducing new topics, conveying an unmarked interpretation, establishing comparisons, effecting discourse coherence and discourse resonance, are equally enabled by negatives. Such evidence attesting to some functional affinity between negative and affirmative interpretations can only be explained by processing mechanisms that do not operate obligatorily but are instead sensitive to global discourse considerations. (Giora 2006, pp. 1009–1010)

By analogy with the case of infinitesimals, paraconsistent negations can be seen as extending the classical one by generalizing some of its features. And this makes sense in terms of the above landscape of many negations: the more properties a negation operator has, the more restricted and specific the operator is. Thus, paraconsistent negations are neither logically "wrong" nor "impossible", but they are part of an enhanced and more general logical scenario, in the same way that infinitesimals are legitimate numbers in a wider sense.

It will be worthwhile to examine for a few moments the model theory of propositional quantified logic, and again to compare this model theory with the underlying model theory of the algebraic extension of fields (from the real to the complex numbers).

Recall that if $\mathfrak{A}$ and $\mathfrak{B}$ are two first-order structures for the same language, such that $\mathfrak{A}$ is a substructure of $\mathfrak{B}$ and $\varphi(x)$ is a formula of that language with just x as free variable, and where only the logical operators $\exists$, $\wedge$ and $\vee$ occur in $\varphi(x)$, then

$$\mathfrak{A} \models \exists x \varphi \text{ implies } \mathfrak{B} \models \exists x \varphi.$$

In particular, taking the language of fields consisting of symbols $+, \cdot, 0, 1$, and $\mathfrak{A}$ and $\mathfrak{B}$ as being the structures of real numbers and complex numbers over that language, respectively, then

$$\mathbb{R} \models \exists x \varphi \text{ implies } \mathbb{C} \models \exists x \varphi$$

for any $\varphi(x)$ as defined above. As is well-known, the converse implication is not valid because $\mathbb{C}$ is an algebraic extension of $\mathbb{R}$, and so it satisfies an existential sentence that the latter does not satisfy. For instance,

$$\mathbb{C} \models \exists x(x.x + 1 = 0) \text{ but } \mathbb{R} \not\models \exists x(x.x + 1 = 0)$$

This means that $\mathbb{R}$ is not an elementary substructure of $\mathbb{C}$.

Consider now **PCL** and $\mathbf{C}_1$, propositional classical logic and da Costa's paraconsistent logic over the signature $\wedge, \vee, \rightarrow, \neg$, respectively. Consider a fixed set $\mathcal{V}$ of propositional letters, and let *For* be the algebra of formulas generated over that signature from $\mathcal{V}$. Let $V_{\mathbf{PCL}}$ and $V_{\mathbf{C}_1}$ be the sets of bivaluations characterizing **PCL** and $\mathbf{C}_1$, respectively. Then $PCL = \langle For, V_{\mathbf{PCL}} \rangle$ can be conceived as a semantic structure interpreting quantified propositional logic in an obvious way, representing **PCL**. Similarly, $C_1 = \langle For, V_{\mathbf{C}_1} \rangle$ can be seen as a structure for that language representing $\mathbf{C}_1$. Since $V_{\mathbf{PCL}} \subseteq V_{\mathbf{C}_1}$ then

$$PCL \models \exists p_1 \ldots \exists p_n \varphi \text{ implies } C_1 \models \exists p_1 \ldots \exists p_n \varphi$$

for any $\varphi(p_1, \ldots, p_n)$ depending on the propositional letters $p_1, \ldots, p_n$ without quantifiers. In a sense, PCL is a "substructure" of C_1. But the converse implication does not hold, and so PCL is not an "elementary substructure" of C_1. In fact,

$$C_1 \models \exists p(p \wedge \neg p) \text{ but } PCL \not\models \exists p(p \wedge \neg p).$$

By analogy with the extension $\mathbb{C}$ of $\mathbb{R}$, which adds new objects outside the classical (real) scope satisfying unexpected (or "absurd") properties, the paraconsistent model-theoretic extension of classical logic by C_1 adds new valuations v (or new formulas p) satisfying "exotic" or "absurd" properties such as $v(p) = v(\neg p) = 1$. Recalling that the inconsistency operator $\bullet$ of *LFI*s allows or guarantees such "exotic" properties (see next section), then, in some sense, the inconsistency operator $\bullet$ or, more precisely, inconsistent formulas of the form $\bullet\varphi$, play a very similar role to that of non-real complex numbers, that is, complex numbers $a + b\mathbf{i}$ such that $b \neq 0$.

C. Dutilh-Novaes convincingly argues that "there is no real negation" and that paraconsistent negation is as "real" as any other (Dutilh-Novaes 2008, p. 470). So, how much is one denying in "real negation"? The comparison is analogous to the joke about naive tourists who keep asking how much is the price of this and that in "real money": the problem only arises if you insist that your money, or your negation, is unique and *is* the real one.

3.3 Can One Sustain a Consistent Contradiction?

But more pointed criticisms to paraconsistency explicitly concern the *LFI*s of Carnielli and Marcos (2002) and Carnielli et al. (2007). The main feature of the approach in these works is that sets $\bigcirc(p)$ of formulas (depending only on p) are used to convey the idea that $\bigcirc(\alpha)$ expresses the fact (or the information, or even the hypothesis) that "α is consistent". Thus $\Gamma \vdash \alpha$ and $\Gamma \vdash \neg\alpha$ do not ensure that Γ is trivial. Instead, Γ is logically explosive iff

$$\Gamma \vdash \alpha \quad \text{and} \quad \Gamma \vdash \neg\alpha \quad \text{and} \quad \Gamma \vdash \bigcirc(\alpha).$$

In most *LFI*s the set $\bigcirc(p)$ is a singleton, defining a consistency connective (primitive or not) denoted by $\circ$. Thus $\circ\alpha$ means "α is consistent", and the usual "Classical Explosion Principle":

$$\text{(exp)} \quad \alpha \Rightarrow (\neg\alpha \Rightarrow \beta)$$

is replaced by a weaker version, the "Gentle Explosion Principle":

$$\text{(bc)} \quad \circ\alpha \Rightarrow (\alpha \Rightarrow (\neg\alpha \Rightarrow \beta)).$$

Therefore, a contradiction (involving α) *plus* the information that α is consistent produces a trivial set. The *inconsistency* of a sentence α can be expressed by a sentence of the form $\bullet\alpha$, where $\bullet$ is an *inconsistency operator*. In most *LFI*s, both operators are related as follows: $\bullet\alpha \equiv \neg\circ\alpha$ and $\circ\alpha \equiv \neg\bullet\alpha$. Let us go a bit further in order to appreciate the importance of this approach to logic and reasoning.

A well-known (by now) twelfth century example of a derivation by Petrus Abelardus in his *Dialectica* is recalled by W. Kneale and M. Kneale (Kneale and Kneale 1985, p. 217): the conclusion *si Socrates est lapis, est asinus* ("if Socrates is a stone, he is an ass") as a consequence of the validity of the Disjunctive Syllogism $\alpha \vee \beta, \neg\alpha \vdash \beta$. Indeed, from the hypothesis *Socrates est lapis* one derives *Socrates est lapis* or *Socrates est asinus*. But surely *Socrates non est lapis, ergo Socrates est asinus*.

W. Kneale and M. Kneale recognize this "very interesting contention" of Abelard as the beginning of the long Medieval debate on paradoxes of implication:

> On the other hand, he thinks that we have departed from the highest standard of rigour as soon as we put forward a consequentia which involves the assumption of two distinct substances ··· For he says that in such a case the sense of the consequent is not contained in the sense of the antecedent and that the truth of the whole can be established only by special knowledge of nature (*'posterius ex naturae discretione et proprietatis naturae cognitione'*).

They recognize that "it is difficult to find any satisfactory interpretation for this passage". But the reason, as Abelard himself explains (cf. Petrus 1970, p. 284), is that the nature of man and stone are incomparable:

> *quod nouimus natura ita hominem et lapidem esse disparata.*

Now, it is an immediate theorem of *LFI*s that the Disjunctive Syllogism is not to be held unrestrictedly, but only for situations in which additional assumptions (or proofs) of certainty or consistency are present, i.e.:

$$\alpha \vee \beta, \neg\alpha \nvdash \beta$$

Here, the Disjunctive Syllogism does not hold in general, but does hold in a controlled form:

$$\circ\alpha, \alpha \vee \beta, \neg\alpha \vdash \beta$$

is a valid rule, where $\circ$ is the consistency connective (in the *LFI*s).

This property can be easily proved in almost all systems developed in Carnielli et al. (2007), and in fact, aside from its conceptual interest, it is essential for the development of effective and natural "paraconsistent logic programming", because it is precisely the appropriate generalization of the famous resolution rule.

With regard to what concerns Abelard and the allegation that the truth involved in a reasoning such as that which proves that *Socrates est asinus* can only be established by special knowledge of nature, our additional hypothesis $\circ\alpha$ in the controlled form of the Disjunctive Syllogism above is perfectly able to express the proviso that the nature of man and of stone cannot be disparate in order to perform a reasoning of this sort. Indeed, just take $\circ\alpha$ to be true, in this case, if you are prepared to defend any reasoned connection between the nature of Socrates and the nature of a stone. If so, you may derive *Socrates est asinus*; if not, you know where your mistake is. Thus the *LFI*s, basic and simple as they are, not only are coherent with Abelard's advice but may also help to *express* this restriction.

Let us examine the criticisms of this approach. There seem to be, to start with, several cases of misunderstanding, misapprehension or pure disregard[3] on what paraconsistent logic(s) is, or rather *are*: for instance, R. Sorensen manifests his firm (and erroneous) belief that paraconsistent logics (in the plural!) must reject weakening (the inference rule from p to $p \vee q$):

---

[3]For some strange reason, this seems to be more acute among writers in certain groups. Although the phenomenon of "relative own-language preference" in citations is well-known, in this case it seems to be one of "relative own-group preference". As an attitude, it is reminiscent of a whimsical statement by the Brazilian writer Oswald de Andrade, who famously said of a book by a rival: "I didn't read it, and I didn't like it".

> Paraconsistent logics are designed to safely confine the explosion. For instance, they reject the inference rule 'p, therefore, p or q' on the grounds that a valid argument must have premises that are relevant to the conclusion. They extend this relevance requirement to conditionals in an effort to head off the paradoxes of implication. (Sorensen 2003, p. 114)

He blames dialetheistic logics for this inability, and in a hasty generalization continues on the same page:

> Dialetheists portray themselves as friends of contradiction. They remind me of ranchers who present themselves as friends of the horses they castrate. A gelding is not just a tamer sort of stallion; it is not a stallion at all. The dialetheist's 'contradiction' may look like contradictions and sound like contradictions, but they cannot perform a role essential to being a contradiction; they cannot serve as the decisive endpoint of a *reductio ad absurdum*. At best they can be the q in a *modus tollens* argument: If p then q; not q, therefore, not p. So in the end, I think Priest falls into Antisthenes' skepticism about contradictions. (Sorensen 2003, p. 114)

The quotation refers to Antisthenes of Athens (445-360 B.C.), a student of Socrates previously trained under the Sophists. Antisthenes thinks there are no contradictions, and Sorensen puts Antisthenes and the dialetheists (and by his reduction, all paraconsistentists) in the same class: their contradictions are not contradictions as such. We cannot respond for dialetheists, but perhaps Sorensen would be happy to learn that *LFI*s neither reject the weakening rule, nor pose themselves as friends of geldings or stallions: they merely separate geldings from stallions, as we argued above. Consistent contradictions *do* indeed serve as the decisive endpoints of *reductio ad absurdum* reasoning: it is proved in Carnielli et al. (2007) that the following *reductio* rule holds in most *LFI*s:

$$\text{If } \Gamma \vdash \circ\alpha, \quad \Delta, \beta \vdash \alpha \text{ and } \Lambda, \beta \vdash \neg\alpha \quad \text{then} \quad \Gamma, \Delta, \Lambda \vdash \neg\beta$$

This illustrates an instance of a more general phenomenon: Any classical rule can be recovered within a class of *LFI*s known as **C**-systems, if a sufficient number of 'consistency assumptions' are assumed (see the "Derivability Adjustment Theorem" in Carnielli et al. 2007, Remark 21).

Some authors consider, however, that this perspective has inherent problems. Specifically, the following passage on p. 27 of Carnielli and Marcos (2002) has been criticized:

> So one may conjecture that consistency is exactly what a contradiction might be lacking to become explosive—if it was not explosive from the start. Roughly speaking, we are going to suppose that a 'consistent contradiction' is likely to explode, even if a 'regular' contradiction is not.

One of the main criticisms comes from F. Berto, who asks:

> How could we have α, $\neg\alpha$ *and* keep claiming that α is consistent? (Berto 2007, p. 162)

In fact you *cannot*, and that is precisely what is meant! In the realm of classical logic, the corresponding objection is: How could we have α *and* $\neg\alpha$ in our theory?

Of course, a classical logician simply *cannot*, and this is exactly what the "Classical Explosion Principle" says:

$$\alpha, \neg\alpha \vdash \beta$$

for any β; that is, if you have a contradiction, you are in trouble.

This principle, as explained above, is generalized in the *LFI*s by the "Gentle Explosion Principle":

$$\alpha, \neg\alpha, \circ\alpha \vdash \beta$$

for any β; that is, if you have a *consistent* contradiction, then you have trouble!

The situation is rather similar to chess: no piece can capture the King, but the King can be under the threat of being captured, that is, in check. In this contradictory situation (the King cannot be captured and is being threatened with capture) the King has to be protected, and there may be several moves to protect him. However, if there is no move which can put the King out of check (that is, if the threat is fatal), this is checkmate and the game is over. This point seems to have been overlooked by Berto, however, who abandons the game:

> These difficulties seem to speak against the philosophical import of the Brazilian approach to paraconsistency (which is why it has been dealt quickly in this Chapter). (Berto 2007, p. 162)

It is perhaps M. Bremer, however, who persuades Berto to resign so quickly:

> introducing 'consistent contradictions' [...] awaits epistemic elucidation: If we have A and $\neg A$, then we should take $\circ A$ as false, shouldn't we? And how can we take A to be consistent and have A and $\neg A$ at the same time? (Bremer 2005, p. 117)

Indeed, Bremer had previously stated his intention to abandon what has called "da Costa systems" from a certain point on, due to what seemed to him insurmountable philosophical difficulties:

> Da Costa-Systeme werden deshalb hier nicht weiter betrachtet.[4] (Bremer 1998, p. 53)

Nevertheless, as observed above, this criticism can be similarly posed to classical logicians as well: "If we have A, then we should take $\neg A$, as false, shouldn't we? And how can we take A, and have A and have $\neg A$ at the same time?"

Obviously you *cannot*, of course, and that is precisely what the "Classical Explosion Principle" says!

Bremer also claims:

> Die da Costa-Negation ist überhaupt nich rekursiv!.[5] (Bremer 1998, p. 50)

The reason, he says, is that the truth-value of $\neg A$ cannot be computed from the truth-value of A. But this is just non-functionality, and if this were a valid philosophical criticism, it could be posed to several other logics, including intuitionistic and

[4]"da Costa' systems are, consequently, not treated here from this point on".

[5]"da Costa's negation is absolutely non-recursive!".

almost all modal logics. It is hard to see the point behind his criticism. The fact is that da Costa (and *LFI*) negations are *bounded non-deterministic* functions, and there is no purpose in disqualifying them for such reasons. It happens that all *LFI*s (including da Costa calculi in the hierarchy C_n) are decidable (and thus, recursive). This is accomplished by the original valuation semantics in the case of C_n, and by the possible-translations semantics (cf. Carnielli et al. 2007) in the general case of *LFI*s.

This is amply confirmed by the non-deterministic semantics of A. Avron and his collaborators (cf. e.g. Avron and Lev 2005). But specifically for da Costa calculi C_n, decidability results by means of the procedure of quasi-matrices have been known for three decades (cf. Alves 1976 and also da Costa and Alves 1977, Sect. 2.2.2—some errors being corrected in Marcos 1999). Disqualifying non-determinism in logic matters thus does not seem to be so easy. Non-deterministic Turing machines are equivalent to deterministic Turing machines: issues on complexity apart, they are the same. Even modal logics of non-deterministic partial recursive functions, which are extensions of the classical propositional logic, are studied in Naumov (2005), and moreover proved to be decidable. Confusions of this sort just add to the difficulty of appraising the real ideas behind *LFI*s. There is no better epistemic elucidation of something, and no better way of assessing its "philosophical import", than by examining it from a fair perspective.

3.4 From Contradictoriness to Buddhism, and Back

It was apparently not a paraconsistent logician who first noted that some contradictions in Buddhist reasoning would possibly made Nāgārjuna, a Buddhist thinker of the II century, to endorse paraconsistent logic; Garfield and Priest (2002, 2003) credit this observation (with which they also concur) to Tillemans (1999).

Garfield and Priest take a decisive step, however, by studying the possibility that these contradictions may be seen as structurally analogous to those arising in the Western tradition. By a penetrating analysis of *catuskoti* or tetralemma, the four-cornered negations of Nāgārjuna, they aptly conclude that Nāgārjuna is not

> ··· an irrationalist, a simple mystic, or crazy; on the contrary: he is prepared to go exactly where reason takes him: to the transconsistent.

There is not doubt that Nāgārjuna's reasoning involves contradictions, and J. Garfield and G. Priest claim to partake what they call the "dialetheist's comfort" (that of admitting the possibility of true contradictions) when they state that Nāgārjuna is "indeed a highly rational thinker". A similar argument that contradictions in Buddhism are essentially dialetheist is found in Deguchi et al. (2008).

The reasoning involved in a tetralemma is shown in the following examples. An interlocutor poses four questions to the Buddha about the re-birth of an *arhat* (illuminated person):

- Is the arhat reborn?
- Is the arhat not reborn?
- Is the arhat both reborn and not reborn?
- Is the arhat neither reborn nor not reborn?

The Buddha replies to each question saying that one cannot say that this is so.

Now, looking at the three first questions, we might conclude that the Buddah would be forced to accept true contradictions—but is that really the case? According to M. Siderits:

> [. . .] the third possibility involves equivocation on 'existent': that the arhat does exist when 'existent' is taken in one sense, but does not exist when it is taken in some other sense. For when the Buddha rejects both of the first two lemmas, this generates an apparent contradiction. And one way of seeking to resolve this contradiction is to suppose that there is equivocation at work. (Siderits 2008)

Siderits then concludes:

> To consider this possibility is not to envision that there might be true contradictions. It is a way of trying to avoid attributing to the speaker the view that a contradiction holds. (Siderits 2008)

Our point is this: if indeed contradictions in Buddhist reasoning are structurally analogous to those we refer to in the Western tradition, and can be put under the scope of a paraconsistent formalism, the task would be to specify which kind of paraconsistent logic could better reflect the logic of Nāgārjuna reasoning.

A conducive strategy would be to get some inspiration on regard to how Buddhist thinkers see negation. And once more, in concord with what has been defended for natural language, we find that negation in this tradition is not restricted to a single interpretation. B. Galloway argues in Galloway (1989) that the *prasajya* negation of the Madhyamaka school of Buddhist philosophy (members of a Buddhist school founded by Nāgārjuna) is not the same as that of the other schools. Would it appease the Madhyamikas to conceive of them as endorsing dialetheism and swallowing contradictions as real? This does not seem to be so – again quoting Siderits:

> Madhyamikas say that only mad people accept contradictions. (Siderits 2008, p. 132)

This seems to exclude that possibility that members of the Nāgārjuna school would embrace dialetheism, the view that there are dialetheias (i.e., sentences which are both true and false). This difficulty is in line with another, pointed out by P. Gottlieb (see also Carnielli and Coniglio 2008):

> While Aristotle is clearly not a dialetheist, it is not clear where he stands on the issue of paraconsistency. Although Aristotle does argue that if his opponent rejects PNC across the board, she is committed to a world in which anything goes, he never argues that if (per impossibile) his opponent is committed to one contradiction, she is committed to anything, and he even considers that the opponent's view might apply to some statements but not to others (Metaph IV 4 1008a10-12). (Gottlieb 2007)

The Shōbōgenzō, a masterpiece that records the teachings of the Japanese Soto Zen Master Eihei Dogen of the thirteenth century, contains several Zen Koan stories that are sometimes disconcertingly contradictory. The text is full of contradictions of several dimensions: contradictions between chapters, contradictions between paragraphs, contradictions between sentences, and even contradictions *within* a sentence (see Nearman 2007 for a careful translation from Japanese). For instance, the discourse by Master Dogen "On Buddha Nature" (Nearman 2007, p. 244) contains two ostensibly contradictory statements, namely, that all sentient beings have a Buddha Nature and that all sentient beings lack a Buddha Nature.

Such a contradiction, however, is not anything like a dialetheia. Buddha Nature is not the existence of something: on the one hand, Buddha Nature is encountered everywhere, and on the other hand sentient beings do not readily find an easy or pleasant way to encounter Buddha Nature (see Nearman 2007, Chap. 21, for a long discussion).

Gudo Wafu Nishijima, a Japanese Zen Buddhist priest and teacher, in trying to explain why the Shōbōgenzō is difficult to understand because of this intricate weave of contradictions, makes clear the following point in Nishijima (1992):

> At this point I want to make a very fundamental point about the nature of contradiction itself. We feel in the intellectual area that something called contradiction exists; that something can be illogical. But in reality, there is no such thing as a contradiction. It is just a characteristic of the real state of things. It is only with our intellect that we can detect the existence of something called contradiction.

It thus seems that both the Buddhist and Aristotelian traditions see perfect coherence in the distinction between *reasoning* in the presence of contradictions and *accepting* them. The former position expands the latter, and any appeal to the principle of rational accommodation would support the following conclusion: If by 'accepting a contradiction' we mean 'considering it as consistent', then both traditions would agree with our vision of paraconsistentism. Infidelity, but only at a bare minimum.

Acknowledgements Both authors acknowledge support from FAPESP—Fundação de Amparo à Pesquisa do Estado de São Paulo, Brazil, Thematic Research Project grant LogCons 2010/51038-0, and from CNPq Brazil Research Grants. The first author has been additionally supported by the Fonds National de la Recherche, Luxembourg.

References

Almeida, D.F., and G.G. Joseph. 2007. Kerala mathematics and its possible transmission to Europe. *Philosophy of Mathematics Education Journal* 20. http://people.exeter.ac.uk/PErnest/pome20/.

Alves, E. 1976. *Logic and inconsistency* (in Portuguese). Ph.D. thesis, University of São Paulo, Brazil.

Avron, A., and I. Lev. 2005. Non-deterministic multi-valued structures. *Journal of Logic and Computation*, 15: 241–261.

Berto, F. 2007. *How to sell a contradiction: The logic and metaphysics of inconsistency*. Studies in logic. London: College Publications.

Berkeley, G. 1734. In *The Analyst; or, a discourse addressed to an infidel mathematician.* ed. D.R. Wilkins. Dublin: School of Mathematics, Trinity College. http://www.maths.tcd.ie/pub/HistMath/People/Berkeley/Analyst/Analyst.html.

Bremer, M. 1998. *Wahre Widersprüche. Einführung in die parakonsistente Logik*. Sankt Augustin: Academia.
Bremer, M. 2005. *An introduction to paraconsistent logics*. Frankfurt: Peter Lang.
Carnielli, W.A., and J. Marcos. 2002. A taxonomy of **C**-systems. In *Paraconsistency: The logical way to the inconsistent*. Lecture notes in pure and applied mathematics, vol. 228, ed. W.A. Carnielli, M.E. Coniglio, and I.M.L. D'Ottaviano, 1–94. New York: Marcel Dekker.
Carnielli, W.A., and M.E. Coniglio. 2008. Aristóteles, Paraconsistentismo e a Tradição Budista. *O que nos faz pensar*, 23: 163–175.
Carnielli, W.A., M.E. Coniglio, and J. Marcos. 2007. Logics of formal inconsistency. In *Handbook of philosophical logic*, vol. 14, ed. D. Gabbay and F. Guenthner, 1–93. Amsterdam: Springer.
Conway, J.H. 1976. *On numbers and games*. New York: Academic.
da Costa, N.C.A., and E. Alves. 1977. A semantical analysis of the calculi C_n. *Notre Dame Journal of Formal Logic* 18(4): 621–630.
Deguchi, Y., J.L. Garfield, and G. Priest. 2008. The way of the dialetheist: Contradictions in Buddhism. *Philosophy East and West* 58(3): 395–402.
Derbyshire, J. 2003. *Prime obsession: Bernhard Riemann and the greatest unsolved problem in mathematics*. New York: Plume/Penguin Group.
Dutilh-Novaes, C. 2008. Contradiction: The real philosophical challenge for paraconsistent logic. In *Handbook of Paraconsistency*, ed. J.-Y. Béziau, W.A. Carnielli, and D. Gabbay, 465–480. Amsterdam: Elsevier.
Galloway, B. 1989. Some logical issues in Madhyamaka thought. *Journal of Indian Philosophy* 17(1): 1–35.
Garfield, J.L., and G. Priest. 2002. Nāgārjuna and the limits of thought. In *Beyond the limits of thought*, ed. G. Priest, 249–270. Oxford: Oxford University Press.
Garfield, J.L., and G. Priest. 2003. Nāgārjuna and the limits of thought. *Philosophy East and West* 53(1): 1–21.
Giora, R. 2006. Anything negatives can do affirmatives can do just as well, except for some metaphors. *Journal of Pragmatics* 38: 981–1014.
Gottlieb, P. 2007. Aristotle on non-contradiction. Stanford encyclopedia of philosophy, Feb 2007. http://plato.stanford.edu/entries/aristotle-noncontradiction/#1.
Kneale, W., and M. Kneale. 1985. *The development of logic*. Oxford: Oxford University Press.
Knuth, D. 1974. *Surreal numbers: How two ex-students turned on to pure mathematics and found total happiness*. Reading: Addison-Wesley.
Lawvere, F.W. 1998. Outline of synthetic differential geometry. Available at http://www.acsu.buffalo.edu/~wlawvere/SDG_Outline.pdf. Accessed Feb 1998.
Marcos, J. 1999. Semânticas de Traduções Possíveis (Possible translations semantics, in Portuguese). Master's thesis, IFCH-UNICAMP, Campinas, Brazil. http://www.cle.unicamp.br/pub/thesis/J.Marcos/.
Naumov, P. 2005. On modal logics of partial recursive functions. *Studia Logica* 81(3): 295–309.
Nearman, H. 2007. *Shōbōgenzō: The treasure house of the eye of the true teaching. A trainee's translation of great master Dogen's spiritual masterpiece*. Mount Shasta: Shasta Abbey Press.
Nelson, E. 1977. Internal set theory: A new approach to nonstandard analysis. *Bulletin of the American Mathematical Society* 83(6): 1165–1198.
Newton, I. 1726. Philosophiae Naturalis Principia Mathematica, third edition. Translated in: Isaac Newton's Philosophiae Naturalis Principia Mathematica, the Third Edition with Variant Readings, ed. A. Koyré and I. B. Cohen, 2 vols., Cambridge: Harvard University Press and Cambridge: Cambridge University Press, 1972.
Nishijima, G.W. 1992. *Understanding the Shōbōgenzō*. London: Windbell Publications.
Petrus, A. 1970. *Dialectica*, ed. L.M. De Rijk, Assen: Van Gorcum. http://individual.utoronto.ca/pking/resources/abelard/Dialectica.txt.
Robinson, A. 1966. *Non-standard analysis*. Princeton: Princeton University Press.
Sandifer, E. 2006. *Infinitely many primes*. MAA online – How Euler did it. Available at http://www.maa.org/news/columns.html. Accessed Mar 2006.
Siderits, M. 2008. Contradiction in Buddhist argumentation. *Argumentation* 22(1): 125–133.

Slater, B.H. 1995. Paraconsistent logics? *Journal of Philosophical Logic* 24: 233–254.
Sorensen, R. 2003. *A brief history of the paradox*. Oxford: Oxford University Press.
Tillemans, T. 1999. Is Buddhist logic non-classical or deviant? In *Scripture, Logic, Language*, ed. T. Tillemans, 187–205. Boston: Wisdom Publications.

Chapter 4
Information, Negation, and Paraconsistency

Edwin D. Mares

4.1 Introduction

The following inference is a well-known fallacy of relevance:

$$\frac{p}{\therefore \; q \vee \neg q}$$

Classical logic holds that this inference is valid because the conclusion is always true. Relevant logicians criticize the argument for having a conclusion that has nothing to do with any of its premises. This criticism is not that the conclusion fails to be true in some circumstances in which the premises are true, but rather its necessary truth is not sufficient for its following from an arbitrary proposition.

In the standard interpretation of the Routley-Meyer semantics, a connection is made between the notion of relevance and the standard interpretation of validity in terms of truth preservation. A one premise argument is valid if and only if in every index (world, set up, situation) in which the premise is true the conclusion is also true. But I suggest this use of the notion of truth is rather forced in this context. What is wrong with this interpretation of the semantics is that it disregards the relevant criticism that the universal truth of a statement is not sufficient for the validity of any inference with it as a conclusion. Rather, this interpretation just claims that no statement is universally true.[1]

[1] Well, this isn't strictly speaking true. Some relevant logicians, especially Meyer and Restall, include the so-called Church constant T in their language. This constant is true at every index in every model.

E.D. Mares (✉)
Philosophy and Centre for Logic, Language, and Computation, Victoria University of Wellington, Wellington, New Zealand
e-mail: Edwin.Mares@vuw.ac.nz

K. Tanaka et al. (eds.), *Paraconsistency: Logic and Applications*, Logic, Epistemology, and the Unity of Science 26, DOI 10.1007/978-94-007-4438-7_4,

For this reason, I think that it is more natural to interpret the semantics for relevant logic in informational terms. On this reading, the above argument is fallacious because the premise does not carry the information that the conclusion is true. I do not propose a new formal semantics, but rather a re-interpretation of the Routley-Meyer semantics (or at least one version of their semantics). I suggest that we understand the indices in the Routley-Meyer semantics as situations, in the sense of Jon Barwise and John Perry's situation semantics.[2] A situation, in the sense that I am using it here, is an abstract representation of a world. It may not be a representation of a complete world. Moreover, it may not accurately represent any possible world—it may include, for example, representations of impossibilities. We should interpret the satisfaction relation between a situation and a formula, as in $s \models A$, as saying that the situation carries the information that the formula is true. As we shall see, thinking in terms of information, rather than truth, changes the way in which we view the clauses that inductively define satisfaction relations. In short, we should adopt Jon Barwise's slogan "information conditions not truth conditions"[3] when it comes to understanding the semantics of relevant logic.

I have set out the informational interpretation of this semantics elsewhere.[4] The point of the present paper is to examine the freedom that the distinction between truth conditions and information conditions gives us. Although truth and information are closely connected, I claim that when we distinguish between truth conditions and information conditions (and associate logical validity with the former) we gain a lot of freedom to adopt a theory of truth that is relatively independent of our logical theory. This may sound incoherent, since we normally think of a theory of truth as a logical theory, but the separation will become clearer when we examine the alternatives. For example, we may, I will argue, adopt a very classical approach to truth (accept the principles of bivalence and consistency) at the same time as accepting a relevant logic. But we are not forced by our logic to accept a classical account of truth. For example, as we shall see, we may adopt a dialetheic view of truth at the same time as accepting a "two-valued" semantics for information. I speculate that this combination could provide the basis of an interesting variant of the naïve theory of truth. And there are other uses of the separation of truth and information: at the end of the paper I explore the use of a theory of information in developing a semantics for vague terms and predicates.

[2]See Barwise and Perry (1983). The suggestion of the link between relevant logic and situation semantics was first made by John Perry (in Barwise and Perry 1985 and in conversation with me in 1992) and Barwise (1993). Their ideas have been developed in Restall (1996), Beall and Restall (2006) and Mares (1996, 2004).

[3]Quoted in Seligman and Moss (1997, p. 288).

[4]I still used the notion of truth conditions in Mares (2004), but laid the foundations for the informational interpretation there. I made my real informational turn in Mares (2009).

4.2 Information Versus Truth

The standard treatment of information takes information to be a syntactic entity of a certain kind. It takes a piece of information to be a well-formed and meaningful piece of *data* (Floridi 2004, p. 46). Luciano Floridi has modified this standard notion to hold that all information must also be true (misinformation, on this view, is not real information). There is, however, another sense of the word 'information' that we will use here. This is the sense of that word for which it is meaningful to talk about finding information in one's surroundings. This is the notion of information that situation semantics attempts to analyze and which I adopt here. The two notions, however, are quite closely related. The notion of objective information is one of *potential data*—features of the environment that can be understood (in the right way) to give us information in Floridi's subjective sense.

This notion of potential data is open to further interpretation. A piece of potential data is some fact that is *available* in a situation. But in what sense of availability? Logic, as a basis for norms about reasoning, uses idealizations. Intuitionists, for instance, talk of what can be proven in a way that far outstrips the ability of any person or even any physical machine. Similarly, we can talk about the availability of information in a situation in a very idealized way. We might think include as information facts that are physically inaccessible to us are nevertheless available in the salient sense, or we might not. How restrictive we should be in this regard is not an issue that I will decide now, in part because I have not yet decided on my own position.[5]

An information condition for a statement is a condition under the environment enables us to extract the relevant data from it. This does not mean that any condition under which we can come truthfully to believe something (or even to know it) can count as an information condition. For example, the fact that the sun has risen every morning this week is an indication that it will always rise, but it is not the case that the particular facts that held in the last week together carry the information that the sun will always rise. Following Dummett, I hold that we need to distinguish between *canonical* and *non-canonical* means of gathering (subjective) information. Let us consider negation, which is the main focus of this paper.

There are various ways that we come to know the truth of a negated statement. We may be reliably told that it is true, we may infer its truth from the behaviour of others, and so on. But these means of gathering information can hardly be considered parts of the meaning of a negated statement. The canonical means of finding that a negated statement is true is to find some information in the environment that is *incompatible* with that statement. The information condition can be used to explain not only how one knows the truth of the statement, but also to explain why it is true.[6]

---

[5]I have developed a position since I wrote this paper. See Mares (2010).

[6]The incompatibility (or sometimes "compatibility") interpretation of negation was brought into relevant logic by Dunn (1993), but was originally formulated by Goldblatt (1974) in the context of orthologic (a generalisation of quantum logic).

Note that the notion of information is not supposed to replace the notion of truth in a semantics. Rather, information is parasitic on truth. A piece of information tells us that something is true. But despite the close link between truth and information, we can make a real distinction between truth conditions and information conditions.

As I shall argue later, adopting an informational semantics gives us a fair amount of freedom in our choice of a theory of truth. But for the moment, let us suppose that we are classical with regard to the truth condition for negation. Then we can say that a negated statement $\neg A$ is true at a possible world if and only if A fails to be true at that world, but a situation carries the information that $\neg A$ if and only if that situation carries some information that is incompatible with A. If we adopt this combination, then we can claim that bivalence is true, although the analogous principle for information fails, since there are certainly situations in which no information about the truth of a particular statement or its negation is available.

4.3 Negation and the Metaphysics of Information

Before we go on to discuss the importance of the theory of information for paraconsistency, it is a good idea to pause for a moment to discuss the commitments of the theory. It might seem that the view of negation just outlined is committed to a *metaphysics of incompatibility*. That is to say, it seems that this theory holds that there is a mind-independent relation of incompatibility that stands between certain situations. This is a rather heavy metaphysical commitment.[7]

Whether we accept this commitment depends on our view of the nature of objective information. On Barwise and Perry's theory, all information is carried by situations relative to constraints. These constraints can come from a wide variety of sources. Among these constraints are, on my view, the background incompatibilities that determine the semantics of negation. Moreover, objective information, if it is to be used for the semantics of human languages, is sorted and understood in terms of human concepts and capacities. As we shall see in the section on Information Fade below, information about what colour things are is a good example of this. The way in which shades in a certain range of the spectrum are sorted into, say, red and orange depends on the way in which we make those distinctions. 'Red' and 'orange' are not natural kinds.

We have a choice about the degree of realism we want for our semantic theory. When we make negative judgments, we often rely on what we think are incompatible properties. The evidence of how we think and talk using negation underdetermines whether we should take the salient incompatibilities to be mind independent or ones that we have accepted as a society. Thus, we have no empirical proof for a realist semantic theory, nor do we have enough evidence to support an anti-realist (in this case a social-relativist) one.

---

[7]I am grateful to Mark Colyvan for pressing me on this point.

There are, however, some good (non-empirical) reasons to accept a more realist theory. First, if there are a class of set incompatibilities, then what counts as negative information remains stable through changes in the theories that are accepted by society. Second, as I said above, all information in actual situations must be actually true. If we have incorrect views about what is really incompatible, then a social-relativist semantics may allow an actual situation to contain false negative information.

Neither of these problems is terribly serious, though. The problem of temporal relativism is one with which many forms of anti-realism have to deal. It would not bother many anti-realists if what counts as negative information changes over time. The second problem—the problem of false information—is more difficult. But I don't think it is insurmountable. The anti-realist could moderate his or her view and claim that any class of incompatibilities is admissible as long as they do not entail that actual situations carry any false information. Perhaps a more interesting possibility is to wed an anti-realist theory of information with an anti-realist theory of truth. For example, consider a form of the pragmatist theory of truth, on which a statement is *true* if it is currently accepted, but *True* if society converges on it as something that is accepted for all time.[8] Clearly there will be no conflict between changing negative information and negative truths in this first sense, and in the long run no conflict between information and Truth in the second sense. Thus, we can distinguish also between two senses in which information can be available in situations.

To sum up what I have said in this section, the information theoretic approach to semantics (and negation in particular) does not carry with it any serious commitments to a mind-independent relation of incompatibility between situations. Rather, the theory awaits further theorizing about how realist our semantics should be.

4.4 Information and Inconsistency

Relevant logic is a paraconsistent logic. In the Routley-Meyer semantics, there are models in which there are situations that satisfy contradictory formulas. How are we to interpret inconsistent situations in information theoretic terms?

Let's start with a depiction of an inconsistent situation. We will analyze this depicted situation and then try to generalize from that to a theory of inconsistent information. The depiction I have chosen is extremely well-known (perhaps hackneyed), but this should make things easier. This is M.C. Escher's drawing, *Ascending and Descending*. The drawing is well known. It depicts a tower with a staircase on top. A number of men are trudging along on the staircase, but although some seem always to be ascending, they end up always where they began. Similarly, although some always seem to be descending, they end up where they began. Here

[8]This is a simplification of William James' view.

the contradiction is (supposedly) that half of the little grey men on the staircase on the roof are continually descending and the other half are continually ascending, yet they each pass the same person in the other group over and over again. In other words, you can keep going down to end up in the same place as where you started and likewise you can keep going up to end up in the same place as you started. The contradiction that we will examine here is between ascending continuously from point x to get to point y and descending continuously from point x to get to point y. These two pieces of information are incompatible with each other. But here we have them depicted in one etching. And this etching describes a set of situations in each of which contain both of these pieces of information. These situations, thus, contain incompatible information. From the standpoint of the incompatibility semantics, each of these situations are incompatible with themselves.

4.5 Truth Again

But things are not as straightforward as this. Suppose that we choose to a classical theory of truth. As we saw in the section on Truth above, objective information is always *true*. If we accept a classical theory of truth, from the very meaning of negation we cannot have a sentence and its negation both being true. If negation is failure then it would seem that we cannot have a situation that represents the world as having both a sentence and its negation both as true. Thus, it looks as though there is a real incoherence here.

We can eliminate this incoherence if we distinguish between two ways in which situations represent a world. On one hand, there is an external perspective on a situation—in which we judge the situation from informational relationships that we have in the actual world. On the other hand, there is an internal perspective—in which we describe the situation from the point of view of someone located within it. Our present situation (e.g. the situation that comprises all true the information in our world) supports incompatibility relationships between other situations. According to our situation, for example, being red all over is incompatible with being green all over at the same time. Similarly, it is incompatible to ascend and descend the same staircase at the same time.

But things seem different when viewed from the internal perspective. Consider again the little grey men in Escher's picture. Here people are getting to the same place from the same place by ascending and descending the staircase at the same time. From the perspective of the little grey men, is there a true contradiction? Let's say that they, like us (for the moment, at least) have a classical view of truth. No statement can be true at the same time as its negation. They can coherently hold a classical view of truth (if they realize what they are doing) only if they view the two salient pieces of information as compatible with one another. That they are compatible is shown to the little grey men because they both really obtain (from their perspective). To put this matter a little more technically, we need to distinguish between contexts of utterance and contexts of evaluation in order to understand inconsistent information. The context of utterance determines which incompatibilities determine the truth of negative statements.

Now, even from a classical point of view, we can interpret the little grey men as believing that there are true contradictions. Suppose that some little grey men speak in a story about them and claim that there is a true contradiction. Our circumstance is one of radical interpretation.[9] We could hold them to believe what we do about negation and be mistaken about their situation or we could think that they are attributing a different meaning to 'not' (or some other connective).

Although I think that believing in both the classical theory of truth and a paraconsistent logic of information is coherent, I don't think that the classical theory is our only option. In the Information and Dialetheism section below, we will discuss the relationship between the informational semantics and a dialetheic approach to truth.

4.6 The Dialetheic Peril

Before we turn to the construction of a dialetheic theory of information, let us consider an argument due to Priest (2000). This is a slippery-slope argument for dialetheism (or something close to dialetheism). Many philosophers and semanticists think that worlds that contain contradictions are needed to treat counterpossible conditionals (counterfactuals with "impossible" antecedents), to treat propositional attitudes that seem to have impossible content, and for a variety of other reasons. Most theorists make these inconsistent worlds impossible worlds in some sense. They are supposed to be more distant than worlds that obey all the supposed laws of metaphysics, including the law of consistency. But once we postulate the existence of contradictory worlds, it would seem that we reject the law of consistency. We admit that contradictions can be true in the sense that there are worlds in which they are true. Moreover, now that we have postulated the existence of contradictory worlds, a sceptical problem arises: what proof is there that we are not ourselves in a contradictory world? Neither our semantic nor our metaphysical theories rule out this possibility. We cannot say that our semantics shows that there can be no true contradictions, since we have worlds at which there are. Moreover, our metaphysics allows that inconsistent worlds may exist.

Let us call this argument the *dialetheic peril*, since it shows that what seems to be a reasonable postulate of a semantical theory (i.e. the postulation of inconsistent worlds) might lead us to holding the extremely controversial view that there may be true contradictions.

The dialetheic peril can be avoided by the distinction between information conditions and truth conditions. The question of why can't a contradiction be true can be answered very easily by the classical truth theorist: there can be no

[9] I am grateful to Greg Restall for forcefully arguing that we need not interpret the little grey men classically even if we adopt a classical theory of truth.

true contradictions because the meaning of negation prevents there from being any. There can be inconsistent situations (as viewed from our perspective), but this does not mean that there can be any true contradictions in any vertebrate sense. The slippery slope is stopped because the informational perspective does not accept the analogy between contradictory information in a situation and there being a true contradiction. The distinction between information conditions and truth conditions (and our double indexing) destroys the analogy.

Priest, however, would seem to have a response available to him.[10] He can make the following claim: *If inconsistent situations represent inconsistent worlds, then there are inconsistent worlds for them to represent.* Thus, if there are inconsistent worlds, then the issue of whether we are in one such still arises problem and the move to situations does not help. Of course I deny that the fact that situations sometimes represent contradictions entails that there are inconsistent worlds. There are two senses in which impossible situations represent worlds. The first sense is that *all situations represent all worlds*, but not all situations represent all worlds accurately. Some situations represent one or more worlds accurately, and others do not. Clearly, this sense of representation does not require the existence of inconsistent worlds. The second sort of representation is close to the notion of representation employed by fictionalists in metaphysics. On this view, inconsistent situations represent *as if* there were inconsistent worlds, but there are no inconsistent worlds for them accurately to represent. On either of these senses of representation, therefore, there is no need for inconsistent worlds. Thus, the dialethic peril is blocked. If we do not postulate inconsistent worlds, then we do not get on Priest's slippery slope.

4.7 Information and Dialetheism

My central theme in this paper is to argue that distinguishing between truth and information and associating logic with the latter liberates logic from various concerns that people have about truth. This division not only allows us to accept a classical theory of truth, but if we wish we can accept a non-classical theory of truth without thereby altering the informational analysis of logic.

Suppose for example, that we accept the dialetheist theory of truth according to which every formula is given zero or more values from the set $\{True, False\}$. We could say, as we did before, that formulas get these values in a world, but situations (which may or may not accurately characterize that world) contain or do not contain the information that a given formula is true or false. Here we can adopt the view (as dialetheists do) that a formula A is false if and only if $\neg A$ is true, so we can hold that a situation contains the information that A is false if and only if it contains the information that $\neg A$. We can, moreover, accept the compatibility semantics

[10] He made a similar reply in conversation. I'm not sure I have his reply exactly right, so I do not attribute it to him.

for negative information as before. There are good technical grounds for adopting the incompatibility semantics. It is easier to use than the dialetheic semantics for relevant logic (see, e.g., Restall 1995).

But we have to be careful about what we claim here. One of the central reasons for accepting the dialetheism to construct a naïve theory of truth. If we accept certain versions of the naïve theory of truth and particular theses concerning implication, then we can construct a trivial theory, that is, a theory in which every proposition is contained in every situation. Suppose that we have singular terms in our language for propositions. That is, in a given world, a particular term of this sort may or may not refer to some proposition. In this sort of theory, the Tarskian T-scheme is:

$$T(t) \leftrightarrow A$$

where t is a name that refers to the proposition that A. Also, suppose that we have some sort of diagonalisation or self-referential device such that we can construct a singular term δ that refers to the Curry proposition expressed by a sentence of the form

$$T(\delta) \rightarrow p$$

(where p is some arbitrary formula).

Now suppose that we have a logic, which like the strong relevant logics R and E, contains the postulate of contraction. Contraction is the thesis $(A \rightarrow (A \rightarrow B)) \rightarrow (A \rightarrow B)$. Now we can construct the following argument:

1. $T(\delta) \leftrightarrow (T(\delta) \rightarrow p)$	T-scheme
2. $T(\delta) \rightarrow (T(\delta) \rightarrow p)$	1, simplification
3. $T(\delta) \rightarrow p$	2, contraction
4. $T(\delta)$	1, 3, MP
5. p	3, 4, MP

So, if it would seem that if we accept a naïve theory of truth together with a logic that contains the contraction postulate we end up with a trivial semantics.

The usual approach among dialetheists is to reject the contraction postulate. Under some interpretations of implication, it makes perfect sense to reject contraction. Let's consider one such interpretation due to Priest (1992). On Priest's semantics, a frame consists in a set of indices. One or more of these indices are "normal indices". For each normal index, a, $a \models A \rightarrow B$ if and only if for every index in the frame b, if $b \models A$, then $b \models B$. For non-normal indices there are no satisfaction conditions for implicational formulas. Implicational formulas are satisfied (or not satisfied) by non-normal indices arbitrarily. Normal indices tell us the truth about entailments—if an implicational formula is true at a normal index, then every index is closed under that implication. Non-normal indices can be thought of as "logical fictions". They do not always tell us the truth about the laws of logic, just as science fictions do not always tell us the truth about the laws of physics.

Priest's semantics justifies the rejection of contraction. Adding contraction would be difficult and unnatural. But note that, on an informational understanding of the indices of his semantics, we would also be justified in rejecting the T-scheme. Suppose that n is a name for the proposition that A. A situation can contain the information that A without thereby containing the information that n is true. The latter information is information about a language (as well as being, perhaps, information about extra-linguistic reality). In order to contain the information that n is true, a situation need not only contain the information that A, but it also needs to contain information such as the information that n refers to the proposition that A. This information about our language is clearly distinct from the information that A. This means that we can have situations in which A but not in which $T(n)$. Given Priest's semantics for implication, it also means that the T-scheme is not true at any normal situation; in other words, the T-scheme is invalid.[11]

Having seen that there are informational semantics that justify the rejection of the T-scheme, it becomes plausible that there are non-trivial semantics that also justify the acceptance of contraction and allow for a fairly naïve theory of truth. If the T-scheme in part defines a naïve theory of truth, then it would seem that we might not be able to get a completely naïve theory of truth. But it would seem that we might be able to allow our language to have diagonalizing devices and accept the principle that if a contains the information that t refers to the proposition that A, then $a \models A$ if and only if $a \models T(t)$. This may not be a fully naïve theory of truth, but it may be naïve enough.

Of course, whether we can really have such a theory needs to be proven. We need a clear specification of a formal semantics and a proof that it is not trivial.

4.8 Information Fade

Fuzzy logic provides a fairly straight-forward treatment of sorites and other paradoxes, but there has been a lot of resistance to the acceptance of infinitely many truth values. For anyone who feels the pull of fuzzy logic, but is reluctant to accept its treatment of truth, an informational approach may provide just what he or she needs.

As I said in the section on Truth above, the notion of objective information is that of potential data contained in an environment. On one interpretation, what information is in an environment is relative to human capacities to extract it. Consider, for example, that you are looking at a sunset on a partly cloudy day. The colours in the clouds are a variety of yellows, oranges, browns, and greys. Some are easily identified. Some are not, because they are borderline cases.

There is *information fade* in situations like this. Some information, such as the fact that there is water in the foreground, is clearly present, and other information,

[11]Of course this is not an argument against Priest since he does not subscribe to an informational reading of his semantics. My point is only that on the informational approach to semantics it is not clear that we have to accept the T-scheme even if we have a naive theory of truth.

like the number of people living on the South Island, is clearly absent. But the information about the borderline colours is there to some degree. We could, I think, quantify the degree to which information is present in a situation, using values in the real interval [0, 1]. Formally, we can think of an abstract situation as (at least in part) a fuzzy set of states of affairs, some states of affairs a situation will contain completely, others it will exclude completely, and still others it will contain to some degree.

Then, we can use Priest's formal semantics for fuzzy relevant logic (Priest 2008, Chap. 11) to give information conditions for complex formulas. To see how this works, we first define an operation, $\ominus$. Where n and m are any two real numbers,

$$n \ominus m = \begin{cases} 1, \; if \; n \leq m \\ 1 - (n - m), \; if \; n > m \end{cases}$$

Now we can give the "information values" corresponding to the various connectives:

$$v_s(\neg A) = 1 - \max\{v_x(A) : \; Csx\}$$

$$v_s(A \wedge B) = \min\{v_s(A), v_s(B)\}$$

$$v_s(A \vee B) = \max\{v_s(A), v_s(B)\}$$

$$v_s(A \rightarrow B) = glb\{v_x(A) \ominus v_y(B) : \; Rsxy\}$$

Where R is the Routley-Meyer ternary accessibility relation on situations.

Now, I am not claiming that we *attribute* specific degrees of presence to states of affairs in situations. Rather, I am claiming that we can in our theorizing about situations attribute these degrees of presence relative to how well people can detect those states of affairs when in those situations. The people in those situations need not themselves be able to detect exactly the degree to which a state of affairs is present.

We can this notion of information fade to construct a theory of vagueness. First order vagueness is explained by information fade, that is, by our inability to tell perfectly whether a state of affairs is present. We find a similar view expressed by Stewart Shapiro in the following passage:

> It would defeat much of the purpose of colour-talk if we had to haul out a digital meter every time we need to establish or communicate the colour of something. And, of course, our powers of observation are limited. Our eyes, ears, noses, taste buds, and hands can discriminate only so much. Indeed, even the discriminatory abilities of digital meters are limited. ... this suggests a principle of tolerance for certain predicates that makes them prone to sorites arguments (Shapiro 2006, p. 194).

Shapiro and I agree that vagueness (at least in some cases) results from the imperfections of human sensory organs as detection devices. I think that the natural way to understand this is in informational terms.

Moreover, the inability of people in situations often to tell exactly how well they are detecting states of affairs indicates a way of treating higher order vagueness. Just as the our abilities for finding out about our environments in part determines the information in situations, our abilities to become aware of these first order abilities should in part determine which second order information is available in the situation.[12] Thus, although it might be that the state of affairs that a particular car is red is wholly in a situation (i.e. that state of affairs has the information value 1), it might be that the state of affairs that the car is definitely red gets a value less than 1 because our ability to tell that our detection of this state of affairs is less than perfect, especially if the colour of the car is, say, just a shade into the colour region in which we always discriminate as red.

4.9 Concluding Remarks

This paper has argued for the separation of the notions of truth and information and for the treatment of logic in terms of information. Validity of arguments should be thought of as information preservation rather than truth preservation. This informational turn makes sense of the semantics for relevant logic.[13] The informational interpretation is compatible with a classical theory of truth, and so it avoids the classicist's complaint that our "truth conditions" for the connectives are unintuitive. The informational interpretation also can help other non-classical logicians. Fuzzy logicians have long been the object of similar attacks from classicists. They too can adopt an informational interpretation of their logic and a classical theory of truth.

But the informational interpretation is not committed to being classical about truth. Just as it liberates logic from the tyranny of truth, it may liberate theories of truth from the annoying needs of logic. I have speculated here that an intuitive and fairly naïve theory of truth *may* be possible using the informational approach to logic. But whether this really works awaits a non-triviality proof.

Acknowledgements This paper benefitted from discussions with Rob Goldblatt, Greg Restall, Graham Priest, Joke Meheus, Franz Berto, Francesco Paoli, Francesco Orilia, Peter Schotch, Luciano Floridi, and Patrick Allo. This paper was funded by grant 05-VUW-079 of the Marsden Fund of the Royal Society of New Zealand.

References

Barwise, J. 1993. Constraints, channels, and the flow of information. In *Situation theory and its applications*, vol 3, ed. P. Aczel, Y. Katagiri, and S. Peters. Stanford: CSLI.

Barwise, J., and J. Perry. 1983. *Situations and attitudes*. Cambridge: MIT Press.

Barwise, J., and J. Perry. 1985. Shifting situations and shaken attitudes. *Linguisitics and Philosophy* 8: 105–161.

[12]Using 'second order' in the sense in which it is used in the vagueness literature, i.e. as in 'second order vagueness'.

[13]I have only looked at the treatment of negation here. For the other connectives, see Mares (2004, 2009, 2012, 2010).

Beall, J.C., and G. Restall. 2006. *Logical pluralism*. Oxford: Oxford University Press.
Dunn, J.M. 1993. Star and perp: Two treatments of negation. *Philosophical Perspectives* 7: 331–357.
Floridi, L. 2004. Information. In *The blackwell guide to philosophy of computing and information*, ed. L. Floridi. Oxford: Blackwell.
Goldblatt, R. 1974. Semantical analysis of orthologic. *Journal of Philosophical Logic* 3: 19–35.
Mares, E.D. 1996. Relevant logic and the theory of information. *Synthese* 109: 345–360.
Mares, E.D. 2004. *Relevant logic: A philosophical interpretation*. Cambridge: Cambridge University Press.
Mares, E.D. 2009. General information in relevant logic. *Synthese* 167: 343–362.
Mares, E.D. 2010. The nature of information: A relevant approach. *Synthese 175 Supplement*, 1: 111–132.
Mares, E.D. 2012. Relevance and conjunction. *Journal of Logic and Computation*, 22: 7–21.
Priest, G. 1992. What is a non-normal world? *Logique et analyse* 35: 291–302.
Priest, G. 2000. Motivations for paraconsistency: The slippery slope from classical logic to dialetheism. In *Frontiers of paraconsistent logic*, ed. D. Batens, C. Mortensen, G. Priest, and J.-P. van Bendigem. Baldock: Research Studies Press.
Priest, G. 2008. *Introduction to non-classical logic: From ifs to is*. Cambridge: Cambridge University Press.
Restall, G. 1995. Four-valued semantics for relevant logics (and some of their Rivals). *Journal of Philosophical Logic* 24: 139–160.
Restall, G. 1996. Information flow and relevant logic. In *Logic, language, and computation*, vol 1, ed. J. Seligman and D. Westerståhl. Stanford: CSLI.
Seligman, J., and L.S. Moss. 1997. Situation theory. In *Handbook of logic and language*, ed. J. van Benthem and A. ter Meulen. Amsterdam: Elsevier Science.
Shapiro, S. 2006. *Vagueness in context*. Cambridge: Cambridge University Press.

Chapter 5
Noisy vs. Merely Equivocal Logics

Patrick Allo

5.1 Introduction

To introduce the substructural pluralist's appeal to ambiguity, it suffices to describe in what precise sense C.I. Lewis's so-called independent argument for explosion can be rejected as a mere fallacy of equivocation. Following among others Read (1981), I reconstruct this argument in terms of the ambiguous natural language connective 'OR' in Fig. 5.1.

Give the assumption that such natural language connectives are ambiguous, while our rules of logic are (or at least should) be stated for unambiguous connectives only, the following diagnosis of the argument in Fig. 5.1 easily follows. First, for each individual argument step to be logically valid, it should be valid on at least one unambiguous reading (henceforth, *disambiguation*); second, for the whole argument to be logically valid, both steps should be valid on the same disambiguation. Unsurprisingly, when we fail to comply with the second demand, we commit a fallacy of equivocation. For many (broadly) relevant logicians (see e.g. Anderson and Belnap 1975; Read 1981; Paoli 2007), this is exactly what happens. Namely, whereas the first argument step is valid in virtue of $\sqcup$-introduction or addition for one possible disambiguation of 'OR', the latter is only valid in virtue of $\oplus$-elimination or disjunctive syllogism for another disambiguation of 'OR'. In Table 5.1, the two relevant senses of 'OR' are illustrated in terms of their introduction and elimination-rules.
In any system which lacks the structural rule of weakening $A \sqcup B$ and $A \oplus B$ are no longer equivalent. The lack of the structural rule of contraction additionally makes these two formulae entirely independent from each other (a description of these structural rules follows in the next section). As a consequence, the above proof

P. Allo (✉)
Postdoctoral Fellow of the Science Foundation (FWO), Centre for Logic and Philosophy of Science, Vrije Universiteit Brussel, Brussel, Belgium
e-mail: patrick.allo@vub.ac.be

K. Tanaka et al. (eds.), *Paraconsistency: Logic and Applications*, Logic, Epistemology, and the Unity of Science 26, DOI 10.1007/978-94-007-4438-7_5,

Fig. 5.1 Explosion

$$\dfrac{\dfrac{A}{A \text{ OR } B}\,(ADD^{\text{OR}}) \qquad \text{NOT}\text{–}A}{B}\,(DS^{\text{OR}})$$

Table 5.1 Rules for $\sqcup$ and $\oplus$

$\dfrac{A}{A \sqcup B}\ \dfrac{B}{A \sqcup B}\,(\sqcup I)$	$\dfrac{\begin{matrix}[A]^i \\ \vdots\,\pi_1 \\ C\end{matrix} \quad \begin{matrix}[B]^j \\ \vdots\,\pi_2 \\ C\end{matrix} \quad A \sqcup B}{C}\,(\sqcup E, i, j)$	$\dfrac{\begin{matrix}[\sim A]^i \\ \vdots\,\pi \\ B\end{matrix}}{A \oplus B} \quad \dfrac{\begin{matrix}[\sim B]^i \\ \vdots\,\pi \\ A\end{matrix}}{A \oplus B}\,(\oplus I, i)$	$\dfrac{A \quad \sim A \oplus B}{B}\,(\oplus E)$

for *explosion* is invalid in weakening-free systems, and *a fortiori* also in systems which are both weakening and contraction-free (I shall, following Došen 1993, refer to such systems as *substructural*). If, in addition to this mere formal point, such substructural logics yield the right logical and therefore unambiguous analysis of our natural language connectives, the classical proof of explosion is not only invalid in certain subclassical systems, but also a fallacy of equivocation in a more robust 'system-independent' sense.

The move from invalid in a logical system to fallacious (or invalid *simpliciter*) should rest on a more substantial argument. This is, however, not the place to give such an argument. Even then, the central role of ambiguous connectives in the substructural pluralist's diagnosis of *explosion* as a fallacy of equivocation should, on its own and independently of the specific challenges it faces, be a sufficient reason for an in-depth inquiry into the nature of ambiguous connectives. Concretely, this amounts to the study of the logical properties of natural language connectives as a specific kind of ambiguous connectives. The latter suggestion that ambiguity could be an object of logical inquiry does not sit well with logical orthodoxy. While this is a point that is usually made with regard to ambiguities in the non-logical vocabulary, Lewis's contention that "[t]he recommended remedy [against triviality] is to make sure that everything is fully disambiguated before one applies the methods of logic." (Lewis 1982, p. 439) generalises to ambiguities in the logical vocabulary. Hence, the ensuing view that only pessimists would have to worry about ambiguity remains particularly compelling.

On a more general level, this broad distrust towards ambiguous connectives bears on two distinct objections.

Objection 1. Ambiguous connectives are not a topic for logical theorising; disambiguation as well as other ways of dealing with ambiguity always ought to precede the application of logical methods.

Objection 2. Even if ambiguous connectives are amenable to logical inquiry, they do not count as genuine logical connectives and therefore do not belong to logic proper.

To answer the first objection, I will establish the following conditional claim: if the substantial view about explosion being a mere fallacy of equivocation is correct, then there should be a logic for ambiguous connectives. A defence of this conditional claim roughly amounts to the view that the substructural pluralist should be a pessimist about the disambiguation of natural language connectives. As a reply to the second objection, I will show that there is indeed such a logic, that it is strictly (and substantially) stronger than the empty logic, and that its connectives obey some common criteria for logicality. The first issue is tackled in the next section, the remainder of this paper is devoted to the second issue.

5.2 Why Care About Ambiguous Connectives?

To a first approximation, the argument for a sub-classical logic of ambiguous connectives rests on the fact that we have at least one good reason to be pessimistic about getting the disambiguation of the natural language connectives in our premises systematically right, and that we consequently need a logic to reason in contexts in which we are stuck with recalcitrant ambiguous premises. To get the details of this argument right, we need the following premises and assumptions.

(I) First assumption:
Some substructural logic is either (1) a *prima facie* candidate for being the correct logic, or (2) at least a logic properly called so. All else being equal, this logic is to be preferred above classical logic.

(II) Second assumption:
At least one natural language connective is ambiguous in the sense that some natural language occurrences of a given connective exhibit the logical features of a first class of unambiguous connectives, while other occurrences of that connective exhibit the logical features of a second class of unambiguous connectives (but no occurrence exhibits the logical features of both classes).

(III) Third assumption:
The connectives of a substructural logic provide all the resources we need for, at least in principle, correctly disambiguating all natural language connectives. We refer to $\sqcup$, $\sqcap$, and $\rightsquigarrow$ as, respectively, the lattice disjunction, conjunction, and implication; and to $\oplus$, $\otimes$, and $\rightarrow$ as, respectively, the group-theoretical disjunction, conjunction, and implication (see Table 5.2 for the rules for the remaining connectives).[1]

[1]Unfortunately, due to the usage of Prawitz-style ND, the introduction-rule for the lattice-conjunction is based on a not so natural way of expressing the *shared context* used to derive both conjuncts. In particular, one should be attentive to the fact that both occurrences of $[A]^i$ in $(\sqcap I)$ count as a single assumption.

Also, rather than to specify the rules for the negation, we only need to note that we use a De Morgan negation which satisfies $\sim\sim A \dashv\vdash A$, as well as all usual De Morgan equivalences for both the lattice and the group-theoretical connectives.

Table 5.2 Rules for ⊓, ⊗, ⇝, and →

$$\dfrac{\begin{matrix}[A]^i \\ \vdots\,\pi_1 \\ B\end{matrix} \quad \begin{matrix}[A]^i \\ \vdots\,\pi_2 \\ C\end{matrix} \quad A}{B \sqcap C}\,(\sqcap I, i) \qquad \dfrac{A \sqcap B}{A} \quad \dfrac{A \sqcap B}{B}\,(\sqcap E) \qquad \dfrac{A \quad B}{A \otimes B}\,(\otimes I) \qquad \dfrac{\begin{matrix}[A]^i[B]^j \\ \vdots\,\pi \\ C\end{matrix} \quad A \otimes B}{C}\,(\otimes E, i, j)$$

$$\dfrac{\sim A}{A \rightsquigarrow B} \quad \dfrac{B}{A \rightsquigarrow B}\,(\rightsquigarrow I) \qquad \dfrac{\begin{matrix}[\sim A]^i \\ \vdots\,\pi_1 \\ C\end{matrix} \quad \begin{matrix}[B]^j \\ \vdots\,\pi_2 \\ C\end{matrix} \quad A \rightsquigarrow B}{C}\,(\rightsquigarrow E, i, j) \qquad \dfrac{\begin{matrix}[A]^i \\ \vdots\,\pi \\ B\end{matrix}}{A \to B}\,(\to I, i) \qquad \dfrac{A \quad A \to B}{B}\,(\to E)$$

$$\dfrac{\dfrac{\dfrac{A \oplus B \qquad \dfrac{\dfrac{[\sim (A \sqcup B)]^1}{\sim A \sqcap \sim B}\,(ND^{\sqcup})}{\sim A}\,(\sqcap E)}{B}\,(\oplus E) \qquad \dfrac{\dfrac{[\sim (A \sqcup B)]^2}{\sim A \sqcap \sim B}\,(ND^{\sqcup})}{\sim B}\,(\sqcap E)}{\bot}\,(\sim E)}{A \sqcup B}\,(RAA, 1, 2)$$

Fig. 5.2 Proof of $A \oplus B \vdash A \sqcup B$

Taken together, these three assumptions serve as a characterisation of a generic relevantist paraconsistent position about logical consequence. For starters, the first assumption gets the basics of relevance right since the rejection of the vacuous discharge of assumptions (i.e. *not* effectively using an assumption, which is how the structural rule of *weakening* appears in a simple ND-setting) is instrumental to invalidate $p \to (q \to p)$ and other paradoxes of material and strict implication. In principle, the rejection of multiple discharge (which, in its turn, corresponds to the structural rule of *contraction*) is not required for the purpose of relevance; for instance, the logic of relevant implication **R** requires this structural rule in order to prove $p \sqcup \sim p$. The present omission of contraction is primarily motivated by the need to keep my argument as general as possible by preventing both presumed unambiguous classes of connectives from interfering (I shall, however, briefly return to this issue at a later point). As illustrated in Fig. 5.2 the multiple discharge of $\sim (A \sqcup B)$ (the assumption used for the *reductio*) suffices to prove $A \oplus B \vdash A \sqcup B$. Similarly, contraction also validates $A \to B \vdash A \rightsquigarrow B$, and $A \sqcap B \vdash A \otimes B$.

As one surely should like to avoid the potentially misleading effect of such logical connections, I opt for a system without weakening or contraction; that is, a system where each assumption must be used exactly once. Consequently, my starting point turns out to be closer to that of Paoli (2007), than to that of Anderson and Belnap, Read, and Mares which all adopt the logic **R** as their preferred system.

Of the three above assumptions, the third one is by far the most controversial. Or better, if this third assumption withstands further scrutiny, support for the other two comes almost for free. Namely, to be plausible at all, any defence of relevant logic needs to reconcile the rejection of the *disjunctive syllogism* with a common-sense practice of deductive reasoning that incorporates the exclusion of cases or possibilities. Solving this problem is crucial for the acceptability of the first assumption. On the substructural pluralist's account, the validity of ($\oplus E$) referred to above does just that: it sanctions the instances of (DS^{OR}) we wish to retain. Such a solution recaptures the *disjunctive syllogism* as a valid rather than as a merely *enthymematic* inference form. This is a clear benefit. What the third assumption claims is precisely that this way of saving the disjunctive syllogism is basically on the right track. As such, the third assumption provides incomplete but crucial support for the first one. Also, since the third assumption makes a claim about unambiguous connectives, it already presupposes the ambiguity of natural language connectives posited in the second one.

Provided that I only want to defend the conditional claim that if substructural logics give the right analysis of unambiguous connectives, an account of ambiguous connectives is also required, then this third assumption may be granted for the sake of argument. Still, there is one aspect of the demarcation of lattice and group connectives to which we should pay special attention. This aspect is explicitly exposed by Burgess. He suggests that any differentiation of $A \sqcup B$ and $A \oplus B$ which refers to the presence of an epistemic connection between A and B in the latter case, and the lack of such a connection in the former case, unavoidably reduces "the whole relevantistic movement as a simple case of confusion between the logical notion of implication and the methodological notion of inference." (Burgess 1981, p. 103). The main target, here, is an analysis of having a ground for asserting $A \sqcup B$ which presupposes knowing either A or B as opposed to having a ground for asserting $A \oplus B$ which would then depend on one's ignorance about which disjunct might be true. While the remark that one should not confuse inference with implication is usually meant to imply that no sense can be made at all from the distinction between lattice and group-theoretical connectives, it should now be seen as a restriction on the third assumption. That is, whatever ultimately grounds the difference between lattice and group connectives, it cannot be a mere epistemic feature and can therefore not be reduced to the properties of a reasoner's premise-set. Or, when put in positive terms, the 'correct disambiguation' of a natural language connective is a knowledge independent feature of the premise it occurs in, in the sense that it is a feature which does not supervene on the logical properties of the premise-set of the assertor (let alone, on a hearer's knowledge of the assertor's premises). Thus refined, the following corollary is an immediate consequence of the third assumption.

(IV) Corollary:
It is possible to be ignorant about the correct disambiguation of one's premises in the sense that there is no method to systematically and fully reliably derive the correct disambiguation of a premise from the totality of one's premises.

For now, we need to leave the precise scope of this possibility of having recalcitrant ambiguous premises rather open-ended. The only thing we might want to stress is that (IV) does not imply that a natural language connective automatically renders any premise in which it occurs irreducibly ambiguous. All it states is that when we have a ground for asserting a natural language premise, we are not always in a position to obtain a ground for asserting one precise disambiguation of that premise, and this even though its correct disambiguation may itself be a determinate matter. To get a grip on the implications of this corollary, it is advisable to treat pluralist and monist attitudes towards logic separately.

(Va) Monist Assumption:
Logic allows one to reason from one's premises without being led from truth to falsehood 'come what may.'

(Vb) Pluralist Assumption:
A logic properly called so allows one to reason from one's premises without (1) being led from truth to falsehood in a particular context, or (2) breaching the contextually operative norms for deductive inference.

Roughly, these assumptions are two ways of connecting logic with its canonical application, Viz. deductive reasoning. While the above assumptions are very crude in comparison to the complex issues we face when we try to explain the relation between logic and reasoning, (Va) and (Vb) are, at least as partial principles, good enough for present purposes. According to the monist, this relation should be conceived in the most straightforward and general way. There is a single norm for deductive reasoning, namely that no falsehood should be deducible from truths, and this norm should not be violated in any (logically possible) context or situation. This is the view defended in Priest (2001). According to the pluralist, that relation should be more flexible; with room for different contexts (Batens 1997; da Costa 1997) and different norms (Beall and Restall 2006). The possibility of being in a context where some premises are irreducibly ambiguous together with the need to reason deductively in these contexts, then leads to the following conclusions.

(VI) First Conclusion:
A system which allows one to reason from ambiguous premises is, both by monist and pluralist standards, a better or at least an equally good candidate for being (a) a *prima facie* candidate for being the correct logic, or (b) a logic properly called so.

As logical orthodoxy has it that ambiguity does not sit well with the canons of logic (see Haack 1974, pp. 116–125 and Lewis 1982), this first conclusion could still be disputed. Perhaps I simply forgot to dismiss all contexts in which one would have to deal with ambiguity, and thus misunderstood the monist or pluralist connection between logic and deductive reasoning. This kind of objection does not pose a real threat. If logics for vagueness can become respectable, there is no reason why we couldn't investigate our deductive resources relative to the irreducibly ambiguous. But also, when ambiguity is related to the functioning of our most basic logical connectives, its study should belong to logic proper.

(VII) Second Conclusion:
Classical logic allows one to reason from ambiguous premises, but in doing so it commits a fallacy of equivocation. A classical retreat is therefore (but also in view of our first assumption) out of question. Hence, a logic which

does not conflate different connectives even when confronted with premises containing ambiguous connectives remains to be given.

With these two conclusions, I have now established the required conditional thesis that once we accept that natural language connectives are ambiguous between a lattice and group-theoretical reading, we cannot but supplement that story with an account of (non-classical) ambiguous connectives. More explicitly, I have now shown that, whether a pluralist or a monist, a relevantist who endorses a solution for the problem of the disjunctive syllogism based on $(\oplus E)$ cannot easily dismiss a logic of ambiguous connectives. Of course, this only carries any weight if there is such a logic, and if that logic contains more than, say, the law of identity. In the remainder of this paper, such a logic will be rigorously characterised. What the present argument shows is that (all else being equal) whatever that logic might be, the pluralist will have to add it to her stock of perfectly good logics, whereas the monist will have to revise or at least supplement (in the sense of Haack (1974)) his logic such as to incorporate the ability to reason from ambiguous premises.

5.3 Classical Logic as a Noisy Logic

One of Burgess's objections to the claim that all acceptable instances of the disjunctive syllogism are valid in virtue of $(\oplus E)$ is based on a scenario in which the premises are distributed between different agents.[2] By giving an informational twist to a similar distributed setting, I replace the original diagnosis of explosion as a fallacy of equivocation by a new one: "classical logic is a noisy logic." A detailed example will clarify this bold claim, but the following simple consideration already sheds some light on the issue. By making assertions using natural language connectives, one shares information about the grounds one has to make that assertion (Prawitz 2006). Yet, since natural language connectives are ambiguous, one cannot convey precise information about one's grounds (and the inferences one had to rely on). Sometimes this might be due to our own ignorance about the precise grounds we have, but in many cases this is merely due to the language we use. In the latter case, we might say that our communication about the grounds we have for making assertions is unreliable: there is information available to the assertor which is no longer available to the receiver.

As a constitutive part of the channel over which we convey information about our grounds for assertion, natural language connectives are unreliable. Information-theory tells us that channels can be unreliable in two ways: they can be *equivocal*, or they can be *noisy*. In the former case, content available at the source is no longer available at the destination. In the latter case, some content available at the

[2]Basically, a situation where a first agent uses addition to obtain a disjunctive formula which he then passes on to a second agent who uses the disjunctive syllogism to recover the original disjunct the first agent started from, see Burgess (1981).

destination is not available at the source: it is a mere artefact of the channel. When combined with a classical consequence-relation, natural language connectives lead to such noisy communication. By contrast, the here intended logic of ambiguous connectives is meant to capture the communication about an agent's grounds for assertion which is merely equivocal. To establish the diagnosis that classical logic is noisy (relative to a predefined substructural logic) more firmly, I need more than the above considerations about grounds for assertion. I have to be able to characterise the informational content of an assertion relative to a logic or formal consequence relation.

Let **L** be a logic. Define the notion of **L**-information at a point (the sender/source or the receiver/destination) as the set of all (non-**L**-tautological) **L**-consequences of the messages (premises) available at that point. Next, consider the following setting. Agent_1 has a set Γ of unambiguous premises which he communicates to Agent_2 in a language containing only ambiguous (natural language) connectives. Since Γ contains unambiguous premises, the **L**-information at the source should be determined relative to the correct substructural logic. For obvious reasons, a similar strategy cannot be used to compute the body of **L**-information that is available at the destination: a substructural logic does not say anything about what follows from ambiguous premises. Classical logic seems perfectly suited for this task, but does not yet enable a direct comparison of the substructural **L**-information at the source and the classical **L**-information at the destination.

To be fully accurate, I should now discriminate between a classical notion of $\mathbf{L_c}$-information and a substructural notion of $\mathbf{L_s}$-information, and then come up with a method for comparing them. While this is surely feasible, I do not require this degree of sophistication right now. The only fact I need to consider is that a formula of the classical language classically entails any of its disambiguations in the substructural language, whereas a formula of the substructural language does not substructurally entail its ambiguous counterpart of the classical language (the latter point can be made explicit in a linear logic with exponentials, but for now the intuitive point is all I need). As such, there is a clear sense in which the classical **L**-information at the destination might exceed the substructural **L**-information at the source. This is a phenomenon that can only be diagnosed as communication over a noisy channel.[3]

Since noise is generally considered an unwanted artefact of unreliable communication, the conclusion that classical logic leads to such noise might be read as a knock-down argument against classical logic. This is perhaps a bit too fast. Classical logic is indeed noisy, but a proponent of classical logic (but also pluralists like Beall and Restall) might still reply that since his preferred logic could never lead one from truth to falsehood, the noise in question is entirely harmless. Apart from the denial

[3]The main complication I ignore here bears on the fact that the **L**-information of a message is defined as the *non-tautological* deductive yield of that message. Since classical tautologies can have a tautological and a non-tautological disambiguation, the diagnosis should in fact be that the channel is both equivocal *and* noisy (cfr. the 'p OR NOT–p' example from Allo (2007)). This more refined diagnosis does not affect my contention that the channel is not merely equivocal, but also noisy.

that classical consequence is not even in a very weak sense truth-preserving, the only remaining option is to explain in what sense truthful noise might after all be harmful. Despite the appearances, this isn't a trivial task; sheer appeals to relevance are just not *that* helpful. To see what is ultimately at stake, I should note that while it makes sense to prefer truthful non-noisy communication to equally truthful noisy communication, it is much less obvious to say that such truthful noise will somehow harm one's epistemic position. As the 'relevant' in relevant logic does not refer to what is actually useful, truthful noisy communication is not the best we can have, but also not necessarily something we should try to avoid. Even though this classical objection rests on the controversial assumption that CL is truth-preserving while relevant logic only adds a relevant component (see Read 1988 *contra*, and Beall and Restall 2006 in favour of this assumption), this objection still calls for a further explanation of the harm caused by veridical but noisy content.

One way to expose the problem with noisy communication takes us back to considerations about grounds for assertion. In general, we could say that when Agent_1 communicates A to Agent_2, then Agent_2's own ground for asserting A is, in the first place, the fact that Agent_1 also asserted A. Even though Agent_2 *could* have independent grounds for asserting A, the deference to Agent_1's original assertion is a sufficient ground for that assertion. As a consequence, Agent_2's minimal ground for asserting A can be reduced to what Agent_2 knows about Agent_1's original grounds for A.

By specifying that a hearer's grounds for asserting p (for some p asserted by someone else) could in principle be reduced to the initial assertor's grounds, and thus be as weak as the hearer's knowledge of those grounds, I actually enforce—or rather posit—a form of transitivity that is akin to Dretske's *Xerox-principle* (Dretske 1999). While this assumption seems innocuous in view of Tarskian orthodoxy about logical consequence as applied to the transfer of grounds for assertion (compare with Mares 1997, § 5.1 for a related point), this very assumption only strengthens the previously mentioned objection of Burgess that some instances of the disjunctive syllogism cannot be valid in virtue of $\oplus$-Elim. The example he advances involves someone who first deduces $p \sqcup q$ from p, and then asserts p OR q, but where the hearer deduces for himself that since NOT$-p$, it must be that q. So presented, it should be clear that such an inference can only be explained as an instance of $\oplus$-Elim if the hearer actually accepts $p \oplus q$. As a consequence, a relevantist cannot explain this inference unless he concedes that the hearer's ground for p OR q exceeds his knowledge of the assertor's grounds for p OR q (this is roughly how I understand the reply given in Paoli (2007) to the challenge posed by Burgess). Upon reflection, the application of $\oplus$-Elim in such situations with distributed premises cannot but conflict with a Xerox-like principle, and the most plausible way out is just to concede that, at least in these cases, transitivity breaks down.[4]

[4]Two remarks: (1) Such failures of transitivity are, as far as I can see, only superficially related to what we find in Tennant (2004); (2) As remarked by Elia Zardini (pc), the suggestion that explosion is a fallacy of equivocation is entirely compatible with the suggestion that transitivity is the real

This concession need not be a problem. All I need to sustain is that the formulation of a merely equivocal logic for ambiguous connectives presupposes the just described transfer of grounds for assertion. Indeed, the presumption of transitivity is itself not affected by the sheer possibility of a hearer who might accept a premise which—strictly speaking—would exceed his knowledge of the original assertor's grounds. As a result, the solution for the problem of the disjunctive syllogism is logical in that it treats $\oplus$-Elim as a valid inference, but it has to leave room for a pragmatic explanation of why the hearer ultimately accepts $p \oplus q$ as a premise.

Noisy communication is, assuming a Xerox-like principle, problematic in the sense that it can lead to Agent_2's assertion of a classical consequence of A such that: (a) Agent_1 has no ground for asserting that consequence, and (b) Agent_2 has no grounds for asserting that consequence save for Agent_1's original assertion of A. So presented, we can see how Agent_2 can still be mistaken in asserting a classical consequence of A which not only happens to be true, but could (by classical standards) not even have been false. Admittedly, this might not convince our classical logician, for he would simply hold that Agent_1 was already in a position to assert that consequence. What this argument does show is that a diagnosis of classical logic as a noisy logic duly reconstructs the relevantists objection to classical logic (in particular the view that natural language connectives can have the logical properties of both lattice and group connectives, but not simultaneously; and the pluralist account of how classical and relevant logic disagree). For now, this is all I need.

While there is a sense in which this informational twist does not substantially improve our insight beyond that of the initial considerations about the grounds one might have for asserting a certain formula, the proposed informational outlook on logic and in particular the usage of the noise and equivocation terminology is not just a rhetorical move. There is a well-defined sense in which natural language connectives *encode* the more refined connectives of the language of substructural logic; and what I have just shown is that this *coding* is not entirely reliable. This phenomenon is even in the strictest sense what information-theory is about. More importantly, an informational perspective also reveals what a logic of ambiguous connectives might look like, and it even helps to individuate it. Put simply: a logic of ambiguous connectives which is free of fallacies of equivocation should lead to a channel which is—as opposed to the channel I identified with a classical consequence relation—merely equivocal. More precisely, such a logic should account for a receiver's ignorance about a sender's precise grounds for assertion.

---

culprit. As a consequence, the proposal set out in the next section can only be motivated if the *Xerox-principle* is already presupposed.

5.4 What Follows from Ambiguous Premises?

The basic idea for a logic of ambiguous connectives is both simple and powerful. Its functioning can be explained in fairly informal terms, and it is, at least at first blush, not unlike a supervaluationist approach to vagueness (Fine 1975). On an informal level, the idea is that one should always keep track of the deductive yield of every possible way to disambiguate one's premises, in the same sense as the deductive yield of vague terms is relative to all its refinements. I prefer not to exploit the analogy with vagueness at this point, but will instead rely on basic information-theoretical considerations to show that this method captures the intended merely equivocal interpretation.

Basic Idea:
Given a set of ambiguous premises Γ, one can (without making any further assumption about that set) deduce an ambiguous formula A from this set iff *all* possible disambiguations of Γ entail *some* disambiguation of A.

As a specification of a derived consequence-relation, this basic idea is incomplete. It only tells us how to obtain a consequence-relation over an ambiguous language from a previously defined entailment-relation over an unambiguous language. Previous assumptions already fix this entailment-relation, and as such this incompleteness does not pose a problem. Above all, it is the way we quantify over possible disambiguations that is crucial to the functioning of this principle, and an argument for its tenability therefore reduces to an argument in favour of a particular choice of quantifiers. For now, I only need to show that the *all/some* quantifier-combination that is introduced in the basic idea duly captures the intended merely equivocal reading of ambiguous connectives. The quantification over *all* possible disambiguations of the premises is hardly objectionable. A universal quantification is perfectly in line with the presumed ignorance about the correct disambiguation, and the need to check separately each possibility which derives from that ignorance. Given these considerations about one's ignorance about the correct disambiguation, the *some*-clause in the second half of the principle is more surprising.

An initial consideration which makes this choice more palpable is derived from an informational reading of logical consequence which—when applied to the present problem—states that A is an ambiguous consequence of Γ iff A does not add any information that is not already contained in Γ (see Corcoran 1998 for a detailed analysis of this information-theoretic reading of logical consequence). When combined with the commonly used link between information and surprise-value, it then follows that A is an ambiguous consequence of Γ iff after having received all of Γ, it would not be surprising to learn that the sender could also send A. Yet, a sender can do so in virtue of a single disambiguation of A.

To adhere to the *all/all* version is, in other words, putting the threshold for what is surprising too low. And since being surprised when one shouldn't is a kind of equivocation, while not being surprised when one should have is the result of noise, it follows that if the *all/some* combination is merely equivocal, the *all/all* version is exceedingly equivocal. This concludes the informal defence of the basic idea beyond a logic of ambiguous connectives (but see van Deemter 1996, §§ 5.2 and 5.6, for a comparison of all four quantifier combinations as well as further refinements of these combinations).

A precise characterisation of a logic of ambiguous connectives requires a translation-function which maps each unambiguous formula on its ambiguous counterpart. Where FORM and $\mathsf{FORM}^{\mathsf{Amb}}$ are the sets of all wffs respectively using the lattice and group connectives, or the ambiguous natural language connectives, *tr*: FORM $\mapsto$ $\mathsf{FORM}^{\mathsf{Amb}}$ is recursively defined by:

$$tr(p) = p$$
$$tr(\sim A) = \text{NOT–}tr(A)$$
$$tr(A \sqcap B) = tr(A \otimes B) = tr(A) \text{ AND } tr(B)$$
$$tr(A \sqcup B) = tr(A \oplus B) = tr(A) \text{ OR } tr(B)$$
$$tr(A \rightsquigarrow B) = tr(A \rightarrow B) = tr(A) \text{ IMPLIES } tr(B)$$

More generally, where $\bigstar$ is an ambiguous n-ary connective which can encode a finite number of unambiguous $\bigstar_i$, we have:

$$tr(\bigstar_i A_1, \ldots, A_n) = \bigstar tr(A_1), \ldots, tr(A_n) \qquad \text{(tr)}$$

This translation-function captures the weakest sense in which ambiguous connectives may be said to encode unambiguous ones. For notational convenience, $\star$ will henceforth stand for a generic ambiguous binary rather than an n-ary connective, and I will write $tr(\Gamma)$ as shorthand for the set of ambiguous wffs obtained by translating all wffs in Γ.

One should also keep in mind that whereas the $\star$-formulation of (tr) remains neutral in several respects, the $\{\sqcup, \sqcap, \rightsquigarrow, \oplus, \otimes, \rightarrow\}$ version I started from is more specific in that it enumerates the connectives that are actually ambiguous, and specifies what their disambiguations are. It is important to note that the basic analysis of ambiguous connectives only depends on the $\star$-version. By contrast, since the latter fails to generate a unique logic of ambiguous connectives, the actual proposal considered in the present paper is based on the former enumeration of the connectives.

When applied to a substructural entailment-relation, the basic idea for a logic of ambiguous connectives can, by using (tr), be formulated as:

Definition 5.1 (All / Some Interpretation: $\vdash_{\forall/\exists}$). Where $\vdash_{\forall/\exists}$ is a relation between sets of ambiguous formulae and a single ambiguous formula, we write $\Gamma \vdash_{\forall/\exists} A$ iff for each smallest $\Gamma_i \subset$ FORM such that $tr(\Gamma_i) = \Gamma$ there is an A_i in $\{B \in \mathsf{FORM} : B = tr(A)\}$ such that $\Gamma_i \vdash A_i$.

This definition precisely captures the intended consequence-relation, and does so by proposing a set of (substructurally) valid entailments which act as necessary

$$\frac{tr^{-1}(A)}{A}\,(tr\ I) \qquad \frac{\begin{matrix}[A_1]^1 & [A_i]^i & [A_n]^n & \\ \vdots\,\pi_1 & \vdots\,\pi_i & \vdots\,\pi_n & \\ B_1 & B_i & B_n & tr(A_i)\end{matrix}}{B}\,(tr\ E, 1\ldots n)$$

Fig. 5.3 tr-Intro and tr-Elim

Fig. 5.4 Rules for $\star$

$$\frac{(p \star_1 q) \sqcup (p \star_2 q)}{p \star q}\,(\star I) \qquad \frac{p \star q}{(p \star_1 q) \sqcup (p \star_2 q)}\,(\star E)$$

Fig. 5.5 Analysis of tr-Intro

$$\dfrac{\dfrac{p \star_i q \quad i \in \{1, 2\}}{(p \star_1 q) \sqcup (p \star_2 q)}\,(\sqcup I)}{p \star q}\,(\star I)$$

and jointly sufficient conditions for a valid argument from ambiguous premises to an ambiguous conclusion. A further step is to capture the meaning of ambiguous connectives in a more direct fashion. To that end, it is advisable to consider the translation of unambiguous formulae into ambiguous formulae as a kind of introduction-rule, and an instance of Definition 5.1 (with $\Gamma_i = \{A_i\}$) as a kind of elimination-rule (Fig. 5.3).

Thus, the introduction-rule states that to have a ground for asserting an ambiguous formula A is to have a ground for asserting at least one disambiguation of A, and the elimination-rule states that when we have a ground for asserting an ambiguous formula $A = tr(A_1) = \ldots = tr(A_i) = \ldots = tr(A_n)$, we also have a ground for asserting another ambiguous formula B if no matter how we disambiguate A, we can derive some B_i such that $tr(B_i) = B$. Given this basic story about what it takes to make ambiguous assertions, an appeal to proof-theoretical harmony suffices to show that the meaning of ambiguous connectives are (from an inferential perspective) coherently defined.

Following Read, an intro/elim-pair is *harmonious* iff where "$\{\Pi_i\}$ denotes the grounds for introducing a formula A (introducing an occurrence of a connective δ in A), then the elimination-rule for δ should permit inference to an arbitrary formula C only if $\{\Pi_i\}$ themselves entail C." (Read 2000, p. 130). Thus, since we have that $B = tr(B_1) = \ldots = tr(B_i) = \ldots = tr(B_n)$, the above pair is in perfect agreement with this criterion.

Given the overt similarity with the standard rules for the disjunction, tr-Intro can be considered as a form of *addition*: only one disambiguation of A is required to produce the ambiguous formula A; and tr-Elim can be considered as a an instance of *proof by cases*: each disambiguation should be checked separately. Consider now the rules given in Fig. 5.4 for the generic ambiguous binary connective $\star$.

Using these, we can unpack the functioning of tr-Intro as illustrated in Fig. 5.5, and the functioning of tr-Elim as illustrated in Fig. 5.6.

$$\dfrac{\dfrac{\begin{array}{c}[p \star_1 q]^1 \\ \vdots\, \pi_1 \\ tr^{-1}(A)\end{array}}{A}\,(tr\ I) \quad \dfrac{\begin{array}{c}[p \star_2 q]^2 \\ \vdots\, \pi_2 \\ tr^{-1}(A)\end{array}}{A}\,(tr\ I) \quad \dfrac{p \star q}{(p \star_1 q) \sqcup (p \star_2 q)}\,(\star E)}{A}\,(\sqcup\, E, 1, 2)$$

Fig. 5.6 Analysis of tr-Elim

A first consequence of this ability to capture ambiguous formulae as the lattice-disjunction of their disambiguations is that from the indirect definition of ambiguous consequence we can derive a more direct version:

Proposition 5.1. *$\Gamma \vdash_{\forall/\exists} A$ iff $\Gamma_1^{\otimes} \sqcup \ldots \sqcup \Gamma_n^{\otimes} \vdash A_1 \sqcup \ldots \sqcup A_m$ where $\Gamma_i^{\otimes}$ is the closure of Γ_i under $\otimes$, and $\{\Gamma_1, \ldots \Gamma_n\}$, $\{A_1, \ldots, A_m\}$ are all A_i and smallest Γ_i such that $tr(A_i) = A$ and $tr(\Gamma_i) = \Gamma$.*

By distributing of $\otimes$ over $\sqcup$, this can be rewritten as:

Proposition 5.2. *$\Gamma \vdash_{\forall/\exists} A$ iff $\bigsqcup(\mathscr{B}_1) \otimes \ldots \otimes \bigsqcup(\mathscr{B}_n) \vdash A_1 \sqcup \ldots \sqcup A_m$ where $\Gamma = \{B_1, \ldots, B_n\}$ and $\mathscr{B}_i$ is the set of all C such that $tr(C) = B_i$.*

The former considerations suggest that the correct analysis of ambiguous connectives is a disjunctive one, and I shall indeed refer to the present proposal as a disjunctive analysis of ambiguous connectives. But there is also something misleading in the use of that terminology. To be precise, Propositions 5.1 and 5.2 give, respectively, a disjunctive analysis of *sets of ambiguous premises*, and a disjunctive analysis of *ambiguous premises*. Crucially, the latter only amounts to a direct disjunctive analysis of *ambiguous connectives* when premises have the form $A \star B$ with A and B literals or otherwise unambiguous formulae. Here, but also in the proposal from van Eijck and Jaspars (1996), the fact that we may interpret $p \star q$ as $(p \star_1 q) \sqcup (p \star_2 q)$ does not guarantee that we may substitute any ambiguous subformula by the disjunction of its disambiguations. For instance, NOT–(p AND q) should be unpacked as $(\sim (p \otimes q)) \sqcup (\sim (p \sqcap q))$, instead of $\sim ((p \otimes q) \sqcup (p \sqcap q))$. This is because the negation is intended to interact directly with the conjunction rather than with the disjunction used to analyse the ambiguous conjunction.

In view of the recursive definition of (tr) this is unavoidable: tr gives a step by step procedure for turning unambiguous formulae into ambiguous formulae, but the result of that procedure cannot be achieved by repeated applications of ($\sqcup$ I). When A contains $n > 1$ connectives, $tr(A)$ cannot be identified with the result of n applications of an addition-like rule as an introduction-rule for ambiguous connectives, but only with a single application of such a rule which then serves as an introduction-rule for an ambiguous expression. As a consequence, any calculus intended to deal with ambiguous connectives should have introduction and elimination-rules for ambiguous expressions—*not* for ambiguous connectives.

Even if the last proviso precludes the definability of ambiguous connectives in the language of substructural logic, it does not affect the possibility of defining ambiguous expressions in that language. This is still largely sufficient to ensure that the result of adding these new connectives to the original language yields a conservative extension of the substructural logic. A consequence of the conservativeness is that it becomes quite hard to deny that ambiguous connectives are genuine logical connectives; a consequence of the definability of ambiguous expressions is that when the argument set out earlier in this paper forces a monist to supplement his preferred logic with a set of ambiguous connectives, this particular move turns out to be entirely cost-free.

It is worthwhile to contrast the disjunctive analysis of ambiguity with the analysis of propositional ambiguity given in van Eijck and Jaspars (1996) as well as with Fine's analysis of ambiguity based on his supervaluationist take on vagueness as a form of hyper-ambiguity.

A first point of disagreement derives from Fine's rejection of a disjunctive and/or conjunctive analysis of ambiguity (Fine 1975) fn. 9. As he has it, asserting an ambiguous expression reduces to the multiple assertion of all of its disambiguations. Since such an expression is true (resp. false) if each of its disambiguations are true (resp. false), neither a disjunctive or a conjunctive analysis can recover these truth-conditions. Irrespectively of the relative merits of this proposal as an analysis of ambiguous non-logical terms, we should still reject it as an analysis of ambiguity at the level of the logical vocabulary. When restricted to logical connectives, the target sense of ambiguity is one which lets particular occurrences of natural language connectives have some of their classical properties, but just not all at once. In particular, the substructural pluralist's appeal to ambiguity is not such that it only recognises uses of ambiguous connectives as truthful if asserted on classical grounds; but this is exactly what the multiple assertion account leads to. This shows the truth-conditions stipulated by Fine cannot be used to elucidate what is meant by the claim that, for instance, 'OR' is ambiguous between its lattice and group-theoretical reading.

This reply largely generalises to worries that might arise from a comparison with the analysis of ambiguity in van Deemter (1996) and van Eijck and Jaspars (1996), where truth-conditions similar to the just described ones (i.e. based on a multiple assertion account) are prescribed for ambiguous expressions. Yet, the proof-theoretical analysis of ambiguous expressions advanced by van Eijck and Jaspars further clarifies the disagreement between a multiple assertion account of ambiguity and the simpler disjunctive analysis presented here. That is, unlike for the latter, the rules for ambiguous expressions of the former exhibit a plain failure of harmony.[5]

When ambiguous expressions are hard to introduce, one expects them to be easy to eliminate. On a multiple assertion account, however, ambiguous expressions

[5]Given the usage of a sequent-calculus in van Eijck and Jaspars (1996), this surfaces as a failure of the principal cuts.

come with an inferential profile that is dual to that of Prior's *Tonk*; the introduction-rules are conjunction-like, and the elimination-rules are disjunction-like. This is not the place to evaluate the relevance of proof-theoretical criteria for the logical analysis of ambiguity, but as a tentative conclusion I would advance the following. Given that both the multiple assertion and the disjunctive analysis get things right in their respective contexts of application, it appears that a logical analysis of ambiguity is feasible for both. Still, only the simple disjunctive account yields genuine logical expressions, and therefore only the ambiguous connectives that are so obtained should be regarded as genuine logical connectives. Put differently, had I used the multiple assertion account to make sense of ambiguous connective, I would have been unable to characterise each ambiguous connective as a single logical connective.

If we only consider the elimination-rule for ambiguous expressions, there is another point of disagreement that is worth highlighting. The rules given in van Eijck and Jaspars (1996) are *context-dependent* ones, which makes the elimination-rule at least as strong to the one defended here. Yet, in the presence of weakening and contraction this leads to an analysis of ambiguity that can only be reformulated by means of classical disjunctions. As there is at least one reason to find this classical analysis plausible, the use of a lattice disjunction in Propositions 5.1 and 5.2 needs further support. The main argument for such a classical analysis is once more a variation on Burgess's objection.

Assume that Agent_1 deduces $(p\star_1 q)\sqcup(p\star_2 q)$ from $p\star_1 q$, and then sends '$p\star q$' to Agent_2. The latter, then, discovers $\sim(p\star_2 q)$ for herself, and concludes that $p\star_1 q$. But if Agent_2 is allowed to reason along these lines, then $\star$-Elim should encompass more than the proof by cases suggested by the definition of $\star$ in terms of $\sqcup$. As this is just an higher-order version of the initial objection, I can offer the same reply as before. Given the usage of a substructural logic, Agent_1's usage of addition can only warrant the lattice-disjunction of both disambiguations. Next, assuming the previously motivated Xerox-principle, the reasoning of Agent_2 is to be explained by her having a ground for asserting $p\star q$ which exceeds her knowledge of Agent_1's own grounds. But if that is the case, the objection has lost its force. Apart from that, I'm not even convinced that this variant has the force as the original objection. I find it much less clear that it is actually a case where some form of *disjunctive syllogism* is applied, and am therefore inclined to leave room for alternative explanations of the hearer's reasoning in this scenario.

A Final Note on the Disjunctive Analysis:
From a logical perspective, a disjunctive analysis of ambiguous terms is undoubtedly natural. Considerations of proof-theoretical harmony are, however, in conflict with the semantics of ambiguity in natural language (as for instance in van Eijck and Jaspars 1996). One way to deal with this tension is to suggest that logical connectives can only be called ambiguous in a loose or non-technical sense. In fact, the OR's and AND's we've been using are more aptly designated as equivocal. While this move presumably explains the difference in analysis, it cannot be used to dispel the objections against a disjunctive analysis of ambiguity put forward in van Deemter (1996) (which I only discovered at a later stage).

A first objection only applies to cases where one disambiguation is logically stronger than the other, and raises the issue that when A is ambiguous between A_1 and A_2 (with $A_1 \vdash A_2$) the disjunctive analysis cannot discriminate between the ambiguous A and the weakest of its disambiguations (i.e. A_2).[6] A second objection points to the fact that the disjunctive analysis confuses the correct analysis of "A is ambiguous between A_1 and A_2" as A either means A_1 or means A_2 with the mistaken analysis A means A_1 or A_2. Each of these deserves a more in depth discussion, but for now a brief response will do.

With regard to the first issue, I should only point out that (a) the problem does not arise for the system presented in this paper, and (b) it isn't even clear that in the presence of contraction the identification of, say, 'OR' with $\sqcup$ is genuinely problematic. As for the second issue, the more cautious reply could be that the mistaken analysis of ambiguity is in fact a correct analysis of equivocation, whereas a more daring reply could be that spelling out the disjunctive analysis in terms of $\sqcup$ suffices to pick out the correct analysis. Further details of these replies have to be left for another occasion.

5.5 The Logical Properties of Ambiguous Connectives

The logical properties of ambiguous connectives and the consequence-relation obtained from Definition 5.1 can be reviewed on two levels: in terms of the general properties of $\vdash_{\forall/\exists}$ and in terms of the valid inferences and theorems of the ambiguous language $\mathsf{FORM}^{\mathsf{Amb}}$. Given the definability of ambiguous expressions, it isn't surprising that $\vdash_{\forall/\exists}$ retains the structural properties of $\vdash$. Considerations about $\vdash_{\forall/\exists}$ being the same consequence relation as $\vdash$ aside, a closer look at how the *all/some* interpretation as it is given in Definition 5.1 enforces the transitivity of $\vdash_{\forall/\exists}$ remains interesting in virtue of its relation to the posited *Xerox-principle*.

Proposition 5.3 (Transitivity of $\vdash_{\forall/\exists}$). *If $\Gamma \vdash_{\forall/\exists} A$ for all $A \in \Pi$, and $\Pi \vdash_{\forall/\exists} B$, then $\Gamma \vdash_{\forall/\exists} B$.*

Proof. Definition 5.1 relates $\vdash_{\forall/\exists}$ to $\vdash$, and $\vdash$ is itself transitive. Hence, it suffices to check that the consequents of the entailments we need to check for all $\Gamma \vdash_{\forall/\exists} A$ to be valid match the antecedents of the entailments we need to check for $\Pi \vdash_{\forall/\exists} B$ to be valid. Since we have to keep track of all the possible disambiguations of Π to establish the validity of $\Pi \vdash_{\forall/\exists} B$, we will automatically also have checked those disambiguations in virtue of which all $\Gamma \vdash_{\forall/\exists} A$ are valid. □

[6]The usage of $\sqcup$ for the disjunctive analysis does not block this argument, whereas the usage of $\oplus$ reduces to the former in the presence of contraction (which is exactly what is required for $A_1 \vdash A_2$).

As opposed to this initial judgement that both consequence-relations coincide, a closer look at valid inferences and theorems reveals a serious gap between the logical properties of ambiguous and unambiguous connectives. To begin with, consider the following valid and invalid inferences:

$$p \vdash p \text{ OR } q \qquad \text{(ADD)}$$

$$p, q \vdash p \text{ AND } q \qquad \text{(ADJ)}$$

which hold in virtue of, respectively, addition for $\sqcup$ and adjunction for $\otimes$, and

$$q \vdash p \text{ IMPLIES } q \qquad \text{(IRR)}$$

$$\text{NOT-}p \vdash p \text{ IMPLIES } q \qquad \text{(IRR')}$$

which both hold in virtue of the irrelevance of $\rightsquigarrow$.

$$p, \text{NOT-}p \text{ OR } q \nvdash q \qquad \text{(DS)}$$

$$p \text{ AND NOT-}p \nvdash q \qquad \text{(ECQ)}$$

which fail in virtue of the failure of extensional disjunctive syllogism, and

$$p, p \text{ IMPLIES } q \nvdash q \qquad \text{(MP)}$$

which fails because $\rightsquigarrow$ is a non-detachable implication.

In the light of these features, it appears that as a relation between ambiguous formulae, $\vdash$ is paraconsistent, but not relevant. It is also a very weak consequence-relation, but given the constraints set by the *all/some* interpretation this is all but surprising. Yet, since in addition to the already mentioned valid inferences, all De Morgan equivalences are also valid:

$$\text{NOT-}(p \text{ OR } q) \dashv\vdash (\text{NOT-}p \text{ AND NOT-}q) \qquad \text{(ND)}$$

$$\text{NOT-}(p \text{ AND } q) \dashv\vdash (\text{NOT-}p \text{ OR NOT-}q) \qquad \text{(NC)}$$

$$\text{NOT-}(p \text{ IMPLIES } q) \dashv\vdash (p \text{ AND NOT-}q) \qquad \text{(NI)}$$

$$\text{NOT-NOT-}p \dashv\vdash p \qquad \text{(DN)}$$

I have, as required, characterised a logic which allows one to reason from ambiguous premises, avoids the classical fallacy of equivocation, and is substantially stronger than the empty logic. Next, consider some of the theorems of the logic of ambiguous connectives:

$$\vdash p \text{ OR NOT-}p \qquad \text{(EM)}$$

$$\dfrac{\dfrac{[p\supset(q\supset r)]^1 \quad \dfrac{\dfrac{[\sim p]^2}{p\supset r}(\rightsquigarrow I)}{(p\supset q)\supset(p\supset r)}(\rightsquigarrow I) \quad \dfrac{\dfrac{[p\supset q]^3[q\supset r]^4}{p\supset r}(\rightarrow trans)}{(p\supset q)\supset(p\supset r)}(\rightarrow I,3)}{((p\supset q)\supset(p\supset r))}(\rightsquigarrow E,2,4)}{(p\supset(q\supset r))\supset((p\supset q)\supset(p\supset r))}(\rightarrow I,1)$$

Fig. 5.7 Strong composition

which holds in virtue of $\vdash p \oplus \sim p$,

$$\vdash p \text{ IMPLIES } (q \text{ IMPLIES } p) \tag{W}$$

which holds in virtue of $\vdash p \rightarrow (q \rightsquigarrow p)$,

$$\vdash (p \text{ IMPLIES } (p \text{ IMPLIES } q)) \text{ IMPLIES } (p \text{ IMPLIES } q) \tag{C}$$

which holds in virtue of $\vdash ((p \rightsquigarrow (p \rightsquigarrow q)) \rightarrow (p \rightsquigarrow q)$, and

$$\vdash ((p \text{ IMPLIES } q) \text{ IMPLIES } p) \text{ IMPLIES } p \tag{P}$$

which holds in virtue of $\vdash ((p \rightsquigarrow q) \rightsquigarrow p) \rightarrow p$.

What is surprising is that even though the structural rules of weakening and contraction were dropped, the corresponding formulae are retained as theorems of the ambiguous language. In particular, the presence of (W) and (C) shows that *qua* theorems the resulting logic of ambiguous connectives is a close cousin of intuitionistic logic, whereas the vacuous ambiguity of 'NOT' warrants (DN) and suffices to bridge the final gap to classical logic. This suggests, and the presence of (EM) and (P) (or Peirce's Law) make this even more plausible, a full recovery of the classical theorems within the ambiguous language.

The following counterexample shows that the present account of ambiguous connectives does not yield all classical theorems, and that in spite of the recovery of weakening and contraction, the impression of full classicality at the level of theorems is misleading. Consider, for that purpose, a proof of the classical axiom of *strong composition*[7] $(p \supset (q \supset r)) \supset ((p \supset q) \supset (p \supset r))$ where, instead of structural rules, the rules for $\rightsquigarrow$ and $\rightarrow$ are both used as rules for the classical implication connective $\supset$ (Fig. 5.7).

[7]The naming convention is inherited from combinatory logic, see Mares and Meyer (2001).

The important thing to notice is that the two final implications in this classical axiom do not get a uniform treatment throughout the proof. In one branch they are understood as lattice implications (cfr. the double application of $\rightsquigarrow$-Intro); in another they are understood as group-theoretical implications (cfr. the application of transitivity and of $\rightarrow$-Intro to discharge the third assumption). As a consequence, there is no unique disambiguation in virtue of which the ambiguous counterpart of this classical axiom could be valid (and there is no way to 'fix' this proof such as to obtain one). The disjunctive account of ambiguous connectives is so formulated that it requires a determinate valid disambiguation for an ambiguous theorem. This 'constructive' demand is precisely what makes it too weak for a full classical recapture at the level of the theorems.

Rather than to fully describe a way to circumvent the need for a determinate disambiguation, I will restrict myself to a few brief comments on this negative result and some promising ways to fix it. First, one should keep in mind that the failure to recapture all classical theorems does not indicate the falsity of the present disjunctive analysis of ambiguity. At best, it reveals that there is an aspect of the ambiguity in classical logic that has not yet been recovered in a merely equivocal sense. Still, this negative result shows that there is room for an introduction-rule for ambiguous expressions that is strictly stronger than the one described above; one that leaves room for ambiguity inside the proofs.

The most plausible strengthening is based on the insight that even though there is no valid disambiguation of *strong composition*, there are a number of intuitively related valid formulae; such as:

$$(p \rightsquigarrow (q \rightarrow r)) \rightarrow (((p \rightarrow q) \rightarrow (p \rightarrow r)) \sqcup ((p \rightsquigarrow q) \rightsquigarrow (p \rightsquigarrow r)))$$

Strictly speaking, this is not a proper disambiguation of the ambiguous version of *strong composition*, but we might still consider it as a proper ground for asserting the ambiguous counterpart of the classical axiom in question. Such a move is an effective strengthening of the introduction-rule for ambiguous expressions; one which includes *indeterminate* grounds for the assertion of an ambiguous expression. Even apart from the question of how this strengthening affects the proper elimination-rule, the incorporation of a rule for indeterminate grounds isn't a trivial matter. The main difficulty is that, for reasons we have stressed before (cfr. the distinction between rules for ambiguous connectives and rules for ambiguous expressions), this strengthening cannot be identified with the full intersubstitutivity of ambiguous subformulae with the disjunction of their disambiguations.

Finally, some distribution principles appear to have an effect that is akin to that of allowing indeterminate grounds for ambiguous expressions. This is revealed by the fact that the substructurally invalid distribution principles $((p \sqcup q) \otimes (p \sqcup r)) \rightarrow (p \sqcup (q \oplus r))$ and $((p \oplus q) \otimes (p \oplus r)) \rightarrow (p \oplus (q \oplus r))$ respectively entail $(p \rightsquigarrow (q \rightarrow r)) \rightarrow ((p \rightsquigarrow q) \rightarrow (p \rightsquigarrow r))$ and $(p \rightarrow (q \rightarrow r)) \rightarrow ((p \rightarrow q) \rightarrow (p \rightarrow r))$. Or, more generally, by the fact that (where A_1/A_2 and B_1/B_2 are disambiguations of A and B) the formula $((A_1 \sqcap A_2) \rightarrow (B_1 \sqcup B_2)) \sqcup ((A_1 \sqcap A_2) \rightsquigarrow (B_1 \sqcup B_2))$ is neither a proper disambiguation of A IMPLIES B nor a sufficient ground for

$(A_1 \to B_1) \sqcup (A_1 \to B_2) \sqcup (A_2 \to B_1) \sqcup (A_2 \to B_2) \sqcup (A_1 \rightsquigarrow B_1) \sqcup (A_1 \rightsquigarrow B_2) \sqcup (A_2 \rightsquigarrow B_1) \sqcup (A_2 \rightsquigarrow B_2)$.[8] I have to leave the role of distribution and the effect of having indeterminate grounds for introducing ambiguous expressions as well as the relation between them at this largely intuitive level. The above two remarks still accomplish the more modest task of identifying the limits of a simple disjunctive analysis of ambiguous connectives, and revealing the specificity of a 'constructive' versus a more 'classical' approach to ambiguity and disambiguation.

5.6 Ambiguity and Disagreement

The main accomplishment of the present paper lies in its showing that the study of ambiguous connectives opens up a new area of logical inquiry. The result that ambiguous connectives can be treated as genuine logical connectives has, however, implications that go beyond the sheer denial that logic cannot properly handle ambiguity. In this final section, I wish to focus on a long standing issue in the philosophy of logic that benefits from this result.

The thesis I would like to defend is that the thorny issue of logical deviance and the disagreement between logics can to some extent be elucidated through the use of ambiguous connectives. That is, by using an ambiguous language as the *locus* to record explicit disagreements, we can understand the rivalry between classical and substructural logic as a genuine disagreement about what follows from what. Logical rivalry surfaces as a disagreement about the correct consequence relation for a shared ambiguous language. At first sight, this might look like a fairly thin sense of disagreement, and certainly one that is only relevant to the specific disagreement between logics with or without the structural rule of weakening. This kind of objection underestimates the importance of the partial recapture of the classical theorems, as well as the wider applicability of the general strategy I've described.

Of course, we have become accustomed to the view that the non-*ad hoc*ness of the translation-function can at best vindicate the intuitive judgement that classical and substructural logics obviously disagree on what follows from ambiguous premises, but that it cannot exclude that ambiguous connectives have distinct classical and substructural meanings (otherwise, the rivalry between classical and intuitionistic logic would never have been called into question). Still, the partial recapture of the classical theorems (and the promise to arrive at a full recapture by more elaborate means), ensures that the ambiguous language does not just give the illusion of a shared subject matter, but a genuine common ground (Azzouni and Armour-Garb 2005). As a result, the disagreement we keep track of within the ambiguous language might be a thin one, but it is still the kind of disagreement

[8] Of course, the use of classical disjunctions as a means to analyse ambiguity would restore the required distribution principles. Given the implicit aim to recover ambiguity within the substructural language, this is not an option I'm inclined to investigate.

which confirms an intuitive understanding of what it means for two logics to present rivalling accounts of consequence. It is the kind of disagreement we should expect in the light of the findings in Humberstone (2005), and leads to a view that is continuous with that in Allo (2007).

Finally, the view that this method is only applicable to logics with more than one conjunction or disjunction largely rests on the misconception that the only ambiguity in classical logic is that between different readings of the connectives. A rather plausible intuitionist critique of classical logic could indeed rest on a similar strategy, and simply claim that classical logic equivocates between p and $\sim\sim p$. From that point on, it is a feasible (though not a trivial) task to construct an 'ambiguous' language to record the disagreement between the classical and intuitionistic accounts of what follows from ambiguous premises. The reference to natural language connectives might be weaker, but the important part of the strategy—a shared language that is genuinely a common ground to keep track of disagreements—remains as effective as before.

Acknowledgements The author wishes to thank the audiences and organisers of the World Congress on Paraconsistency (Melbourne) and the Foundations of Logical Consequence Workshop at Arché (Saint-Andrews), as well as Francesco Paoli and Sebastian Sequoiah-Grayson for correspondence on the material in this paper.

References

Allo, P. 2007. Logical pluralism and semantic information. *Journal of Philosophical Logic* 36(6): 659–694.

Anderson, A.R., and N.D. Belnap. 1975. *Entailment: The logic of relevance and necessity*, vol. 1. Princeton: Princeton University Press.

Azzouni, J., and B. Armour-Garb. 2005. Standing on common ground. *The Journal of Philosophy* 102(10): 532–544.

Batens, D. 1997. Inconsistencies and beyond: A logical-philosophical discussion. *Revue Internationale de Philosophie* 51(2): 259–273.

Beall, J.C., and G. Restall. 2006. *Logical pluralism*. Oxford: Oxford University Press.

Burgess, J.P. 1981. Relevance: A fallacy? *Notre Dame Journal of Formal Logic* 22(2): 97–104.

Corcoran, J. 1998. Information-theoretic logic. In *Truth in Perspective*, ed. C. Martínez, U. Rivas, and L. Villegas-Forero, 113–135. Aldershot: Ashgate.

da Costa, N.C.A. 1997. *Logiques classiques et non classiques: Essai sur les fondements de la logique*. Paris: Masson.

Došen, K. 1993. A historical introduction to substructural logic. In *Substructural logics*, ed. K. Došen and P. Schroeder Heister, 1–30. Oxford: Oxford University Press.

Dretske, F. (1999). *Knowledge and the flow of information*. Stanford: CSLI.

Fine, K. 1975. Vagueness, truth and logic. *Synthese* 30(3): 265–300.

Haack, S. 1974. *Deviant logic. Some philosophical issues*. Cambridge: Cambridge University Press.

Humberstone, I.L. 2005. Logical discrimination. In *Logica universalis*, ed. J.-Y. Béziau, 207–228. Basel: Birkhäuser Verlag.

Lewis, D. (1982). Logic for equivocators. *Noûs* 16: 431–441.

Mares, E. 1997. Relevant logic and the theory of information. *Synthese* 109(3): 345–360.

Mares, E. 2004. *Relevant logic: A philosophical interpretation*. Cambridge: Cambridge University Press.

Mares, E., and R.K. Meyer. 2001. Relevant logics. In *The blackwell guide to philosophical logic*, ed. L. Goble, 280–308. Oxford: Blackwell.

Paoli, F. 2007. Implicational paradoxes and the meaning of logical constants. *Australasian Journal of Philosophy* 85(4): 553–579.

Prawitz, D. 2006. Meaning approached via proofs. *Synthese* 148(3): 507–524.

Priest, G. 2001. Logic: One or many? In *Logical consequence: Rival approaches*, ed. J. Woods and B. Brown, 23–38. Stanmore: Hermes.

Read, S. 1981. What is wrong with disjunctive syllogism? *Analysis* 41: 66–70.

Read, S. 1988. *Relevant logic: A philosophical examination of inference*. Oxford: Basil Blackwell.

Read, S. 2000. Harmony and autonomy in classical logic. *Journal of Philosophical Logic* 29(2): 123–154.

Tennant, N. 2004. Relevance in reasoning. In *Handbook of philosophy of logic and mathematics*, ed. S. Shapiro, 696–726. Oxford: Oxford University Press.

van Deemter, K. 1996. Towards a logic of ambiguous expressions. In *Semantic ambiguity and underspecification*, ed. K. van Deemter and S. Peters, 203–237. Stanford: CSLI.

van Eijck, D.J.N., and J.O.M. Jaspars. 1996. *Ambiguity and reasoning*. Amsterdam: CWI, CS-9616. http://db.cwi.nl/rapporten/abstract.php?abstractnr=568

Chapter 6
Assertion, Denial and Non-classical Theories

Greg Restall

6.1 Assertion, Denial and Sequents

Friends of truth-value GAPS and truth-value GLUTS both must distinguish the *assertion of a negation* (asserting $\ulcorner \neg p \urcorner$) and *denial* (denying $\ulcorner p \urcorner$). If you take there to be a truth-value *glut* at $\ulcorner p \urcorner$ the appropriate claim to make (when asked) is to assert $\ulcorner \neg p \urcorner$ without thereby denying $\ulcorner p \urcorner$. If you take there to be a truth-value *gap* at $\ulcorner p \urcorner$ the appropriate claim to make (when asked) is to deny $\ulcorner p \urcorner$ without thereby asserting $\ulcorner \neg p \urcorner$.

This is why taking $\ulcorner p \urcorner$ to be in a truth-value gap is not the same attitude as *ignorance* or *agnosticism* concerning $\ulcorner p \urcorner$. If I am ignorant of $\ulcorner p \urcorner$, I assert neither $\ulcorner p \urcorner$ nor $\ulcorner \neg p \urcorner$ and neither do I deny them. I am open to the possibilities. Taking $\ulcorner p \urcorner$ to be in truth-value gap involves *denying* it, together with denying its negation. Similarly, this is why a taking $\ulcorner p \urcorner$ to suffer from a truth-value glut is not the same attitude as being *confused* concerning $\ulcorner p \urcorner$. I might mistakenly believe both $\ulcorner p \urcorner$ and $\ulcorner \neg p \urcorner$,[1] but in *that* case I take my assertion (when asked) of $\ulcorner \neg p \urcorner$ to rule out assertion of $\ulcorner p \urcorner$, and I take my assertion of $\ulcorner p \urcorner$ to rule out assertion of $\ulcorner \neg p \urcorner$. I am, alas, in two minds concerning $\ulcorner p \urcorner$. Someone who takes $\ulcorner p \urcorner$ to be genuinely both true and false is not in this state. To take $\ulcorner p \urcorner$ to be both true and false is to be prepared to assert $\ulcorner \neg p \urcorner$ without thereby denying $\ulcorner p \urcorner$.

I think this is important, because logical consequence has something to say not only about assertion but also about denial and the connection between assertion

[1] A nice example of confusion is David Lewis' discussion of the orientation of Nassau Street in Lewis (1982).

G. Restall (✉)
School of Historical and Philosophical Studies, University of Melbourne, Parkville, Australia
e-mail: restall@unimelb.edu.au

K. Tanaka et al. (eds.), *Paraconsistency: Logic and Applications*, Logic, Epistemology, and the Unity of Science 26, DOI 10.1007/978-94-007-4438-7_6,

and denial (see Restall 2005). To take an argument to be valid does not mean that when one asserts the premises one should also assert the conclusion (that way lies madness, or at least, making *many* assertions). No, to take an argument to be valid involves (at least as a part) the commitment to take the assertion of the premises to stand against the denial of the conclusion. In general, we can think of logical consequence as governing *positions* involving statements asserted and those denied. Logical validity governs positions in the following way:

- If $A \vdash B$, then the *position* consisting of asserting $\ulcorner A \urcorner$ and denying $\ulcorner B \urcorner$ *clashes*.

If $\ulcorner B \urcorner$ deductively follows from $\ulcorner A \urcorner$, and I assert $\ulcorner A \urcorner$ and deny $\ulcorner B \urcorner$, I have made a mistake. This generalises in the case of more than one assertion and more than one denial.

- If $\Gamma \vdash \Delta$, a position in which we assert each member of Γ and deny each member of Δ *clashes*.

What can we say about this relation of logical consequence, between collections of premises and conclusions, governing positions involving assertions and denials? At the very least we can say that the following rule (*Id*) holds, meaning that a position is a clash if the same thing is both asserted and denied.

$$\Gamma, A \vdash A, \Delta \quad (Id)$$

Furthermore, if there is no clash in asserting every member of Γ and denying every member of Δ, we can see that together with asserting each member of Γ and denying each member of Δ either there is no clash in asserting $\ulcorner A \urcorner$ or there is no clash in denying $\ulcorner A \urcorner$. In other words, if asserting $\ulcorner A \urcorner$ is ruled out by means of the rules of the game alone, then since $\ulcorner A \urcorner$ is unassertible, its denial is implicit in the assertion of every member of Γ and the denial of every member of Δ, so its explicit denial involves no clash. Contraposing this, if there is a clash in denying $\ulcorner A \urcorner$ (together with asserting every member of Γ and denying every member of Δ) and there is also a clash in asserting $\ulcorner A \urcorner$ (together with asserting every member of every member of Γ and denying every member of Δ), then there is a clash in asserting every member of Γ and denying every member of Δ alone. In other words, we have the rule (*Cut*).

$$\frac{\Gamma \vdash A, \Delta \quad \Gamma, A \vdash \Delta}{\Gamma \vdash \Delta}\ (Cut)$$

We should note three things concerning the structural features of consequence understood in this way. First, Γ and Δ here are *sets* of statements. While the use of more discriminating collections (multisets, lists, etc.) can be very useful from a proof-theoretic point of view, as long as there is no normative difference between a position in which something has been asserted *twice* and where it has been asserted merely *once*, this seems to be a distinction that makes no difference. Second, it is important to notice that implicit in the rule (*Id*) of identity, is the rule of *weakening*. If a position has a clash, this is not alleviated by the addition of *more* assertions or denials. It could well be alleviated by a *retraction* of something formerly asserted or denied, but a retraction is not the same thing as a denial or an assertion. Retracting a claim means moving to a position in which that claim is taken 'off the table' as an assertion (or as a denial), which need not involve any further assertion or denial (of that claim, its negation, or anything else). If I discover that the claim that p has untoward consequences, I can retract an assertion of $\ulcorner p \urcorner$ without being committal to its truth or falsity. Third, the vocabulary of sequents here is, so far, *independent* of the logical vocabulary used in the statements that are themselves asserted and denied. We have only sketched some structural features which quite plausibly govern the practice of making assertions and denials.[2]

Now, let's consider logical vocabulary, and in particular the operator of negation. Gentzen's own sequent rules for negation are simple:

$$\frac{\Gamma \vdash A, \Delta}{\Gamma, \neg A \vdash \Delta}\,(\neg L) \qquad \frac{\Gamma, A \vdash \Delta}{\Gamma \vdash \neg A, \Delta}\,(\neg R)$$

They tell us that asserting $\ulcorner \neg A \urcorner$ has the same status as denying $\ulcorner A \urcorner$. If there is a clash in denying $\ulcorner A \urcorner$ (in the context of asserting every member of Γ and denying every member of Δ), there is a clash in asserting $\ulcorner A \urcorner$ too (in that context). Similarly, denying $\ulcorner \neg A \urcorner$ has the same status as asserting $\ulcorner A \urcorner$.

Clearly, given what we have already said about friends of gaps and gluts, not everyone will find these rules acceptable. Depending on our attitudes to truth-value gaps and gluts, we may find some rules acceptable and others not. ($\neg L$) corresponds to *Ex Contradictione Quodlibet*[3] and ($\neg R$) corresponds to the Law of the Excluded Middle.[4] The four different possibilities seem to be these:

1. *No gaps, no gluts*: both ($\neg L$) and ($\neg R$) are acceptable.
2. *Gaps, no gluts*: ($\neg L$) is acceptable, but ($\neg R$) is not. We can have $\nvdash \neg A, A$.
3. *No gaps, gluts*: ($\neg L$) is not acceptable, but ($\neg R$) is acceptable. We can have $A, \neg A \nvdash$.
4. *Gaps, gluts*: both ($\neg L$) and ($\neg R$) are not acceptable We have both $\nvdash \neg A, A$ and $A, \neg A \nvdash$.

In other words, the interpretation of sequents in terms of assertion and denial gives us a way to characterise different treatments of negation. While there is something to be said taking ($\neg L$) and ($\neg R$) as a *definition* of a concept of negation, this will be

---

[2]I have argued for them in some detail elsewhere, see Restall (2005). I do not take these considerations to be *conclusive*, but on the other hand, I have not seen any rival account of the norms of assertion and denial that is in any way a plausible alternative to this picture. I urge defenders of non-classical logic who take the assumptions I have made to be mistaken to develop an alternative account, explaining when the rules (*Id*) or (*Cut*) might fail, when governing assertion and denial, and what should be put in their place.

[3]Since $p \vdash p, q$, we have by ($\neg L$), $p, \neg p \vdash q$.

[4]Since $p \vdash p$, we have by ($\neg R$), $\vdash p, \neg p$, and by the disjunction rule, $\vdash p \vee {\sim} p$.

a definition that friends of gaps or gluts take to fail to characterise the true concept of negation.

Does that mean there is a *false* concept of negation defined by ($\neg L$) and ($\neg R$), or does it mean that these rules don't define a concept at all? These are subtle matters for the friend of gaps or gluts. Graham Priest, for one, takes the rules ($\neg L$) and ($\neg R$) to not define a concept at all. He takes there to be *no* concept satisfying those two rules, see Priest (1990).

The fact that ($\neg L$) is not acceptable for friends of gaps, and ($\neg R$) is not acceptable for friends of gluts does not mean that friends of gaps or gluts must reject any rules that together *look* like ($\neg L$) and ($\neg R$). For example, rules of this general shape are satisfied by negation in linear logic (see Girard 1987; Restall 2000), without allowing a derivation of either the law of the excluded middle or the law of non-contradiction. You can prove a sequent of the form $\vdash p, \neg p$, but this does not mean that $\vdash p \vee \neg p$, only that $\vdash p + \neg p$, where '$+$' is an *intensional* disjunction, and where $p + p$ does not entail p, and where p does not entail $p + q$ (both *weakening* and *contraction* fail for this disjunction). The disagreement is not over the general shape of the rules, but over the rules ($\neg L$) and ($\neg R$) where we have interpreted the structure (the comma, the turnstile) in just the way we have here, governing assertion and denial. We must be clear on how we are interpreting our vocabulary, *especially* when using non-classical logics, in which common assumptions are questioned or rejected.

Just as there are rules governing negation, there are rules governing other connectives. Different logics, classical and non-classical, have different rules for the connectives such as conjunction, disjunction, the conditional (or many conditionals) and quantifiers. In what follows, these rules will be unimportant, as the topics we are considering can be characterised without reference to those connectives.

Let's consider the position of the dialetheist, or in fact, any proponent of a paraconsistent logic. We will presume (*Id*) and (*Cut*) in what follows. This is not to say that they cannot be resisted by a dialetheist: of course they can. However, to resist them is to open up the question: what is to be held in their place? It seems that maintaining (*Id*) and (*Cut*) are *no* bar to holding a paraconsistent logic, nor even being a dialetheist, who holds that some contradictions are true. We can very well make sense of this position, agree that (*Id*) and (*Cut*) are valid, and simply reject ($\neg L$) as a rule satisfied by a genuine negation operator. That is an understandable position for the dialetheist. Were the dialetheist to go *further*, to reject either of (*Id*) or (*Cut*), they would need to answer further questions. Chief among them is this: how *are* we to constrain assertion and denial? If not (*Id*) and (*Cut*), then what? If we have a valid argument from premises to conclusion, how does this constrain assertion and denial? Do we not take there to be a mistake in asserting the premises of a valid argument and denying the conclusion? Something must be said here, and it is a challenge for the dialetheist who wishes to reject (*Id*) or (*Cut*) to sail between the Scylla of agreeing with (*Id*) and (*Cut*) and the Charybdis of rejecting so much that logic has no evaluative force.

6.2 Theories, Cotheories and Bitheories

With that background on assertion and denial, granting the role of (*Id*) and (*Cut*) constraining assertion and denial, but allowing different accounts of negation and the other logical connectives to vary from logic to logic, it is time to consider the notion of a *theory*. For among many different logics, such as classical logic, constructive logics, logics with truth-value gaps and—especially for our discussion here, logics with truth-value gluts—the notion of a *theory* makes sense, and has a prominent role. Given any of these choices of logic, and given the context of the formalisation of mathematical concepts or the presentation of other 'theories', intuitively understood, it is common to treat a formal *theory* as a collection of statements. Perhaps when presented it is characterised as the consequences of a number of basic axioms. Perhaps instead it is characterised as the application of a number of basic rules. Perhaps, thirdly, it is characterised as the collection of statements true in some class of models. However they are characterised, the result is the following condition: a THEORY is a collection T of sentences closed under logical consequence. That is,

$$T \text{ is a THEORY } \textit{iff } (\forall A)(T \vdash A \;\Rightarrow\; A \in T)$$

If some statement is a consequence of the theory, it is also a part of the theory. So, if you *endorse* the theory, commitment to this theory means that you are making a mistake if you deny any statement in the theory. The consequences of T are *undeniable*, granted commitment to T.[5] What does the theory tell us we *should* deny, or contrapositively, what we *shouldn't* assert? As far as the theory goes, if assertion of a negation does not bring denial along with it (as it doesn't, for friends of gluts), the commitment to the theory itself need carry no consequences concerning what is not to be asserted (or what should be denied). The fact that a theory tells us $\ulcorner \neg A \urcorner$ does not give us guidance on the matter of ruling $\ulcorner A \urcorner$ out, at least if we have countenanced gluts.[6]

What is there to do? It seems that we must not only keep track of what is to be accepted, on the terms of a theory, but we should also keep track of what is to be *rejected*. Dual to a theory is the notion of a COTHEORY, a collection U of sentences closed 'over' logical consequence. That is,

$$U \text{ is a COTHEORY iff } (\forall A)(A \vdash U \Rightarrow A \in U)$$

[5]Since T is closed under consequence, that is not saying much of course. We could strengthen things by noting that the statements of the theory are undeniable, given commitment to the *axioms* of the theory, if the axioms X form a set from which all members of T follow.

[6]It might be thought that commitment to something like $\ulcorner A \rightarrow \bot \urcorner$ would do it, where $\ulcorner \bot \urcorner$ is to be rejected always. Perhaps that will express a feature appropriate for denial, but now the trouble is that it is too strong. In non-classical logics used for the paradoxes, $\ulcorner A \vee (A \rightarrow \bot) \urcorner$ is rejected (and it must be, lest the liar paradox arise for the 'negation' of implying $\ulcorner \bot \urcorner$), so $\ulcorner A \urcorner$ and $\ulcorner A \rightarrow \bot \urcorner$ are to be rejected. But this means we must have some way of rejecting $\ulcorner A \urcorner$ which does not involve accepting $\ulcorner A \rightarrow \bot \urcorner$. So, $\ulcorner A \rightarrow \bot \urcorner$ may express *one* kind of rejection, but it is not enough to express the entirety of the notion.

That is, if some statement has the cotheory as a consequence, it is also a part of the cotheory. So, if you *reject* the cotheory, this rejection means that you are making a mistake if you assert any statement in the cotheory.[7] The statements which have U as a consequence are *unacceptable*, granted commitment to deny U.

A cotheory is the natural dual partner to a theory. However, we don't want to restrict our attention to treating either a theory or a cotheory in isolation, or merely in tandem. Perhaps given the assertion of some members of T and the denial of some other members of U, some other statements are unassertible, or are undeniable. In each case, these statements belong in U or in T respectively. In other words, what we *really* need is a BITHEORY, consisting of both a theory and also a cotheory.

$$\langle T, U\rangle \text{ is a BITHEORY } \textit{iff } (\forall A)(T \vdash A, U \;\Rightarrow\; A \in T \textit{ and } T, A \vdash U \;\Rightarrow\; A \in U)$$

In other words, $\langle T, U\rangle$ gives us direction both on what is to be asserted (T) and what is to be denied (U). And if, in this context, $\ulcorner A \urcorner$ is undeniable, it also belongs in T, and if $\ulcorner A \urcorner$ is unassertible, it belongs in U.

If we were merely to consider logics in which ($\neg L$) and ($\neg R$) held in their generality, we would not need to consider either cotheories or bitheories. In that case $\ulcorner T, A \vdash U \urcorner$ is equivalent to $\ulcorner T \vdash \neg A, U \urcorner$, so membership in U is decided by membership (of the negation) in T. In cases where we do not have a negation connective with such an intimate connection with assertion and denial, this trick will not always work, and hence the need to explicitly consider both components of a bitheory.

Does this distinction actually matter in practice? I think it does. In the rest of this paper I'll look at three non-classical theories as bitheories: they are theories of Numbers, Classes, and Truth. We will see that attention to considerations of assertion and denial—considering these theories as *bi*theories—will provide a range of insights obscured when we consider presentation in the guise of theories alone.

6.3 Numbers, Classes and Truth

We will start with a simple case. Numbers: theories of arithmetic.

6.3.1 Numbers

Axiomatic presentations of theories of arithmetic typically involve many connectives: axioms take the form of conditional statements such as $x' = y' \rightarrow x = y$, and so on. It is noticeable, however, that the details of the logic of the conditional

[7]Note, to reject the cotheory is not to reject the conjunction of its members, since the cotheory marks what is to be rejected. To reject it is to reject each member, just as to accept a theory is to accept each member of the theory.

in question often does not matter very much (see Meyer and Restall 1999; Restall 2009; Slaney et al. 1996). Now that we have the machinery of sequents, and their interpretation in terms of assertion and denial, it turns out that we can strip the extra logical vocabulary away from the core of the presentation of arithmetic. The language for our statements of arithmetic will involve the following items:

$$= \qquad 0 \qquad ' \qquad + \qquad \times$$

Identity is a two-place predicate, $\ulcorner 0 \urcorner$ is a constant, successor is a one-place function, and addition and multiplication are binary two-place functions. For addition and multiplication, the salient requirements in the theory are simple recursive *equations*. We endorse the following:

$$\vdash x + 0 = x \qquad \vdash x + y' = (x + y)'$$

$$\vdash x \times 0 = 0 \qquad \vdash x \times y' = x \times y + x$$

The use of free variables indicates at least the commitment to each *instance* for any choice of terms to fill in for $\ulcorner x \urcorner$ and for $\ulcorner y \urcorner$, but also commitment to each further instance whenever we extend our language to contain more terms of the same type. If you wish to consider quantificational statements, then the logic of the universal quantifier should dictate that not only do we endorse $\ulcorner x + 0 = x \urcorner$ but its generalisation $\ulcorner (\forall x)(x + 0 = x) \urcorner$.[8]

These recursive equations are items to be asserted. They say nothing about what is to be denied.[9] More interesting are the rules governing identity and the successor function. These involve denial:

$$x' = y' \vdash x = y \qquad 0 = x' \vdash$$

The first of the rules is an axiomatic sequent: indicating that successor is a one-to-one function. It pairs an assertion and a denial, dictating that it would be a clash to assert $\ulcorner x' = y' \urcorner$ but to deny $\ulcorner x = y \urcorner$. Given a position in which we have asserted $\ulcorner x' = y' \urcorner$, the only option for $\ulcorner x = y \urcorner$ is to assert it, as it is undeniable. This seems quite plausible: to take $\ulcorner x' = y' \urcorner$ to hold but to reject $\ulcorner x = y \urcorner$ seems to involve a mistaken conception of numbers or of the successor function. Conversely, if we reject $\ulcorner x = y \urcorner$ then the only option for $\ulcorner x' = y' \urcorner$ is to reject it, too.

Similarly, the claim that 0 is not a successor, often formalised as an *axiom*, of the form $\ulcorner 0 \neq x' \urcorner$, is better formulated as a denial. While it is interesting to observe that there are models of arithmetic that get arithmetical truths correct while also including a claim of the form $\ulcorner 0 = n' \urcorner$ for some n, while also committing us to

[8] In sequent presentations of logic, that would be a direct consequence of $\ulcorner x + 0 = x \urcorner$ by $(\forall R)$, since x occurs nowhere else in the sequent: it is *arbitrary*.

[9] To accept that $\vdash x + 0 = x$ is to take $\ulcorner x + 0 = x \urcorner$ to be undeniable, so it tells us about what is *not* to be denied. For positive advice on what is to be denied, however, we need to look elsewhere.

$\ulcorner 0 \neq x' \urcorner$ in general,[10] there is no doubt that these models get something *wrong*. They endorse something that is to be rejected, by the lights of the concepts of arithmetic. They may endorse everything that arithmetic tells us is to be endorsed, but that is not enough to be a model of the *bi*theory of arithmetic, and only a bi-theoretical perspective is enough to draw out this fact.

From the rules so far, we may derive simple statements, such as these:

$$\vdash 0 = 0 \qquad \vdash 0'' + 0'' = 0'' \times 0''$$

$$0 = 0' \vdash \qquad 0' \times 0'' = 0''' \vdash$$

using the recursive equations and (*Cut*). However, it is harder to prove things in *generality*. For this, we need principles of induction. It seems harder to do away with logical vocabulary when it comes to induction, for an induction axiom is typically formulated with a thicket of connectives and quantifiers:

$$\phi(0) \to ((\forall x)(\phi(x) \to \phi(x')) \to (\forall x)\phi(x))$$

Here, the logic truly makes a difference.[11] However, it seems like the logic of the choice of this or that conditional used in the formulation of an induction *axiom* should not make a difference. Induction is a least number principle. It tells us that when a property fails to hold of all numbers and it holds of 0, there is a number for which it holds where for the *next* number it fails. Contrapositively, it tells us that when a property holds of some numbers, and it doesn't hold of 0, there is a number for which it fails, but where it holds at the *next* number. In other words, we have the following two principles of *ascent* and *descent*.

$$\frac{\Gamma \vdash \phi(0), \Delta \quad \Gamma, \phi(x) \vdash \phi(x'), \Delta}{\Gamma \vdash \phi(x), \Delta} \ (\textit{Ascent})$$

$$\frac{\Gamma, \phi(x') \vdash \phi(x), \Delta \quad \Gamma, \phi(0) \vdash \Delta}{\Gamma, \phi(x) \vdash \Delta} \ (\textit{Descent})$$

Reading these principles from bottom-to-top, *ascent* tells us that if we have denied $\ulcorner \phi(x) \urcorner$ we should either be prepared to deny $\ulcorner \phi(0) \urcorner$, or we should be prepared to (for some term x) assert $\ulcorner \phi(x) \urcorner$ and deny $\ulcorner \phi(x') \urcorner$. Or from top-to-bottom, if we have claimed $\ulcorner \phi(0) \urcorner$ and if $\ulcorner \phi(x) \urcorner$ brings with it $\ulcorner \phi(x') \urcorner$, then we have claimed $\ulcorner \phi(x) \urcorner$ in general, ascending the tower of numbers.

[10] These are the so-called 'mod' models of arithmetic, over the integers modulo n, see Meyer (1976) and Meyer and Mortensen (1984).

[11] In the absence of contraction or weakening as structural rules governing the conditional, it makes a difference as to whether the induction scheme is formulated as above, or as $\phi(0) \to ((\forall x)(\phi(x) \to \phi(x')) \to (\forall x)\phi(x))$ or as $(\forall x)(\phi(x) \to \phi(x')) \to (\phi(0) \to (\forall x)\phi(x))$ or as a myriad of other formulations, each subtly different.

The *descent* principle is the dual. If we have asserted $\ulcorner\phi(x)\urcorner$ we should either be prepared to assert $\ulcorner\phi(0)\urcorner$, or we should be prepared to deny $\ulcorner\phi(x)\urcorner$ while asserting $\ulcorner\phi(x')\urcorner$ (for some term x). Or from top-to-bottom, if we have denied $\ulcorner\phi(0)\urcorner$ and if $\ulcorner\phi(x')\urcorner$ brings with it $\ulcorner\phi(x)\urcorner$, then we had better deny $\ulcorner\phi(x)\urcorner$ in general, lest we be able to descend the tower of numbers to 0, from wherever we started.

Clearly, no conditional in the object-language is required for this formulation of induction principles, but it is just as clear that we have not managed to rid the induction conditions of all conditionality entirely—the turnstile of consequence expresses a conditional connection: there is no escaping that. However, we have been able to formulate induction in such a way that patterns of assertion and denial of statements—themselves not containing conditionals—are enough for us to judge whether induction has been violated or not. This is an advance, for now we can formulate bitheories of arithmetic in which the induction principle is present, yet where we need make no choice over what kind of object-language conditional is present in the theory. We can get some way with arithmetic without having to make that choice at all.

Induction here has split into two rules because in the absence of a negation satisfying both ($\neg L$) an ($\neg R$), we have no way, in general, to get from one principle to the other, yet it seems that both are equally appropriate rules governing arithmetic. Anyone prepared to endorse the premises of either rule, without endorsing the appropriate conclusion would seem to thereby have a non-standard understanding of the concept of *number*, so they seem appropriate to countenance as axiomatic principles. They are essentially bi-theoretic, governing both assertion and denial. Better still, they apply in the absence of other connectives, so we can examine a great deal of the theory of arithmetic without deciding between a logic for gaps or gluts.

6.3.2 Classes

The aim of this paper is to introduce a new line of inquiry, not to pursue any of those directions to any length. So, instead of exploring arithmetic further, let's now consider class theories. Non-classical logics, of gaps or of gluts, are often proposed as the right means for a 'solution' to the paradoxes confronting Frege's general conception of classes. Frege's axiom (V), the general principle of comprehension for classes, has this form

$$a \in \{x : \phi(x)\} \textit{ if and only if } \phi(a) \quad \text{(V)}$$

Membership in the class $\{x : \phi(x)\}$ is found by way of the membership condition. An object a is in that class if and only if the defining condition $\ulcorner\phi(a)\urcorner$ holds. Again, considered as a single *axiom* it features a biconditional, so the question must be raised: which conditional, what logic? Non-classical logics for axiom (V) differ in their choice at this point (see Brady 2006; Gilmore 1974, 1986; Priest 2006;

Restall 1992).[12] Independent of concerns over conditionality, there is a central core to commitment to axiom (V): $\ulcorner\phi(a)\urcorner$ and $\ulcorner a \in \{x : \phi(x)\}\urcorner$ stand and fall together. The assertion of $\ulcorner\phi(a)\urcorner$ has the same upshot as the assertion of $\ulcorner a \in \{x : \phi(x)\}\urcorner$; a denial of $\ulcorner\phi(a)\urcorner$ has the same upshot as a denial of $\ulcorner a \in \{x : \phi(x)\}\urcorner$. Anyone prepared to *assert* $\ulcorner\phi(a)\urcorner$ but to *deny* $\ulcorner a \in \{x : \phi(x)\}\urcorner$ rejects condition (V). Similarly, anyone prepared to *deny* $\ulcorner\phi(a)\urcorner$ but to *assert* $\ulcorner a \in \{x : \phi(x)\}\urcorner$ also rejects condition (V). We have the following two introduction rules for membership in classes:

$$\frac{\Gamma, \phi(a) \vdash \Delta}{\Gamma, a \in \{x : \phi(x)\} \vdash \Delta}\,(\in L) \qquad \frac{\Gamma \vdash \phi(a), \Delta}{\Gamma \vdash a \in \{x : \phi(x)\}, \Delta}\,(\in R)$$

Now, what makes $\ulcorner\{x : \phi(x)\}\urcorner$ an expression denoting a *class* and not a *property* is the commitment to extensionality, the commitment that

Classes with *the same members are the same.*

For this, we require some means to express the binary relation of identity. It is traditional, again, to express this as an axiom involving quantification and conditionals (actually, a conditional and a *bi*conditional) something like this:

$$(\forall x)(\forall y)((\forall z)(z \in x \leftrightarrow z \in y) \rightarrow x = y)$$

and again, there are many debates concerning the appropriate formulation of this condition.[13] Again, we can avoid such baroque discussions by expressing the commitment to extensionality as, at root, commitment to an inference rule in which conditionality is eliminated altogether from the object language.[14]

---

[12]This is not an idle worry: Curry's paradox wreaks havoc with axiom (V), so in the presence of a conditional, the inference from $\ulcorner p \rightarrow (p \rightarrow q)\urcorner$ to $\ulcorner p \rightarrow q\urcorner$ is to be rejected (see Meyer et al. 1979). But then, $\ulcorner p \rightarrow (p \rightarrow q)\urcorner$ and $\ulcorner p \rightarrow q\urcorner$ express different conditional connections between $\ulcorner p\urcorner$ and $\ulcorner q\urcorner$. For the first, *two* instances of *modus ponens* are required to get from $\ulcorner p\urcorner$ to $\ulcorner q\urcorner$, for the second, one suffices. Which of these conditional notions is to be used in the statement of axiom (V)? This is a genuinely hard problem. Suppose I write $\ulcorner p \rightarrow (p \rightarrow q)\urcorner$ as $\ulcorner p \rightarrow_2 q\urcorner$, and replace my theory expressed in terms of $\ulcorner\rightarrow\urcorner$ with one expressed in terms of $\ulcorner\rightarrow_2\urcorner$. *Modus ponens* holds for $\ulcorner\rightarrow_2\urcorner$ as much as it does for $\ulcorner\rightarrow\urcorner$. What changes? Which is the *real* conditional? In the absence of a wider semantic story, the difference is vacuous. Yet for the proponent of a non-classical logic, the difference is important, for if there was no difference, the theory is trivial. So, a wider semantic story of some kind must be told.

[13]Not only are there debates concerning contraction: there are also debates over *relevance*. Should the main conditional be taken to express a *relevant* connection, or should we weaken the condition to involve a prophylactic $\ulcorner t\urcorner$?

$$(\forall x)(\forall y)((\forall z)(z \in x \leftrightarrow z \in y) \wedge t \rightarrow x = y)$$

Furthermore, does the identity $\ulcorner x = y\urcorner$ *entail* $\ulcorner(\forall z)(x \in z \leftrightarrow y \in z)\urcorner$ or is the connection here not relevance preserving? Options abound.

[14]In the rule ($Ext_\in$) we have the side condition that x is absent from Γ and Δ.

$$\frac{\Gamma, x \in a \vdash x \in b, \Delta \qquad \Gamma, x \in b \vdash x \in a, \Delta}{\Gamma \vdash a = b, \Delta}\ (Ext_{\in})$$

The rule ($Ext_{\in}$) tells us that if we are prepared to deny $\ulcorner a = b \urcorner$, then we must be prepared either to assert $\ulcorner x \in a \urcorner$ and deny $\ulcorner x \in b \urcorner$, or *vice versa*.[15] I cannot see how anyone prepared to reject any of ($\in L$), ($\in R$) or ($Ext_{\in}$) truly accepts an extensional theory of classes satisfying law (V) in its intended meaning.[16] This is problematic, since we shall see that, independently of any fancy footwork concerning the logic of the propositional connectives, ($\in L$), ($\in R$) and ($Ext_{\in}$) are already very strong. To explain the untoward consequences of these rules, we need to explain how to understand the logic of identity in this context. I take it that the appropriate notion of identity is one in which the following three rules are satisfied.

$$\frac{\Gamma, \phi(a) \vdash \Delta}{\Gamma, a = b, \phi(b) \vdash \Delta}\ (=L_l) \qquad \frac{\Gamma \vdash \phi(a), \Delta}{\Gamma, a = b \vdash \phi(b), \Delta}\ (=L_r)$$

$$\frac{\Gamma, Xa \vdash Xb, \Delta \quad \Gamma, Xb \vdash Xa, \Delta}{\Gamma \vdash a = b, \Delta}\ (=R)$$

Identity is, at its heart, a second-order notion.[17] If I assert $\ulcorner a = b \urcorner$ and $\ulcorner \phi(b) \urcorner$, then I am thereby committed to $\ulcorner \phi(a) \urcorner$. After all, if I were to assert $\ulcorner \phi(a) \urcorner$ and deny $\ulcorner \phi(b) \urcorner$, what more evidence do I need to the effect that $a \neq b$?[18] Similarly, if I assert $\ulcorner a = b \urcorner$ and I *deny* $\ulcorner \phi(b) \urcorner$: I am thereby committed to denying $\ulcorner \phi(a) \urcorner$. This motivates the left identity rules. For the right identity rule ($=R$), if I *deny* $\ulcorner a = b \urcorner$ I must be prepared to countenance something (perhaps not in the vocabulary I already have: it may be a schematic 'property' not expressible in my own vocabulary) holding of a but not of b, or *vice versa*. The second-order nature of the identity

---

[15]The terms $\ulcorner a \urcorner$ and $\ulcorner b \urcorner$ denote classes, nothing else here. There is no implicit commitment to the effect that different *numbers*, *electrons* or *tables* must have different $\ulcorner \in \urcorner$-members.

[16]I have myself explored theories and models in which a kind of 'naïve comprehension' holds but in which ($\in L$), ($\in R$) fail. The simple LP-models of naïve comprehension (see Restall 1992) validate 'extensionality' in the weak form

$$a \in \{x : \phi(x)\} \equiv \phi(a)$$

where $\ulcorner \equiv \urcorner$ is a material conditional. Here, models do not truly validate ($\in L$) and ($\in R$), for a class B in which everything both *is* and *isn't* a member validates that material biconditional, doing the job for $\{x : \phi(x)\}$ for *any* predicate $\ulcorner \phi \urcorner$. A material biconditional with one side 'both' true and false is, at least, true. In models in which B does the job of the empty set $\{x : \bot\}$, we have $\ulcorner a \in B \equiv \bot \urcorner$ materially true, but we are prepared to *assert* $\ulcorner a \in \{x : \bot\} \urcorner$ but at the same time *deny* $\bot$. Here ($\in L$) fails. The case for B standing in for the universal set is dual.

[17]This is clearly articulated by Read (2004).

[18]Yes, we must be careful of the nature of the context $\ulcorner \phi(\cdot) \urcorner$ and the terms $\ulcorner a \urcorner$ and $\ulcorner b \urcorner$. Here there will be no such opaque contexts or non-rigid designators.

rule $(=R)$ may seem worrying. In what follows we need not worry at all. For our purposes we need only appeal to $(=L_l)$, and for that rule we need only the case where $\ulcorner \phi(x) \urcorner$ is $\ulcorner t \in x \urcorner$. Nothing more is required.

Now there is a puzzle. Friends of Frege's Law (V) have long worried about Russell's paradox, involving class $\mathfrak{R}$ defined as $\{x : x \notin x\}$. For us, this is not the main concern, for nothing *we* have said involves negation, at least in the object-language.[19] Russell's paradox, if it is a paradox at all, is meant as a problem for naïve theories of classes, and as we have seen, we can express these as bitheories governed by the rules $(\in L)$, $(\in R)$, $(Ext_{\in})$, and the rules of identity. Can we express the core idea of Russell's paradox in the absence of negation or other propositional connectives in the object-language used to define conditions on classes? It turns out that we can. Using an idea from Hinnion and Libert (2003), we can express the paradox using class abstraction, membership and identity alone: using only the concepts we have used in the rules $(\in L)$, $(\in R)$, $(Ext_{\in})$ and nothing else. Therefore, we avoid all of the argument concerning the design of the logical vocabulary governing the predicates $\ulcorner \phi \urcorner$. We cut across all discussion of truth-value gaps or truth-value gluts, contraction, intensional connectives, or anything else. Hinnion and Libert give the following definition (Hinnion and Libert 2003, p.831),[20] which I will call the *Hinnion* class:

$$\mathfrak{H} =_{\mathrm{df}} \{x : \{y : x \in x\} = \{y : p\}\}$$

Notice, the vocabulary is what is given in the statements of comprehension and extensionality. There is no negation, conditionality, quantifiers, in the definition. It turns out that using only the rules $(\in L)$, $(\in R)$, $(Ext_{\in})$, $(=L_l)$, (Cut) and (Id) we can derive $\ulcorner p \urcorner$. The derivation is agnostic concerning gaps, gluts and any detail other than these rules.

Here is the first part of the derivation. Call it δ_1.

$$\dfrac{\dfrac{\dfrac{\mathfrak{H} \in \mathfrak{H} \vdash \mathfrak{H} \in \mathfrak{H}}{\mathfrak{H} \in \mathfrak{H} \vdash x \in \{y : \mathfrak{H} \in \mathfrak{H}\}}\,(\in R) \qquad \dfrac{\dfrac{p \vdash p}{x \in \{y : p\} \vdash p}\,(\in L)}{x \in \{y : \mathfrak{H} \in \mathfrak{H}\}, \{y : \mathfrak{H} \in \mathfrak{H}\} = \{y : p\} \vdash p}\,(=L_l)}{\mathfrak{H} \in \mathfrak{H}, \{y : \mathfrak{H} \in \mathfrak{H}\} = \{y : p\} \vdash p}\,(Cut)}{\mathfrak{H} \in \mathfrak{H} \vdash p}\,(\in L)$$

[19] Yes, we have kept track of assertion and denial. We have not committed ourselves to any particular theory of negation, or even the claim that our language has a single concept of negation. Just as we may be able to express a range of conditional notions, why not a range of negative notions? To think that there is one Russell set is to think that there is one negation.

[20] Actually, they use $\ulcorner \bot \urcorner$, not an arbitrary $\ulcorner p \urcorner$ used here. Nothing hangs on this, except the formulation here is slightly more general, designed to apply even in the case where we have no special statement taken to entail all others.

Now consider the next part. Call it δ_2.

$$\dfrac{\dfrac{\dfrac{\dfrac{\delta_1}{\vdots}}{\mathfrak{H}\in\mathfrak{H}\vdash p}}{x\in\{y:\mathfrak{H}\in\mathfrak{H}\}\vdash x\in\{y:p\},p}(\in L)\qquad \dfrac{p\vdash x\in\{y:\mathfrak{H}\in\mathfrak{H}\},p}{x\in\{y:p\}\vdash x\in\{y:\mathfrak{H}\in\mathfrak{H}\},p}(\in L)}{\dfrac{\vdash\{y:\mathfrak{H}\in\mathfrak{H}\}=\{y:p\},p}{\vdash\mathfrak{H}\in\mathfrak{H},p}(\in R)}(Ext_{\in})$$

Finally, we paste the two pieces together, to conclude $\ulcorner p\urcorner$.

$$\dfrac{\begin{matrix}\delta_2\\ \vdots\\ \vdash\mathfrak{H}\in\mathfrak{H},p\end{matrix}\qquad\begin{matrix}\delta_1\\ \vdots\\ \mathfrak{H}\in\mathfrak{H}\vdash p\end{matrix}}{\vdash p}(Cut)$$

Given that $\ulcorner p\urcorner$ is to be denied (for some $\ulcorner p\urcorner$ or other), *everyone* has to reject one of the rules $(\in L)$, $(\in R)$, $(Ext_{\in})$, $(=L_l)$, (*Cut*) and (*Id*). At some stage the derivation of $\ulcorner p\urcorner$ is to break down, but where? Orthodoxy tells us that the rules to reject (at least where $\ulcorner\in\urcorner$ expresses class membership) are $(\in L)$ or $(\in R)$, and the underlying assumption that every predicate determines a set: to reject Law (V). For defenders of Law (V), however, some other move must be rejected. For defenders of Law (V) concerning *classes*, the pickings seem extremely thin: either defend Law (V) despite rejecting $(\in L)$ or $(\in R)$—in the face of criticism that to reject $(\in L)$ or $(\in R)$ is to reject what we meant by Law (V) in the first place—or reject $(Ext_{\in})$ in the face that this was what we meant by *extensionality* in the first place—or finally, find fault in $(=L)$, (*Cut*) or (*Id*).

What option can the defender of Law (V) take? The bitheoretical perspective seems to constrain the options for non-classical theories of classes much more stark. Evading this paradox will, at least, help clarify what is at stake in taking a non-classical position on classes in defence of Law (V).

6.3.3 Truth

In the last section, I will see to what extent these results apply to theories of truth defending Tarski's T-scheme in the face of paradoxes like the liar. The structure is similar to Russell's paradox, but at face value there seems to be nothing playing the role of extensionality. Tarski's T-schema is often presented as a biconditional

$$T\{\!|A|\!\}\ \textit{if and only if}\ A$$

where $\ulcorner\{\!|\cdot|\!\}\urcorner$ is some quotation device, and there is some biconditional connecting $\ulcorner T\{\!|A|\!\}\urcorner$ with $\ulcorner A\urcorner$. But as is probably obvious by now, we will not take that route. Instead, we will notice that anyone prepared to assert $\ulcorner T\{\!|A|\!\}\urcorner$ and deny $\ulcorner A\urcorner$ or to deny $\ulcorner T\{\!|A|\!\}\urcorner$ and assert $\ulcorner A\urcorner$ is rejecting the equivalence. We have the following two rules, governing the bitheory of truth, governing expressions of the form $\ulcorner T\{\!|A|\!\}\urcorner$ in positions of assertion and of denial respectively.

$$\frac{\Gamma, A \vdash \Delta}{\Gamma, T\{\!|A|\!\} \vdash \Delta}\,(TL) \qquad \frac{\Gamma \vdash A, \Delta}{\Gamma \vdash T\{\!|A|\!\}, \Delta}\,(TR)$$

To understand the significance of these rules, we need to ask ourselves this: what kind of object is $\{\!|A|\!\}$? In particular, when is $\{\!|A|\!\}$ equal to $\{\!|B|\!\}$? One option is to think of $\{\!|A|\!\}$ as the *truth value* of $\ulcorner A\urcorner$. If that were the case we would have *extensionality* for the term $\ulcorner\{\!|A|\!\}\urcorner$.

$$\frac{\Gamma, T\{\!|A|\!\} \vdash T\{\!|B|\!\}, \Delta \qquad \Gamma, T\{\!|B|\!\} \vdash T\{\!|A|\!\}, \Delta}{\Gamma \vdash \{\!|A|\!\} = \{\!|B|\!\}, \Delta}\,(Ext_{T\{\!|\ |\!\}})$$

Anyone prepared to deny $\ulcorner\{\!|A|\!\} = \{\!|B|\!\}\urcorner$ must either be prepared to assert $\ulcorner T\{\!|A|\!\}\urcorner$ and deny $\ulcorner T\{\!|B|\!\}\urcorner$ or vice versa. Given extensionality for $\ulcorner\{\!|\ |\!\}\urcorner$ in this form, we have the analogue of the Hinnion–Libert paradox. We let the term $\ulcorner\mathfrak{L}\urcorner$ name the truth value of the expression $\ulcorner\{\!|T\mathfrak{L}|\!\} = \{\!|p|\!\}\urcorner$.[21] So we have the following definition

$$\mathfrak{L} =_{\mathrm{df}} \{\!|\{\!|T\mathfrak{L}|\!\} = \{\!|p|\!\}|\!\}$$

With this in place, we can form the following derivation. First, δ_3:

$$\frac{\dfrac{T\mathfrak{L} \vdash T\mathfrak{L}}{T\mathfrak{L} \vdash T\{\!|T\mathfrak{L}|\!\}}\,(TR) \qquad \dfrac{\dfrac{p \vdash p}{T\{\!|p|\!\} \vdash p}\,(TL)}{T\{\!|T\mathfrak{L}|\!\}, \{\!|T\mathfrak{L}|\!\} = \{\!|p|\!\} \vdash p}\,(=L_l)}{\dfrac{T\mathfrak{L}, \{\!|T\mathfrak{L}|\!\} = \{\!|p|\!\} \vdash p}{T\mathfrak{L} \vdash p}\,(TL)}\,(Cut)$$

[21] Diagonalisation, demonstratives, or other devices give you 'self-reference' enough for this.

Then using δ_3, we form δ_4:

$$\dfrac{\dfrac{\dfrac{\dfrac{\begin{array}{c}\delta_3\\ \vdots\\ T\mathfrak{L}\vdash p\end{array}}{T\{\!|T\mathfrak{L}|\!\}\vdash p}\,(TL)}{T\{\!|T\mathfrak{L}|\!\}\vdash T\{\!|p|\!\},p}\,\textit{Weakening}\qquad \dfrac{p\vdash T\{\!|T\mathfrak{L}|\!\},p}{T\{\!|p|\!\}\vdash T\{\!|T\mathfrak{L}|\!\},p}\,(TL)}{\vdash \{\!|T\mathfrak{L}|\!\}=\{\!|p|\!\},p}\,(\mathit{Ext}_{T\{\!|\;|\!\}})}{\vdash T\mathfrak{L},p}\,(TR)$$

Then δ_3 and δ_4 give us:

$$\dfrac{\begin{array}{c}\delta_4\\ \vdots\\ \vdash T\mathfrak{L},p\end{array}\qquad \begin{array}{c}\delta_3\\ \vdots\\ T\mathfrak{L}\vdash p\end{array}}{\vdash p}\,(\mathit{Cut})$$

It follows that *everyone* who rejects some proposition $\ulcorner p\urcorner$ has to reject one of (TL), (TR), $(\mathit{Ext}_{T\{\!|\;|\!\}})$, $(=L_l)$, (Cut) and (Id). Here the problem does not seem to be so stark, as the commitment to truth *values* in the form required by $(\mathit{Ext}_{T\{\!|\;|\!\}})$ seems rather strong for the defender of a non-classical logic with gaps or gluts.[22]

However, the problem does not go away. Instead of focussing on *truth values* perhaps we should consider *propositions*. We can replace the appeal to $(\mathit{Ext}_{T\{\!|\;|\!\}})$ by appeal to identity of *co-entailing* propositions. Think of $[\![A]\!]$ as the *proposition* to the effect *that* A. Here the criterion of intensional identity is that denying

[22]How many truth values are there? Using $(\mathit{Ext}_{T\{\!|\;|\!\}})$ it seems there are only *two*, since we can derive $\vdash \{\!|A|\!\}=\{\!|B|\!\},\{\!|B|\!\}=\{\!|C|\!\},\{\!|C|\!\}=\{\!|A|\!\}$. Using the form of $(\mathit{Ext}_{T\{\!|\;|\!\}})$ with weakening built in:

$$\dfrac{\Gamma,T\{\!|A|\!\}\vdash T\{\!|B|\!\},\Delta\quad \Gamma',T\{\!|B|\!\}\vdash T\{\!|A|\!\},\Delta'}{\Gamma,\Gamma'\vdash \{\!|A|\!\}=\{\!|B|\!\},\Delta,\Delta'}\,[\mathit{Ext}]$$

simply to make the proofs *narrow* enough to fit on the page, we have

$$\dfrac{\dfrac{T\{\!|B|\!\},T\{\!|C|\!\}\vdash T\{\!|C|\!\},T\{\!|A|\!\}\quad T\{\!|A|\!\}\vdash T\{\!|C|\!\},T\{\!|A|\!\}}{T\{\!|B|\!\}\vdash T\{\!|C|\!\},T\{\!|A|\!\},\{\!|C|\!\}=\{\!|A|\!\}}\,[\mathit{Ext}]\qquad T\{\!|B|\!\},T\{\!|C|\!\}\vdash T\{\!|B|\!\}}{T\{\!|B|\!\}\vdash T\{\!|A|\!\},\{\!|B|\!\}=\{\!|C|\!\},\{\!|C|\!\}=\{\!|A|\!\}}\,[\mathit{Ext}]$$

and similarly, we can prove $T\{\!|A|\!\}\vdash T\{\!|B|\!\},\{\!|B|\!\}=\{\!|C|\!\},\{\!|C|\!\}=\{\!|A|\!\}$, which together give us

$$\dfrac{T\{\!|A|\!\}\vdash T\{\!|B|\!\},\{\!|B|\!\}=\{\!|C|\!\},\{\!|C|\!\}=\{\!|A|\!\}\qquad T\{\!|B|\!\}\vdash T\{\!|A|\!\},\{\!|B|\!\}=\{\!|C|\!\},\{\!|C|\!\}=\{\!|A|\!\}}{\vdash \{\!|A|\!\}=\{\!|B|\!\},\{\!|B|\!\}=\{\!|C|\!\},\{\!|C|\!\}=\{\!|A|\!\}}\,[\mathit{Ext}]$$

In other words, of any three truth values, two are equal. To prove that there are *at least two* truth values, more must be done. I suggest finding sentences $\top$ and $\bot$ such that $\{\!|\top|\!\}=\{\!|\bot|\!\}\vdash$.

$\ulcorner\llbracket A\rrbracket = \llbracket B\rrbracket\urcorner$ involves either asserting $\ulcorner T\llbracket A\rrbracket\urcorner$ and denying $\ulcorner T\llbracket B\rrbracket\urcorner$ or vice versa, now no longer keeping other assumptions as side-conditions. Reading the rule from top-to-bottom, it means merely that if $\ulcorner A\urcorner$ *entails* $\ulcorner B\urcorner$ and $\ulcorner B\urcorner$ *entails* $\ulcorner A\urcorner$ then the propositions $\llbracket A\rrbracket$ and $\llbracket B\rrbracket$ are identical.

$$\dfrac{T\llbracket A\rrbracket \vdash T\llbracket B\rrbracket \qquad T\llbracket B\rrbracket \vdash T\llbracket A\rrbracket}{\vdash \llbracket A\rrbracket = \llbracket B\rrbracket}\,(Int_{T\llbracket\ \rrbracket})$$

This is, as with $(Ext_{T\{\!|\ |\!\}})$, a substantial commitment. However, it is a commitment to the heart of the model-theory of many non-classical logics. In these models if the entailment from $\ulcorner A\urcorner$ to $\ulcorner B\urcorner$ fails, there is some *point* (world, situation, whatever) where $\ulcorner A\urcorner$ holds and $\ulcorner B\urcorner$ does not. Construing a proposition as a set of points (perhaps satisfying some kind of closure or coherence condition), if $\ulcorner A\urcorner$ and $\ulcorner B\urcorner$ are co-entailing, they hold at the same points, and so correspond to the same propositions, construed as sets of points. So the condition $(Int_{T\llbracket\ \rrbracket})$ is not a foreign idea.

However, $(Int_{T\llbracket\ \rrbracket})$ causes nearly as much trouble as $(Ext_{T\{\!|\ |\!\}})$. If we have a statement $\bot$, from which anything follows (which is always unassertible), we can replace $(Ext_{T\{\!|\ |\!\}})$ by $(Int_{T\llbracket\ \rrbracket})$ in our problematic derivation. Using fixed-points, define our problematic proposition term $\ulcorner\mathfrak{P}\urcorner$ by setting

$$\mathfrak{P} =_{\mathsf{df}} \llbracket\llbracket T\mathfrak{P}\rrbracket = \llbracket\bot\rrbracket\rrbracket$$

Then we have δ_5:

$$\dfrac{\dfrac{\dfrac{\dfrac{\dfrac{}{\bot\vdash}\,(\bot L)}{T\llbracket\bot\rrbracket\vdash}\,(TL)}{}\!\!\!\!\!\quad}{}}{}$$

$$\dfrac{\dfrac{\dfrac{T\mathfrak{P}\vdash T\mathfrak{P}}{T\mathfrak{P}\vdash T\llbracket T\mathfrak{P}\rrbracket}\,(TR) \qquad \dfrac{\dfrac{\dfrac{}{\bot\vdash}\,(\bot L)}{T\llbracket\bot\rrbracket\vdash}\,(TL)}{T\llbracket T\mathfrak{P}\rrbracket, \llbracket T\mathfrak{P}\rrbracket = \llbracket\bot\rrbracket\vdash}\,(=L_l)}{T\mathfrak{P}, \llbracket T\mathfrak{P}\rrbracket = \llbracket\bot\rrbracket\vdash}\,(Cut)}{T\mathfrak{P}\vdash}\,(TL)$$

Using δ_5 we can construct δ_6:

$$\dfrac{\dfrac{\dfrac{\dfrac{\dfrac{\begin{matrix}\delta_5\\ \vdots\\ T\mathfrak{P}\vdash\end{matrix}}{T\mathfrak{P}\vdash\bot}\,(\bot R)}{T\llbracket T\mathfrak{P}\rrbracket\vdash\bot}\,(TL)}{T\llbracket T\mathfrak{P}\rrbracket\vdash T\llbracket\bot\rrbracket}\,(TR) \qquad \dfrac{\bot\vdash T\llbracket T\mathfrak{P}\rrbracket}{T\llbracket\bot\rrbracket\vdash T\llbracket T\mathfrak{P}\rrbracket}\,(TL)}{\dfrac{\vdash\llbracket T\mathfrak{P}\rrbracket = \llbracket\bot\rrbracket}{\vdash T\mathfrak{P}}\,(TR)}\,(Int_{T\llbracket\ \rrbracket})$$

Together, δ_5 and δ_6 give us

$$\dfrac{\begin{array}{c}\delta_6\\ \vdots\\ \vdash T\mathfrak{P}\end{array}\qquad \begin{array}{c}\delta_5\\ \vdots\\ T\mathfrak{P}\vdash\end{array}}{\vdash}\;(Cut)$$

which is a problematic conclusion, since $\Gamma \vdash \Delta$ follows for every Γ, Δ. This tells us that there is a clash in every position.

It follows that *everyone* has to reject one of (TL), (TR), $(Int_{T\llbracket\,\rrbracket})$, $(=L_l)$, $(\bot L)$, $(\bot R)$, (Cut) and (Id). If, in particular, the defender of (TL) and (TR) wishes to reject $(Int_{T\llbracket\,\rrbracket})$, the onus is on her or him to give an account of the semantics of the non-classical logic in use in such a way as to not allow for a definition of propositions which motivates $(Int_{T\llbracket\,\rrbracket})$. Given the widespread use of world-like semantics, this seems to be a significant challenge.

6.4 Conclusion

Here is the moral of the story so far: bitheories (and sequent rules) give us a way to specify natural conditions on concepts, such as

$$numbers \quad \in \quad \{:\} \quad T \quad \{\!|\ |\!\} \quad \llbracket\ \rrbracket$$

in a way that abstracts away from debates over this or that logic. The results here apply to logics with gaps, with gluts, with any number of different connectives. Attending instead to the way entailment constrains assertion and denial allows us to avoid stepping in to those difficult debates, to uncover common structure underlying many different theories in many different logics.

Let me end with some homework for everyone interested in these issues.

1. *For everyone*: Use *bitheories*. The bitheoretical formulation of theories of numbers, classes and truth has proved to be clarifying. You do not need to be a partisan in favour of non-classical logics to be interested in a formulation of arithmetic which allows for the arithmetic rules to be independent from the connectives and quantifiers. In this way, we have a natural account of *positive* arithmetic (arithmetic without negation), of the shared core between classical Peano arithmetic and intuitionist Heyting arithmetic, and many other connections may be explored.
2. *For friends of gaps or gluts*: Articulate and defend your commitments connecting consequence, negation, assertion and denial. I have attempted to sketch what I take to be those connections. Perhaps the story told here is wrong. Regardless, it is certainly incomplete. Friends of gaps and gluts should not merely present *theories* of things they take to be true. Given a gap or a glut at the boundary

between truth and falsity, the presentation of a theory is more complicated, and the connections between logical consequence, assertion and denial—and the role of the concept (or concepts?) of negation must be articulated. The role of (*Cut*) and (*Id*) sketched here (and defended in Restall 2005) are crucial in everything we have done. If there was some way to live without (*Cut*) or (*Id*), that would open up more space for strong non-classical theories of classes and truth. But what can we leave in their place? What is the connection between assertion and denial, consequence and negation if not the one sketched here? What story can be told?

3. Finally, *for friends of strong theories such as* ($\in L/R$) *or* (TL/R): articulate and defend your response to these paradoxes. In particular, a friend of Law (V) for classes, or the T-scheme for truth must explain which of (*Id*), (*Cut*), ($=L_l$) and (*Ext & Int*) are to be rejected. In particular, the defender of these theories needs to isolate a point in each derivation where it breaks down: a rule cannot fail merely because it does not satisfy this or that strong constraint on validity, but rather, in the cases in question we must find a spot in the derivation where we are prepared to grant the premises of a rule but reject the conclusion.

   Answering this challenge will involve work. (In particular, it will involve giving an answer to Homework Task 2.) No matter how this challenge is met, an answer will help us understand non-classical theories of classes and truth much more.

Acknowledgements Comments on this paper are very welcome. Please check the webpage http://consequently.org/writing/adnct to post comments and to read comments left by others. Thanks to the Logic Seminar at the University of Melbourne and the audience at WCP4 (including Jc Beall, Patrick Girard and Jerry Seligman), and to Michael De, for comments on this material, and to an anonymous referee, who helpfully pressed on certain points, and asked me to write up the little proof (in footnote 22) that there are at most two truth values. This research is supported by the Australian Research Council, through grant DP0343388, and Max Richter's *24 Postcards in Full Colour.*

References

Brady, R. 2006. *Universal logic*. Stanford: CSLI.

Gilmore, P.C. 1974. The consistency of partial set theory without extensionality. In *Axiomatic set theory. Proceedings of symposia in pure mathematics*, vol. 13, ed. T.J. Jech, 147–153. Providence: American Mathematical Society.

Gilmore, P.C. 1986. Natural deduction based set theories: A new Resolution of the old paradoxes. *Journal of Symbolic Logic* 51: 393–411.

Girard, J-Y. 1987. Linear logic. *Theoretical Computer Science* 50: 1–101.

Hinnion, R., and T. Libert. 2003. Positive abstraction and extensionality. *The Journal of Symbolic Logic* 68(3): 828–836.

Lewis, D. 1982. Logic for equivocators. *Noûs* 16(3): 431–441.

Meyer, R.K. 1976. Relevant arithmetic. *Bulletin of the Section of Logic* 5: 133–137.

Meyer, R.K., and C. Mortensen. 1984. Inconsistent models for relevant arithmetics. *Journal of Symbolic Logic* 49: 917–929.

Meyer, R.K., and G. Restall. 1999. 'Strenge' arithmetic. *Logique et Analyse* 42: 205–220.

Meyer, R.K., R. Routley, and J.M. Dunn. 1979. Curry's paradox. *Analysis* 39: 124–128.
Priest, G. 1990. Boolean negation and all that. *Journal of Philosophical Logic* 19(2): 201–215.
Priest, G. 2006. *In contradiction: A study of the transconsistent*. Oxford: Oxford University Press.
Restall, G. 1992. A note on naïve set theory in *LP*. *Notre Dame Journal of Formal Logic* 33(3): 422–432.
Restall, G. 2000. *An introduction to substructural logics*. London/New York:Routledge.
Read, S. 2004. Identity and harmony. *Analysis* 64(2): 113–115.
Restall, G. 2005. Multiple conclusions. In *Logic, methodology and philosophy of science: Proceedings of the twelfth international congress*, ed. P. Hájek, L. Valdés-Villanueva, and D. Westerståhl, 189–205. London: KCL Publications.
Restall, G. 2009. Models for substructural arithmetics. In *Miscellanea logica*, vol. VII, ed. M. Bilková, 1–20. Charles University, Prague. See also http://consequently.org/writing/mfsa
Slaney, J.K., R.K. Meyer, and G. Restall. 1996. Linear arithmetic desecsed. *Logique et Analyse* 39: 379–388.

Chapter 7
New Arguments for Adaptive Logics as Unifying Frame for the Defeasible Handling of Inconsistency

Diderik Batens

7.1 Introduction

A variety of formats is used to present defeasible logics. More often than not, the format is typical for the logic and derives from the accidental way in which the logic was discovered. Not only the object level description, but also the proof techniques needed for metatheorems vary with those formats. Unifying this domain seems highly useful if not necessary.

As soon as a standard format for adaptive logics was devised,[1] it seemed to offer an attractive means for unification. Today nearly all (first order) defeasible logics have been characterised by adaptive logics. Moreover, the unification is a strong one. If an adaptive logic is in standard format, the format itself defines the logic's proof theory and semantics. Moreover, most of the metatheory has been proved in terms of the standard logic alone. This includes soundness and completeness and a host of properties.

The standard format of adaptive logics may still prove not to be the right unifying frame. New defeasible logics may be discovered and may require that the format is modified or replaced. Or another format may turn out superior in the end. Nevertheless, especially in terms of the new arguments presented below, it is certainly worthwhile to continue the unification in terms of the standard format.

In the present paper, four new arguments are presented in favour of characterizing defeasible reasoning forms by adaptive logics in standard format. The arguments are diverse in nature, but all point in the same direction.

---

[1]The first steps were taken in Batens (2001), but later the matter was refined. The best published formulation appears in Batens (2007). The most reliable reference on adaptive logics is Batens (201+), of which the central chapters are available on the web.

D. Batens (✉)
Centre for Logic and Philosophy of Science, Ghent University, Ghent, Belgium
e-mail: diderik.batens@ugent.be

K. Tanaka et al. (eds.), *Paraconsistency: Logic and Applications*, Logic, Epistemology, and the Unity of Science 26, DOI 10.1007/978-94-007-4438-7_7,

For the first two arguments, more technical papers are in preparation. This is why I shall consider them briefly, pointing out the results and commenting on their significance while referring, for technical matters, to the forthcoming papers. The third and fourth argument are presented a bit more at length.

7.2 Preliminaries

In order to make the paper minimally self contained, I shall first briefly summarise the standard format of adaptive logics. First, however, I need to introduce some logics.

Where **CL** is classical logic, let **CLuN** be the full positive fragment of **CL** together with the axiom $A \vee \neg A$.[2] **CLuN** is just like **CL** except that it allows for gluts with respect to negation (whence its name). So it is a paraconsistent logic and actually (with respect to **CL**) the most basic paraconsistent logic that is not also paracomplete. **CLuNs**, studied at length in Batens and De Clercq (2004), is the paraconsistent logic obtained by extending **CLuN** with double negation (in both directions) De Morgan axioms, axioms expressing the standard classical behaviour of negations of implications, negations of equivalences, and negations of the quantifiers, and Replacement of Identicals—its name refers to Schütte who first described its propositional fragment in Schütte (1960). **LP** is a fragment of **CLuNs**: all logical symbols have the same meaning as in **CLuNs** except for implication and equivalence, which are explicitly defined by $A \supset B =_{df} \neg A \vee B$ and $A \equiv B =_{df} (A \wedge B) \vee (\neg A \wedge \neg B)$ and hence are not detachable.

The sequel of this section may be skipped by people familiar with adaptive logics. An adaptive logic **AL** is defined by a triple:

1. A *lower limit logic* **LLL**: a reflexive, transitive, monotonic, and compact logic for which there is a positive test.
2. A *set of abnormalities* Ω: a set of **LLL**-contingent formulas, characterised by a (possibly restricted) logical form F which contains at least one logical symbol.
3. An *adaptive strategy*: Reliability, Minimal Abnormality, ...

The lower limit logic is the stable part of the adaptive logic; anything that follows from the premises by **LLL** will never be revoked. For technical reasons, all classical symbols are added to the lower limit logic, whence this extends **CL**. In the present context, this means that classical negation, $\check{\neg}$, is added next to the standard negation, $\neg$, which is paraconsistent. In standard applications, $\check{\neg}$ does not occur in the premises or in the conclusion. Its function is technical and metatheoretical. Abnormalities are supposed to be false "unless and until proven otherwise". Strategies are ways to handle derivable disjunctionsof abnormalities:

[2]Replacement of Identicals is not derivable in **CLuN** but can be added.

an adaptive strategy picks one specific way to interpret the premises as normally as possible. To keep the discussion with bounds, I shall only consider the Minimal Abnormality strategy—see below—in the present paper.

From now on, I shall take "adaptive logic" to mean adaptive logic in standard format. Inconsistency-adaptive logics are adaptive logics the lower limit of which has a paraconsistent standard negation.

Let us review some examples of inconsistency-adaptive logics. $\mathbf{CLuN}^m$ has **CLuN** as its lower limit logic, $\Omega = \{\exists(A \wedge \neg A) \mid A \in \mathscr{F}\}$, and Minimal Abnormality as its strategy—$\mathscr{F}$ is the set of open and closed formulas and $\exists(A \wedge \neg A)$ is the existential closure of $A \wedge \neg A$. $\mathbf{CLuNs}^m$ is similar, except that **CLuNs** is its lower limit and its set of abnormalities is $\Omega^a = \{\exists(A \wedge \neg A) \mid A \in \mathscr{F}^a\}$, in which $\mathscr{F}^a$ is the set of open and closed *primitive* formulas (those that contain no logical symbol except possibly for identity). $\mathbf{LP}^m$ is exactly like $\mathbf{CLuNs}^m$ except that **LP** is its lower limit.

If the lower limit logic is extended with an axiom by which all abnormalities entail triviality, one obtains the *upper limit logic* **ULL**. The upper limit logic of $\mathbf{CLuN}^m$, of $\mathbf{CLuNs}^m$, and of $\mathbf{LP}^m$ is **CL**. If a premise set Γ does not require that any abnormalities are true, the **AL**-consequences of Γ are identical to its **ULL**-consequences. In the opposite case, the **AL**-consequence set of Γ will in general be a superset of its **LLL**-consequences.

In the expression $Dab(\Delta)$, Δ is a finite subset of Ω and $Dab(\Delta)$ denotes the *classical* disjunction of the members of Δ. $Dab(\Delta)$ is called a *Dab-formula*. $Dab(\Delta)$ is a *minimal Dab-consequence of* Γ iff $\Gamma \vdash_{\mathbf{LLL}} Dab(\Delta)$ whereas $\Gamma \nvdash_{\mathbf{LLL}} Dab(\Delta')$ for all $\Delta' \subset \Delta$. Where $Dab(\Delta_1)$, $Dab(\Delta_2)$, ... are the minimal *Dab*-consequences of Γ, $\Phi(\Gamma)$ comprises the minimal choice sets of $\{\Delta_1, \Delta_2, \ldots\}$. Where M is a **LLL**-model, $Ab(M)$ is the set of abnormalities verified by M.

Definition 7.1. A **LLL**-model M of Γ is *minimally abnormal* iff there is no **LLL**-model M' of Γ such that $Ab(M') \subset Ab(M)$.

Definition 7.2. $\Gamma \vDash_{\mathbf{AL}^m} A$ iff A is verified by all minimally abnormal models of Γ.

It was proved in Batens (2007) that a **LLL**-model M of Γ is *minimally abnormal* iff $Ab(M) \in \Phi(\Gamma)$.

Adaptive logics have also a dynamic proof theory, which is defined by rules of inference and by a marking definition. An annotated **AL**-proof consists of lines that have four elements: a line number, a formula, a justification and a condition. Where

$$A \qquad \Delta$$

abbreviates that A occurs in the proof as the formula of a line that has Δ as its condition, the (generic) inference rules are:

PREM If $A \in \Gamma$:

$$\frac{\begin{array}{ll} \ldots & \ldots \end{array}}{\begin{array}{ll} A & \emptyset \end{array}}$$

RU If $A_1, \ldots, A_n \vdash_{\mathbf{LLL}} B$:

$$\frac{\begin{array}{ll} A_1 & \Delta_1 \\ \ldots & \ldots \\ A_n & \Delta_n \end{array}}{\begin{array}{ll} B & \Delta_1 \cup \ldots \cup \Delta_n \end{array}}$$

RC If $A_1, \ldots, A_n \vdash_{\mathbf{LLL}} B \check{\vee} Dab(\Theta)$

$$\frac{\begin{array}{ll} A_1 & \Delta_1 \\ \ldots & \ldots \\ A_n & \Delta_n \end{array}}{\begin{array}{ll} B & \Delta_1 \cup \ldots \cup \Delta_n \cup \Theta \end{array}}$$

In RU, $\check{\vee}$ abbreviates classical disjunction. By applying the above rules, one moves from one stage of a proof to another. A *stage* is a list of lines—stage 0 of any proof is the empty list. Stage s' is an *extension* of s iff all lines that occur in s occur in the same order in s'. A dynamic proof is a chain of stages.

That A is derivable on the condition Δ from the premise set Γ may be interpreted as follows: it follows from Γ that A or one of the members of Δ is true. As the members of Δ, which are abnormalities, are supposed to be false, A is considered as derived, unless and until the supposition cannot be upheld. The precise meaning of this depends on the strategy, which determines the marking definition (see below) and hence determines which lines are marked at a stage. If a line is marked at a stage, its formula is considered as not derived at that stage.

$Dab(\Delta)$ is a *minimal Dab-formula* at stage s of an **AL**-proof iff, at stage s, $Dab(\Delta)$ is derived on the condition $\emptyset$ and there is no $\Delta' \subset \Delta$ for which $Dab(\Delta')$ is derived on the condition $\emptyset$. Where $Dab(\Delta_1), \ldots, Dab(\Delta_n)$ are the minimal *Dab*-formulas at stage s of a proof from Γ, $\Phi_s(\Gamma)$ is the set of minimal choice sets of $\{\Delta_1, \ldots, \Delta_n\}$.

Definition 7.3. Marking for Minimal Abnormality: Line l is marked at stage s iff, where A is derived on the condition Δ at line l, (1) there is no $\varphi \in \Phi_s(\Gamma)$ such that $\varphi \cap \Delta = \emptyset$, or (2) for some $\varphi \in \Phi_s(\Gamma)$, there is no line on which A is derived on a condition Θ for which $\varphi \cap \Theta = \emptyset$.

This reads more easily: where A is derived on the condition Δ at line l, line l is *unmarked* at stage s iff (1) there is a $\varphi \in \Phi_s(\Gamma)$ for which $\varphi \cap \Delta = \emptyset$ and (2) for every $\varphi \in \Phi_s(\Gamma)$, there is a line at which A is derived on a condition Θ for which $\varphi \cap \Theta = \emptyset$.

Definition 7.4. A is *finally derived* from Γ at line l of a stage s iff (1) A is the second element of line l, (2) line l is not marked at stage s, and (3) every extension of the stage in which line l is marked may be further extended in such a way that line l is unmarked.

Definition 7.5. $\Gamma \vdash_{\mathbf{AL}} A$ (A is *finally* **AL**-*derivable* from Γ) iff A is finally derived at a line of a proof from Γ.

As announced, most of the metatheory is provable in terms of the standard format, including that $\Gamma \vdash_{\mathbf{AL}} A$ iff $\Gamma \vDash_{\mathbf{AL}} A$.

7.3 Equivalent Premise Sets

This section reports on joint work with Peter Verdée and Christian Straßer (see Batens et al. 2009b). It is often important to determine whether two premise sets, Γ and Γ', are equivalent with respect to a logic **L**, i.e. $Cn_{\mathbf{L}}(\Gamma) = Cn_{\mathbf{L}}(\Gamma')$. Thus, two theories may be 'identical' or not and two people may or may not share the same view on some topic. Determining whether two premise sets are identical by computing the sets $Cn_{\mathbf{L}}(\Gamma)$ and $Cn_{\mathbf{L}}(\Gamma')$ is obviously an impossible task. Fortunately certain criteria may be applied if the underlying logic is a Tarski logic (a reflexive, transitive, monotonic consequence relation), which is the common type of logics.[3]

Let $\mathbf{L}'$ be *weaker than* **L** iff $Cn_{\mathbf{L}'}(\Gamma) \subset Cn_{\mathbf{L}}(\Gamma)$ for some Γ and $Cn_{\mathbf{L}'}(\Gamma) \subseteq Cn_{\mathbf{L}}(\Gamma)$ for all Γ. The three most straightforward criteria are C1–C3 below. C1 is a direct criterion; the other criteria refer to a different logic. C2 and C3 are especially handy if **L** is a complicated logic.

C1 If $\Gamma' \subseteq Cn_{\mathbf{L}}(\Gamma)$ and $\Gamma \subseteq Cn_{\mathbf{L}}(\Gamma')$, then Γ and Γ' are **L**-equivalent.
C2 If $\mathbf{L}'$ is a Tarski logic weaker than **L**, and Γ and Γ' are $\mathbf{L}'$-equivalent, then Γ and Γ' are **L**-equivalent.
C3 If every $Cn_{\mathbf{L}}(\Delta)$ is closed under a Tarski logic $\mathbf{L}'$ (viz. $Cn_{\mathbf{L}'}(Cn_{\mathbf{L}}(\Delta)) = Cn_{\mathbf{L}}(\Delta)$ for all Δ), and Γ and Γ' are $\mathbf{L}'$-equivalent, then Γ and Γ' are **L**-equivalent.

For most defeasible logics, as formulated in the literature, one or more of the criteria break down. Easy examples are the Strong (or inevitable) consequence relation ($\Gamma \vdash_{Strong} A$ iff $\Gamma' \vdash_{\mathbf{CL}} A$ for every maximal consistent subset of Γ' of Γ) and the Weak consequence relation ($\Gamma \vdash_{Weak} A$ iff $\Gamma' \vdash_{\mathbf{CL}} A$ for some maximal consistent subset of Γ' of Γ)—see Rescher and Manor (1970) and Benferhat et al. (1997). Note that C1 does not hold for the Weak consequence relation and that C3 fails for the Strong consequence relation. The way in which some defeasible logics are presented causes the situation even to be worse. Thus criteria C1–3 require heavy reformulation before they even make a chance to be applicable to the many kinds of default logics or to the very transparent pivotal-assumption consequence relations defined in Makinson (2005).

The situation is completely different for adaptive logics: criteria C1–C3 provably hold for all of them. The proofs (in Batens et al. 2009b) rely on the fact that all adaptive logics have the following properties: reflexivity, fixed point ($Cn_{\mathbf{AL}}(Cn_{\mathbf{AL}}(\Gamma)) = Cn_{\mathbf{AL}}(\Gamma)$), cumulative monotonicity (if $\Gamma' \subseteq Cn_{\mathbf{AL}}(\Gamma)$, then

[3]Tarski logics that are compact and semi-recursive may be characterised as logics that have static proofs, whereas defeasible logics have dynamic proofs. A first version of the theoretical analysis of such notions is presented in Batens (2009a).

$Cn_{\mathbf{AL}}(\Gamma) \subseteq Cn_{\mathbf{AL}}(\Gamma \cup \Gamma')$), and cumulative transitivity (if $\Gamma' \subseteq Cn_{\mathbf{AL}}(\Gamma)$ then $Cn_{\mathbf{AL}}(\Gamma \cup \Gamma') \subseteq Cn_{\mathbf{AL}}(\Gamma)$)—note that these properties are provable from the standard format. So, for adaptive logics, we have handy criteria for determining the equivalence of premise sets (and the identity of theories) and these criteria are the same as for Tarski logics.

Some will wonder how this is possible, given the claim that all defeasible first-order logics can be characterised by an adaptive logic. The reason is that the characterization often proceeds under a translation. An example might clarify this. Let the premises be formulated with classical negation, $\check{\neg}$. Let $\Gamma^{\neg\check{\neg}} = \{\neg\check{\neg}A \mid A \in \Gamma\}$ and let $\mathcal{W}^{\not\neg}$ be the set of closed formulas that do not contain $\neg$ (but may contain $\check{\neg}$). It was proved in Batens (2000) that $Cn_{Strong}(\Gamma) = Cn_{\mathbf{CLuN}^m}(\Gamma^{\neg\check{\neg}}) \cap \mathcal{W}^{\not\neg}$. So while C3 does not hold for the Strong consequence relation, C3 applies once the two premise sets are so translated and the 'logic' *Strong* is replaced by $\mathbf{CLuN}^m$.

There is a further result on extending premise sets. For every Tarski logic **L**, $\Gamma \cup \Delta$ and $\Gamma' \cup \Delta$ are **L**-equivalent if Γ and Γ' are. This does not hold for defeasible logics, not even for adaptive ones. However, for adaptive logics there is (apart from a specific criterion) a very close approximation: If **L** is a Tarski logic weaker than **AL** and Γ and Γ' are **L**-equivalent, then $\Gamma \cup \Delta$ and $\Gamma' \cup \Delta$ are **AL**-equivalent for all Δ.

Two other important results are proven in Batens et al. (2009b). Where **AL** is an adaptive logic and **LLL** is its lower limit logic: (1) every monotonic logic **L** that is weaker than **AL** is weaker than **LLL** or identical to it and (2) if $Cn_{\mathbf{AL}}(\Gamma)$ is closed under a monotonic logic **L**, then **L** is weaker than **LLL** or identical to it. This means that the lower limit logic provides very sharp versions of C2 and C3 and of the criterion mentioned in the previous paragraph.

7.4 Reducing Tinkering

Both the structure of the $\mathbf{C}_n$ logics and certain statements of da Costa's seem to suggest that a certain stratagem should be applied to theories that turn out inconsistent. Whether da Costa had this application in mind or not, the stratagem is clearly interesting and suggested by the $\mathbf{C}_n$ logics. It is worthwhile to develop inconsistency-adaptive logics that have the $\mathbf{C}_n$ systems as their lower limit because these enable one to accomplish, in more comfortable circumstances, the task served by the stratagem. The results presented in this section are studied at length in Batens (2009). So I shall be brief here.

7.4.1 The $\mathbf{C}_n$ Logics and the Stratagem

The $\mathbf{C}_n$-logics form a hierarchy. A simple way to describe it—not da Costa's original one—goes as follows. Let $\mathbf{C}_{\overline{\omega}}$ be full positive (predicative) **CL** together with the axioms $A \vee \neg A$ and $\neg\neg A \supset A$ and the rule "if $A \equiv^c B$, then $\vdash A \equiv B$", in which $A \equiv^c B$ iff A and B are congruent in the sense of (Kleene 1952, p. 153) or one is obtained from the other by deleting vacuous quantifiers.[4]

Let A^1 abbreviate $\neg(A \wedge \neg A)$,[5] let A^2 abbreviate $\neg(A^1 \wedge \neg A^1)$, etc., and let $A^{(n)}$ abbreviate $A^1 \wedge A^2 \wedge \ldots \wedge A^n$. The logic $\mathbf{C}_n$ ($n \in \{1, 2, \ldots\}$) is obtained by extending $\mathbf{C}_{\overline{\omega}}$ with the following axioms

$$B^{(n)} \supset ((A \supset B) \supset ((A \supset \neg B) \supset \neg A))$$

$$(A^{(n)} \wedge B^{(n)}) \supset (A \dagger B)^{(n)} \qquad \text{where } \dagger \in \{\vee, \wedge, \supset\}$$

$$\mathsf{Q}x(A(x))^{(n)} \supset (\mathsf{Q}x\, A(x))^{(n)} \qquad \text{where } \mathsf{Q} \in \{\forall, \exists\}$$

A formula of the form $A^{(n)}$ is a *consistency statement* in $\mathbf{C}_n$. It expresses that A behaves consistently—see for example da Costa (1974)—in that $A, \neg A, A^{(n)} \vdash_{\mathbf{C}_n} B$. Incidentally, $\neg^{(n)} A =_{df} \neg A \wedge A^{(n)}$ defines classical negation in $\mathbf{C}_n$.

The $\mathbf{C}_n$ logics form a hierarchy in that $\Gamma \vdash_{\mathbf{C}_n} A$ if $\Gamma \vdash_{\mathbf{C}_m} A$ for some $m > n$. $\mathbf{C}_{\overline{\omega}}$ forms a limit of this hierarchy. As it will be useful to have classical negation available even in $\mathbf{C}_{\overline{\omega}}$, let us extend the language with the symbol $\check{\neg}$ and give it the meaning of classical negation (by introducing the usual axioms)—the standard negation, $\neg$, is still paraconsistent. Note the difference between $\neg^{(n)}$ and $\check{\neg}$. The first is definable within the standard language and behaves like classical negation in all $\mathbf{C}_m$ with $m \leq n$, but is not definable in $\mathbf{C}_{\overline{\omega}}$. The second symbol does not belong to the standard language, and hence does not occur in the premises, but is added to the language for technical reasons.[6]

Two features of the $\mathbf{C}_n$ logics may cause some wonder. First, what is the use of having classical negation, viz. the symbol $\neg^{(n)}$, definable within paraconsistent logics? Next, what is the use of the hierarchy of $\mathbf{C}_n$ logics? The following paragraphs answer these questions, possibly with hindsight.

The paraconsistent $\mathbf{C}_n$ were introduced to replace **CL** in inconsistent contexts. Let $T_0 = \langle \Gamma_0, \mathbf{CL} \rangle$ turn out to be inconsistent. Replacing T_0 by $T_1 = \langle \Gamma_0, \mathbf{C}_1 \rangle$ saves the theory from triviality—I suppose that Γ_0 does not contain any formulas of the form $\neg(A \wedge \neg A)$ because these are **CL**-tautologies. At the same time, however, T_1 is much poorer than is desirable. Suppose that $A \vee B$ and $\neg A$ are $\mathbf{C}_1$-derivable from Γ_0 and that A is not. As Γ_0 was intended to be consistent, one would expect B to be derivable as well. But $A \vee B, \neg A \nvdash_{\mathbf{C}_1} B$. So, if A is not $\mathbf{C}_1$-derivable from Γ_0,

[4]All $\mathbf{C}_n$ logics defined below in the text are identical to da Costa's, except that he introduces $\mathbf{C}_\omega$ as the limit. $\mathbf{C}_\omega$ is like $\mathbf{C}_{\overline{\omega}}$ except that the former has positive intuitionistic logic where the latter has positive classical logic. An interesting study of limits of the hierarchy is presented in Carnielli and Marcos (1999). The logic $\mathbf{C}_{\overline{\omega}}$ is there called $\mathbf{C}_{min}$.

[5]While $\neg A \wedge A$ and $A \wedge \neg A$ are $\mathbf{C}_{\overline{\omega}}$-equivalent, $\neg(\neg A \wedge A)$ and $\neg(A \wedge \neg A)$ are not. Which of both is taken to express the consistency of A is a conventional matter.

[6]The approach is related to, but different from, the one followed in Carnielli et al. (2007), where a consistency operator, $\circ A$, belongs to the standard language and is implicitly defined by, for example, $\circ A \supset ((A \wedge \neg A) \supset B)$.

one might extend Γ_0 with the consistency statement $A^{(1)}$. This delivers the desired result because $A \vee B, \neg A, A^{(1)} \vdash_{\mathbf{C}_1} B$. Exactly the same situation arises if $\neg B \supset A$ and $\neg A$ are $\mathbf{C}_1$-derivable from Γ_0. So the addition of consistency statements to an inconsistent theory has dramatic effects. Within the paraconsistent context, it drastically enriches the theory. Moreover, the so enriched theory approaches the original theory, T_0, as it was originally *intended.*

Adding consistency statements involves a danger. Let $T_1' = \langle \Gamma_1, \mathbf{C} \rangle$ in which Γ_1 is obtained by adding a set of consistency statements of the form $A^{(1)}$ to Γ_0. T_1' may very well be trivial. When this is the case, one may retract some of the added consistency statements. There is, however, another possibility.

The transition from T_0 to T_1 involves the replacement of **CL**, which da Costa also calls $\mathbf{C}_0$, by $\mathbf{C}_1$ in order to avoid triviality. If T_1' turns out trivial, one may replace $\mathbf{C}_1$ by $\mathbf{C}_2$—let the result be T_2. In this way, triviality is avoided again; statements of the form $A^{(1)}$ are not consistency statements in the context of $\mathbf{C}_2$. Moreover, relying on the insights from the failed previous attempt, one may enrich Γ_1 with consistency statements of the form $A^{(2)}$, which have the desired effect in the context of $\mathbf{C}_2$. This process may be repeated. If $T_n' = \langle \Gamma_n, \mathbf{C}_n \rangle$, Γ_n comprising no statements $A^{(m)}$ for which $m > n$,[7] and is trivial, replacing $\mathbf{C}_n$ by $\mathbf{C}_{n+1}$ restores non-triviality because no $A^{(m)}$ occurring in T_n' is a consistency statement with respect to $\mathbf{C}_{n+1}$.

The stratagem demands the presence of classical negation and the $\mathbf{C}_n$ hierarchy and so motivates them. Certain phrases used by da Costa also suggest the stratagem. Thus he states that $\mathbf{C}_n$ logics isolate inconsistencies and he distinguishes between ‘good’ and ‘bad’ theorems of $\mathbf{C}_n$-theories, the bad ones being those whose negation is also a theorem. In order to isolate the bad theorems and to take advantage of the good ones, one needs to add consistency statements to the theory.

7.4.2 The Adaptive Logics

I shall proceed in two steps. First we need adaptive logics that interpret the premise set as consistently as possible with respect to a $\mathbf{C}_n$-logic. Let us call these $\mathbf{C}_n{}^m$ logics. These inconsistency-adaptive logics enrich a premise set with the consistency statements that are justifiable by logical means. The $\mathbf{C}_n{}^m$-logics should have been devised a long time ago, were it only because of the historical significance of the $\mathbf{C}_n$ logics. There was, however, a difficulty. $\mathbf{C}_n$ logics validate relations between contradictions and whenever this is the case there is a possibility that a flip-flop logic results. Flip-flop logics are adaptive logics, but are uninteresting for most application contexts. They behave like the upper limit logic whenever the premise set is normal, which is all right, and behave like the lower limit logic whenever

[7]Just as A^1 is a **CL**-theorem, viz. a $\mathbf{C}_0$-theorem, A^m is a $\mathbf{C}_n$-theorem whenever $m > n$. So one may suppose that no formula of the form A^m or $A^{(m)}$ is $\mathbf{C}_{n+1}$-derivable from the non-logical axioms of a theory that has $\mathbf{C}_n$ as underlying logic.

the premise set is abnormal (requires at least one abnormality to be true), which is not all right. Fortunately, a criterion for flip-flop behaviour, in terms of a specific indeterministic semantics, was developed for the application of the criterion to the $\mathbf{C}_n$ logics. In view of this result, the following logics are not flip-flops. For each n, $\mathbf{C}_n{}^m$ is defined as the triple consisting of (1) $\mathbf{C}_n$, (2) $\Omega = \{\exists(A \wedge \neg A) \mid A \in \mathscr{F}\}$, and (2) Minimal Abnormality—the result generalises to Reliability, which I do not consider for lack of space.

These logics assign as consequences of a premise set Γ all formulas true in the minimally abnormal $\mathbf{C}_n$-models of Γ—this obviously includes all $\mathbf{C}_n$-consequences of Γ.

Applying the adaptive logics has certain advantages over following the stratagem. First of all, the logic itself adds consistency statements that can be added on logical grounds; no tinkering is involved. Next, for some (actually most) premise sets, the consequence set will comprise an *infinite* number of consistency statements as well as all their consequences. Note that this effect cannot be obtained by tinkering. Moreover, it is possible that a *Dab*-formula is derivable, say $(p \wedge \neg p) \vee (q \wedge \neg q)$, of which no disjunct is derivable. In this case, there is no logical justification for either of the two disjuncts. So the logic will not chose between $\neg(p \wedge \neg p)$ and $\neg(q \wedge \neg q)$, but will have the disjunction of the consistency statements, $\neg(p \wedge \neg p) \vee \neg(q \wedge \neg q)$, as a consequence together with all that follows from it.

An interesting fact concerns the choice of a $\mathbf{C}_n{}^m$ logic that is suitable for a set of premises. It turns out that $\mathbf{C}_{\overline{\omega}}{}^m$ is the suitable choice for *all* premise sets. To be more precise, it holds for every $\mathbf{C}_n{}^m$ that $Cn_{\mathbf{C}_n{}^m}(\Gamma)$ is either trivial or identical to $Cn_{\mathbf{C}_{\overline{\omega}}{}^m}\Gamma$.

Now we come to the second step. Following the stratagem has also an advantage over applying the adaptive logic. Consider again a case where $(p \wedge \neg p) \vee (q \wedge \neg q)$ is derivable but none of both disjuncts is. A person following the stratagem is able to chose at this point, for example to consider $p \wedge \neg p$ as false, and hence $q \wedge \neg q$ as true.

It is possible to introduce such 'new premises' within an adaptive framework and it is actually possible to do this in a more elegant way than the stratagem permits. First of all, the minimal *Dab*-formulas that are derived evoke the question which of the disjuncts is true; so they indicate the points at which choices may be made. Next, there cannot be logical reasons for the choices. So the person applying the adaptive logics has to justify the new premises on the basis of *extra-logical* grounds. Moreover, the addition of new premises should proceed in a *defeasible* way in order to avoid possible triviality. Finally, each such new premise is better introduced in a prioritised way. Indeed, the justification of some consistency statements will be stronger than that of others. Given all this, the matter may be handled by a well known combined adaptive logic, which should only be adjusted to the circumstances in that the lower limit of the combining adaptive logics should be $\mathbf{C}_{\overline{\omega}}$. The combined logic *guides* the addition of prioritised consistency statements. To the $\mathbf{C}_{\overline{\omega}}{}^m$-consequences the combined logic first adds as many as possible of the consistency statements with the highest priority; to the result of this it adds as many as possible of the consistency statements with the next highest priority; and so on.

For the details of the combined logic, I refer to Batens (2009). It is interesting, however, to note that, while the hierarchy of $\mathbf{C}_n$ logics proves useless on the present approach, the priorities are expressed by formulas that largely follow da Costa's hierarchy of consistency statements. Thus $\neg\exists(A \wedge \neg A)$ is the least prioritised consistency statement concerning A, $\neg\exists(A \wedge \neg A) \wedge \neg(\exists(A \wedge \neg A) \wedge \neg\exists(A \wedge \neg A))$ is the next stronger consistency statement concerning A, and so on.

7.4.3 Two Comments

The enrichment that will be described in the next section may be introduced within the context of the $\mathbf{C}_n^m$ logics. This departs rather heavily from the stratagem, but is clearly meaningful in the present context.

The second comment concerns decidability. Not taking anything back of what I said about the advantages of the adaptive approach over the stratagem, let me try to avoid a misunderstanding. The adaptive approach clearly cannot make the situation more decidable than it is. For example, if the premise set is (finite and) propositional, the adaptive consequence set is decidable. In this case, an able logician may manage to obtain the right result in terms of the stratagem. Where the premise set is predicative, the stratagem may lead one to the wrong conclusions because one may never find out that an added consistency statement causes triviality. By following the adaptive approach, a similar situation may arise: one takes a conclusion as finally derived while it is not, because one does not manage to derive the required *Dab*-formulas. If matters are undecidable, no approach can repair this—see Horsten and Welch (2007) for a challenge and Batens et al. (2009a) for an answer.

The advantages of the adaptive approach are mainly threefold. First, it *defines* the consequence set in a correct way, even if this set is not recursive or not even semi-recursive. Next, there are proof procedures (see Batens 2005 and Verdée 201+) that, for some Γ and A, lead after finitely many steps to the conclusion that A is or is not a final consequence of Γ. It the answer is decidable, the proof procedure will provide it, and if it provides an answer, the answer is correct. Finally, the adaptive approach rigorously distinguishes between consistency statements that can be added on logical grounds and those that require an extra-logical justification. It guides the addition of the latter by delineating the choices to be made and it handles the added statements according to their priority.

7.5 Variations

The first inconsistency-adaptive logic, dating from around 1980, had the aim to offer a maximally consistent interpretation of premise sets, or theories, that were intended as consistent but had turned out to be inconsistent. So when it was recently found possible to realise the aim in a more efficient way, this came as a shock.

Two other problems are solved at once. Inconsistency-adaptive logics are instruments: formal characterizations of defeasible reasoning forms. We want to have a manifold of them around to suit specific application purposes. While there is a lot of variation with respect to the lower limit logic and the strategy, every lower limit logic seems to determine a unique set of abnormalities[8]—I disregard flip-flop logics (see previous section). In this paper, the limitation is overcome.

The second problem concerns the comparison between different lower limit logics. Stronger paraconsistent logics have in general larger consequence sets than weaker ones, but also spread inconsistencies. While the former property makes more formulas derivable on the empty condition, the latter restricts the number of formulas that are finally derivable but have a non-empty condition. In general, varying the lower limit logic often leads to incomparable adaptive consequence sets. The result presented in this paper changes the picture drastically. By varying the set of abnormalities, adaptive logics with a very weak lower limit logic may be given a very rich consequence set. I shall present comparative results below.

7.5.1 Characterization of the Abnormalities

The idea behind the enriched set of abnormalities is surprisingly simple. When certain complex $\mathbf{CLuN}^m$-abnormalities are derivable, these may have different causes. Thus if $(p \vee q) \wedge \neg(p \vee q)$ is **CLuN**-derivable from the premises, this may be because p is so derivable, or q is, or $p \vee q$ is whereas neither p nor q is. These three cases can be distinguished.

Consider the premise set $\Gamma_1 = \{\neg(p \vee q), q, p \vee r\}$ and let the underlying logic be $\mathbf{CLuN}^m$. Note that $\neg p$ is derivable on the condition $\{(p \vee q) \wedge \neg(p \vee q)\}$ and hence r is derivable on the condition $\{(p \vee q) \wedge \neg(p \vee q), p \wedge \neg p\}$. By the presence of q and $\neg(p \vee q)$, however, $(p \vee q) \wedge \neg(p \vee q)$ is derivable from Γ_1 on the empty condition and so cannot be taken to be false. So neither $\neg p$ nor r are $\mathbf{CLuN}^m$-derivable from Γ_1. At first sight, this seems justified. Note, however, that the derivability of $(p \vee q) \wedge \neg(p \vee q)$ is caused by the presence of q, not by the presence of p.

It is possible to turn this idea in a technically feasible definition? It is. In the presence of $\neg(p \vee q)$, each of $p \vee q$, p, and q may cause the abnormality. The disjunction is derivable from either disjunct. Moreover, any **CLuN**-model verifying $p \vee q$ verifies p or q, but not necessarily both. This suggests that we consider $(p \vee q) \wedge \neg(p \vee q)$, $p \wedge \neg(p \vee q)$, and $q \wedge \neg(p \vee q)$ as separate abnormalities. The gain is clear: as $\neg(p \vee q) \vdash_{\mathbf{CLuN}} \neg p \vee (p \wedge \neg(p \vee q))$, p is derivable from $\neg(p \vee q)$ on the condition $\{p \wedge \neg(p \vee q)\}$ if the member of this singleton counts as an abnormality. Moreover, while $q \wedge \neg(p \vee q)$ is unconditionally derivable from Γ_1, $p \wedge \neg(p \vee q)$ is provably not a disjunct of any minimal *Dab*-consequence of Γ_1. Of course, this is merely an example; the matter requires elaboration.

---

[8]This is typical for inconsistency-adaptive logics, not for other adaptive logics.

Primitive formulas and their negations will be called *atoms*. Formulas that are not atoms are classified as $\mathfrak{a}$-formulas or $\mathfrak{b}$-formulas, varying on a theme from Smullyan (1968). To each of them, two other formulas are assigned according to the following table.

$\mathfrak{a}$	$\mathfrak{a}_1$	$\mathfrak{a}_2$	$\mathfrak{b}$	$\mathfrak{b}_1$	$\mathfrak{b}_2$
$A \wedge B$	A	B	$A \vee B$	A	B
$A \equiv B$	$A \supset B$	$B \supset A$	$A \supset B$	$\check{\neg} A$	B
			$\neg A$	$\check{\neg} A$	$\check{\neg} A$
$\check{\neg}(A \vee B)$	$\check{\neg} A$	$\check{\neg} B$	$\check{\neg}(A \wedge B)$	$\check{\neg} A$	$\check{\neg} B$
$\check{\neg}(A \supset B)$	A	$\check{\neg} B$	$\check{\neg}(A \equiv B)$	$\check{\neg}(A \supset B)$	$\check{\neg}(B \supset A)$

Next, a set $\text{sp}(A)$ of *specifying parts* is assigned to every open or closed formula A as follows:

1. Where A is a conjunction of (one or more) atoms, possibly preceded by a sequence of quantifiers, $\text{sp}(A) = \{A\}$.
2. $\text{sp}(\mathfrak{a}) = \{\mathfrak{a}\} \cup \{\text{sp}(A \wedge B) \mid A \in \text{sp}(\mathfrak{a}_1); B \in \text{sp}(\mathfrak{a}_2)\}$.
3. $\text{sp}(\mathfrak{b}) = \{\mathfrak{b}\} \cup \text{sp}(\mathfrak{b}_1) \cup \text{sp}(\mathfrak{b}_2)$.
4. $\text{sp}(\forall \alpha A) = \{\text{sp}(\forall \alpha B) \mid B \in \text{sp}(A)\}$.
5. $\text{sp}(\exists \alpha A) = \{\text{sp}(\exists \alpha B) \mid B \in \text{sp}(A)\}$.

The adaptive logic $\mathbf{CLuN}_1^m$ is defined by the following triple: (1) lower limit: **CLuN**, (2) set of abnormalities: $\Omega^s = \{\exists(B \wedge \neg A) \mid A \in \mathcal{F}; B \in \text{sp}(A)\}$, and (3) strategy: Minimal Abnormality.

The mechanism is one of *refinement*. Even if $(p \vee q) \wedge \neg(p \vee q)$ is true in some models of a premise set, either $p \wedge \neg(p \vee q)$ or $q \wedge \neg(p \vee q)$ may be false in some of those models and this enables us to rule out some further models as more abnormal than required by the premises.

We have seen that the logic $\mathbf{CLuN}_1^m$ is richer than $\mathbf{CLuN}^m$ with respect to Γ_1. However, the enrichment is not restricted to similar cases. Let me mention two further examples. Consider first $\Gamma_2 = \{p \vee q, \neg(p \vee q), p \vee r, q \vee s\}$. In view of the explicit contradiction between the first two premises, one might expect to obtain no gain in this case. Yet, there is one. It is easily seen that r is derivable from Γ_3 on the condition $\{p \wedge \neg(p \vee q)\}$ and that s is derivable on the condition $\{q \wedge \neg(p \vee q)\}$. So $r \vee s$ is derivable on both conditions. Moreover, the only minimal *Dab*-consequences of Γ_3 are $(p \vee q) \wedge \neg(p \vee q)$ and $(p \wedge \neg(p \vee q)) \vee (q \wedge \neg(p \vee q))$. It follows that $r \vee s$, which is not a $\mathbf{CLuN}^m$-consequence of Γ_3, is a $\mathbf{CLuN}_1^m$-consequence of this premise set.

Another enrichment is illustrated by $\Gamma_3 = \{\neg\neg(p \wedge q), \neg p, \neg q \vee r\}$. Neither q nor r is a $\mathbf{CLuN}^m$-consequence of Γ_2, but both are $\mathbf{CLuN}_1^m$-consequences of it.

7.5.2 A Combined Inconsistency-Adaptive Logic

For all that was said, one might have the impression that $\mathbf{CLuN}_1^m$ offers a net gain over $\mathbf{CLuN}^m$, but this is false. In order to obtain a net gain, we need a combined adaptive logic. To see this, consider $\Gamma_4 = \{\neg(\neg s \vee (\neg p \wedge \neg r)), \neg(\neg p \vee \neg q), \neg(s \vee p)\}$.

The only members of minimal *Dab*-consequence of Γ_4 (with respect to both Ω and Ω^s) are provably 1–9 below. All are members of Ω^s and only 1–3 are members of Ω.

1	$(\neg s \vee (\neg p \wedge \neg r)) \wedge \neg(\neg s \vee (\neg p \wedge \neg r))$
2	$(\neg p \vee \neg q) \wedge \neg(\neg p \vee \neg q)$
3	$(s \vee p) \wedge \neg(s \vee p)$
4	$\neg s \wedge \neg(\neg s \vee (\neg p \wedge \neg r))$
5	$(\neg p \wedge \neg r) \wedge \neg(\neg s \vee (\neg p \wedge \neg r))$
6	$\neg p \wedge \neg(\neg p \vee \neg q)$
7	$\neg q \wedge \neg(\neg p \vee \neg q)$
8	$s \wedge \neg(s \vee p)$
9	$p \wedge \neg(s \vee p)$

It is also provable that we may restrict our attention, in this specific propositional case, to models of Γ_4 that verify the premises together with some of the relevant propositional letters and the classical negation of the others. A survey is displayed in Table 7.1. Unmentioned letters may receive an arbitrary value, provided they are not inconsistent. The numbers in the table refer to the abnormalities listed before. The first row of stars depicts the (kinds of) models that are minimally abnormal with respect to $\mathbf{CLuN}^m$; the second row of stars those that are *moreover* minimally abnormal with respect to $\mathbf{CLuN}_1^m$. The two-step selection is required because the second, fourth, sixth, eight, and ninth models are minimally abnormal with respect to Ω^s-abnormalities, but none of them is minimally abnormal with respect to Ω-abnormalities. The so combined selection delivers the consequences q, $p \vee r$, $s \vee r$, ... on top of those delivered by $\mathbf{CLuN}^m$.

Let us call the combined adaptive logic $\mathbf{CLuN}_c^m$ and let $Cn_{\mathbf{CLuN}_c^m}(\Gamma) = Cn_{\mathbf{CLuN}_1^m}(Cn_{\mathbf{CLuN}^m}(\Gamma))$, which offers the right selection of models. Proof theoretically such logics seem to be disastrous: it seems that one needs to compute $Cn_{\mathbf{CLuN}^m}(\Gamma)$ before one can even start to apply $\mathbf{CLuN}_1^m$. But this is not so. As was spelled out already in Batens (2001), the dynamic proof theory of thus combined adaptive logics is hardly more complex than that of the combining logics.

7.5.3 Some Comparisons

As promised, I shall now show that the combined logic $\mathbf{CLuN}_c^m$ does not only better than $\mathbf{CLuN}^m$, but does also very well in comparison to inconsistency-adaptive

Table 7.1 **CLuN**-models of Γ_4

p	p	p	p	p	p	p	p	$\check{\neg}p$	$\check{\neg}p$	$\check{\neg}p$	$\check{\neg}p$	$\check{\neg}p$	$\check{\neg}p$	$\check{\neg}p$	$\check{\neg}p$
q	q	q	q	$\check{\neg}q$	$\check{\neg}q$	$\check{\neg}q$	$\check{\neg}q$	q	q	q	q	$\check{\neg}q$	$\check{\neg}q$	$\check{\neg}q$	$\check{\neg}q$
r	r	$\check{\neg}r$	$\check{\neg}r$	r	r	$\check{\neg}r$	$\check{\neg}r$	r	r	$\check{\neg}r$	$\check{\neg}r$	r	r	$\check{\neg}r$	$\check{\neg}r$
s	$\check{\neg}s$	s	$\check{\neg}s$	s	$\check{\neg}s$	s	$\check{\neg}s$	s	$\check{\neg}s$	s	$\check{\neg}s$	s	$\check{\neg}s$	s	$\check{\neg}s$
	1		1		1		1		1	1	1		1	1	1
				2	2	2	2	2	2	2	2	2	2	2	2
3	3	3	3	3	3	3	3	3		3		3		3	
*		*							*		*		*		*
	4		4		4		4		4		4		4		4
										5	5			5	5
								6	6	6	6	6	6	6	6
				7	7	7	7					7	7	7	7
8		8		8		8		8		8		8		8	
9	9	9	9	9	9	9	9								
*		*							*						

logics that have a richer lower limit. Below, I consider five premise sets to compare $\mathbf{CLuN}^m_c$ with the corresponding adaptive logics that have as their lower limit logic respectively the maximal paraconsistent logic **CLuNs** and **LP**. I list the results for the latter logics together where they are identical with respect to the formulas that are listed—they differ from each other with respect to formulas that contain implications or equivalences.

$$\Gamma_5 = \{\neg(p \vee q), q \vee r, p, \neg p \vee s\}$$

$\mathbf{CLuN}^m$	$\mathbf{CLuN}^m_c$	$\mathbf{CLuNs}^m/\mathbf{LP}^m$
p	p	p
		$\neg p$
	$\neg q$	$\neg q$
$q \vee r$	r	r
s	s	

$$\Gamma_6 = \{p \vee q, \neg(p \vee q), p \vee r, q \vee s\}$$

$\mathbf{CLuN}^m$	$\mathbf{CLuN}^m_c$	$\mathbf{CLuNs}^m/\mathbf{LP}^m$
		$\neg p$
		$\neg q$
$p \vee q$	$p \vee q$	$p \vee q$
$p \vee r$	$p \vee r$	$p \vee r$
$q \vee s$	$q \vee s$	$q \vee s$
	$\neg p \vee \neg q$	
	$r \vee s$	

$\Gamma_7 = \{p, \neg p \vee q, \neg(p \vee r), \neg\neg p \supset s\}$

$\mathbf{CLuN}^m$	$\mathbf{CLuN}_c^m$	$\mathbf{CLuNs}^m$	$\mathbf{LP}^m$
p	p	p	p
		$\neg p$	$\neg p$
$\neg\neg p$	$\neg\neg p$	$\neg\neg p$	$\neg\neg p$
	$\neg r$	$\neg r$	$\neg r$
q	q		
s	s	s	

$\Gamma_8 = \{p, \neg p \vee q, \neg(p \vee r), \neg\neg p \supset s, \neg q \vee t, r \vee u\}$

$\mathbf{CLuN}^m$	$\mathbf{CLuN}_c^m$	$\mathbf{CLuNs}^m$	$\mathbf{LP}^m$
p	p	p	p
		$\neg p$	$\neg p$
$\neg\neg p$	$\neg\neg p$	$\neg\neg p$	$\neg\neg p$
	$\neg r$	$\neg r$	$\neg r$
q	q		
s	s	s	
t	t		
	u	u	u

It is interesting to study the difference between the consequence sets. In all cases, (1) the $\mathbf{CLuNs}^m$-consequences or $\mathbf{LP}^m$-consequences that are not $\mathbf{CLuN}_c^m$-consequences cause additional inconsistency and (2) some $\mathbf{CLuN}_c^m$-consequences are neither $\mathbf{CLuNs}^m$-consequences nor $\mathbf{LP}^m$-consequences and they do not cause additional inconsistency. I am not claiming, however, that $\mathbf{CLuN}_c^m$ is better than the other adaptive logics. An instrument should be used where it is suitable. The only point I wanted to make is that $\mathbf{CLuN}_c^m$ maximally isolates inconsistencies, just as much as $\mathbf{CLuN}^m$, but nevertheless offers an extremely rich consequence set.

7.6 Parsimonious Axiomatisations

7.6.1 The Problem

Let $\mathscr{L}_A$ be the language of arithmetic (with one constant, 0, and three functions, $'$, $+$, and $\times$). In several places, for example Priest (1994, 1997, 2000, 2006), Graham Priest considers inconsistent models of arithmetic (see also Paris and Pathamanathan 2006; Paris and Sirokofskich 2008). In these models, the logical symbols are interpreted in terms of Priest's **LP**—implication and equivalence are defined and non-detachable. I shall only consider the so-called collapsed models.

$\mathcal{M}_p^n$ denotes the model with the following successor graph:

$$\begin{array}{ccccc} 0 \to 1 \to \ldots \to & n & & \to n+1 \\ & \uparrow & & \downarrow \\ & n+p-1 \leftarrow & & \ldots \end{array}$$

In order to simplify the subsequent argument, let us concentrate on models $\mathcal{M}_1^n$, which have the following successor graph

$$0 \to 1 \to \ldots \to n \circlearrowleft$$

Let us more particularly concentrate on $\mathcal{M}_1^2$. In order to avoid confusion between numbers and numerals, let the domain of the model be $\{\mathfrak{f}, \mathfrak{m}, \mathfrak{a}\}$ and let the interpretation of the successor function be characterised by the following graph:

$$\mathfrak{f} \to \mathfrak{m} \to \mathfrak{a} \circlearrowleft$$

with $v(0) = \mathfrak{f}$, viz. the constant 0 is taken to name $\mathfrak{f}$. So $0'$ names $\mathfrak{m}$, and $0''$, $0'''$, etc. all name $\mathfrak{a}$.

Every $\mathcal{M}_1^n$ can be seen as modelling a specific inconsistent arithmetic $\mathbf{A}_1^n = \{A \mid \mathcal{M}_1^n \Vdash A\}$ (the formulas of $\mathcal{L}_A$ that are verified by $\mathcal{M}_1^n$). As every $\mathcal{M}_1^n$ is a finite model, $\mathbf{A}_1^n$ can be finitely axiomatised with **LP** as the underlying logic. This means that there is a recursive, and actually finite, set of formulas Γ such that $\mathbf{A}_1^n = \{A \mid \Gamma \vdash_{\mathbf{LP}} A\}$.

There is, however, an oddity. Not only $\mathcal{M}_1^2$, but also $\mathcal{M}_1^1$ as well as the trivial model $\mathcal{M}_1^0$ are models of $\mathbf{A}_1^2$. This is related to the fact that $\mathbf{A}_1^0 \supseteq \mathbf{A}_1^1 \supseteq \mathbf{A}_1^2 \supseteq \ldots$. It is also related to the fact that $\mathcal{M}_1^2$ is a model of classical arithmetic,[9] which, provided it is consistent, is a limit of this sequence of sets. The sentences of $\mathcal{L}_A$ that are true in the standard model of arithmetic are also true in the finite and inconsistent model $\mathcal{M}_1^2$. In the same way the sentences of $\mathcal{L}_A$ that are true in $\mathcal{M}_1^2$ are also true in $\mathcal{M}_1^1$ and in $\mathcal{M}_1^0$.

It follows from Gödel's first theorem that no consistent axiomatisation of first-order sentences true in the standard model of arithmetic identifies the standard model. Every such axiomatisation also has non-standard models, the domain of which comprises objects not named by any numeral. So no (first-order) axiomatisation identifies the standard model. The situation is similar for every $\mathbf{A}_1^n$, except that the domains of the non-intended models comprise not more but less objects than the domain of the intended model—the larger n, the greater the number of non-intended models. In many other respects, the situation is dissimilar from the situation of classical arithmetic, but in this sense it is similar.

[9] By "classical arithmetic" I obviously mean the set of formulas true in the standard model and not the theorems of some axiom system.

The failure to identify a single model, say $\mathscr{M}_1^2$, is obviously contingent on the object language and on the underlying logic. Let us first have a look at variant logics.

7.6.2 A **LPm**-*Axiomatisation*

One might hope to identify $\mathscr{M}_1^2$ by presenting an axiomatisation that has **LPm** as its underlying logic rather than **LP** (see Priest 1991, 2006). Indeed, **LPm** selects the 'minimal abnormal' **LP**-models of a premise set—see below for the quotation marks. In $\mathscr{M}_1^2$, the denotation of $0''$ and of all higher numerals are inconsistent with respect to identity (that is $0'' = 0'' \wedge \neg\, 0'' = 0''$ is a theorem), but the denotations of $0'$ and of 0 are consistent with respect to identity. In $\mathscr{M}_1^1$, the denotation of $0'$ is also inconsistent with respect to identity, and in $\mathscr{M}_1^0$ the denotation of every numeral is inconsistent with respect to identity—$\mathscr{M}_1^0$ is a trivial model.

Unfortunately, **LPm** does not provide a solution. The cause lies with the way in which minimal abnormal models are defined in **LPm**. Here are, again, the successor graphs of $\mathscr{M}_1^2$, $\mathscr{M}_1^1$, and $\mathscr{M}_1^0$:

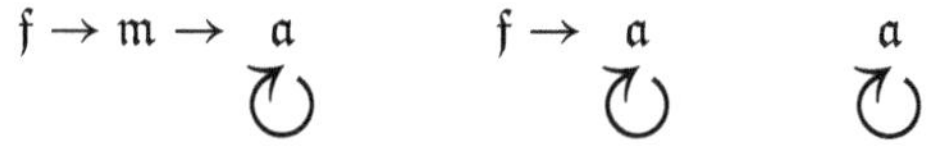

The 'abnormal part' of a model is represented in **LPm** by the atomic inconsistent 'facts' that hold in the model. In other words, for every n-ary predicate R, the n-tuples that belong to both the extension of R, $v^+(R)$, and to the anti-extension of R, $v^-(R)$ (see Priest 2006 for details). The only predicate that matters in the present context is identity and all three models have the same abnormal part, viz. $v^+(=) \cap v^-(=) = \{\langle \mathfrak{a}, \mathfrak{a}\rangle\}$. So all three models are **LPm**-models of $\mathbf{A}_1^2$. This means that no **LPm**-axiomatisation identifies $\mathscr{M}_1^2$ and that the difficulty remains.

Incidentally, we obviously need $v(0) = \mathfrak{a}$ instead of $v(0) = \mathfrak{f}$ in the displayed model $\mathscr{M}_1^0$. Some isomorphic models have $\mathfrak{f}$ as the only element of the domain, and these have exactly the same abnormal part as some models isomorphic with $\mathscr{M}_1^1$ and $\mathscr{M}_1^2$.

7.6.3 $\mathbf{LP}^m$-*Axiomatisation*

Unlike **LPm**, $\mathbf{LP}^m$ is an adaptive logic in standard format; it was described earlier. The difference with **LPm** is that abnormalities are not 'inconsistent' n-tuples of members of the domain, but *formulas*, viz. existentially closed contradictions. The abnormal part of a **LP**-model M, $Ab(M)$, is the set of abnormalities verified by M. So $Ab(\mathscr{M}_1^2)$ comprises all formulas of the form $0^i = 0^i \wedge \neg\, 0^i = 0^i$ in which i is a sequence of two or more names of the successor function, as well as the

LP-consequences of these, for example $\exists x(x = x \wedge \neg x = x)$. The set $Ab(\mathscr{M}_1^1)$ *moreover* comprises $0' = 0' \wedge \neg 0' = 0'$ and $Ab(\mathscr{M}_1^0)$ even comprises $0 = 0 \wedge \neg 0 = 0$. So, of the three considered models, only $\mathscr{M}_1^2$ is a minimally abnormal model of $\mathbf{A}_1^2$. An $\mathbf{LP}^m$-axiomatisation of $\mathbf{A}_1^2$ is obtained, for example by adding the axiom $0''' = 0''$ to the Peano Axioms. Let this set of axioms be called $\mathbf{PA}_1^2$—there are obviously simpler, viz. finite, sets that do exactly the same job. The axiom system $\langle \mathbf{PA}_1^2, \mathbf{LP}^m \rangle$ (the axioms $\mathbf{PA}_1^2$ closed under $\mathbf{LP}^m$) identifies $\mathbf{A}_1^2$.

It is important to realise that the effect results from changing the underlying logic. If this is $\mathbf{LP}^m$, the models of $\mathbf{A}_1^2$ have to be $\mathbf{LP}^m$-models, and the only such model is $\mathscr{M}_1^2$.

Some may wonder whether an axiomatisation with $\mathbf{LP}^m$ as underlying logic is really an axiomatisation. Indeed, a $\mathbf{LP}^m$-proof of A from the premise set Γ requires a list of formulas together with a reasoning in the metalanguage establishing that A is finally derived in the list of formulas (see for example Batens et al. 2009a for details). So this kind of proofs, which are called dynamic, do not form a positive test for (final) derivability. In the present context, however, this complication does not arise. Given the model $\mathscr{M}_1^2$, which is finite, and the language, there are prospective proofs, see for example Batens (2005), that form a decision method for derivability. In other words, $Cn_{\mathbf{LP}^m}(\mathbf{PA}_1^2)$ is a decidable set and the couple $\langle \mathbf{PA}_1^2, \mathbf{LP}^m \rangle$ is a legitimate axiomatisation of $\mathbf{A}_1^2$. For those who are still mistrusting, let $\langle \Delta, \mathbf{LP} \rangle$ be an axiomatisation of $\mathbf{A}_1^2$—so $Cn_{\mathbf{LP}}(\Delta) = \mathbf{A}_1^2$. Next, consider the axiomatisation $\langle \Delta, \mathbf{LP}^m \rangle$ and note that $Cn_{\mathbf{LP}^m}(\Delta) = \mathbf{A}_1^2$.[10] As every $\mathbf{A}_1^2$-theorem is **LP**-derivable from Δ, it is unconditionally $\mathbf{LP}^m$-derivable from Δ. So in view of this metatheoretic fact, there is a positive test for $\mathbf{A}_1^2$-theoremhood.

7.6.4 A Richer Language

Other axiomatisations are possible, even with a Tarski logic as the underlying logic, but they all have the disadvantage that they require replacing $\mathscr{L}_A$ by a richer language.

The first alternative is that one adds classical (or Boolean) negation, $\check{\neg}$. A suitable axiomatisation is obtained by extending $\mathbf{PA}_1^2$ with, for example, $\check{\neg} 0' = 0''$. In the presence of classical negation, $\check{\neg} 0 = 0'$ is derivable from this and, in general, $\check{\neg} A$ is derivable whenever A is "false only" in $\mathscr{M}_1^2$. Apart from requiring an extension of the language, this approach has the further disadvantage that it is opposed to Priest's philosophical views—he has argued against the meaningfulness of classical negation, a point which I shall not discuss here.

Another alternative is to extend the language with a relevant implication, $\rightarrow$, as well as with bottom, $\bot$, and adding to $\mathbf{PA}_1^2$ axioms like $0' = 0'' \rightarrow \bot, 0 = 0' \rightarrow \bot$,

[10] This further clarifies the claim made in the previous paragraph. Although $Cn_{\mathbf{LP}}(\Delta) = Cn_{\mathbf{LP}^m}(\Delta)$, $\langle \mathbf{PA}_1^2, \mathbf{LP}^m \rangle$ identifies $\mathscr{M}_1^2$ whereas $\langle \mathbf{PA}_1^2, \mathbf{LP} \rangle$ does not.

and so on. If the relevant implication is the one from Priest (2006, § 18.3), the "and so on" should not be underestimated; even $0' = 0 \rightarrow \bot$ is not a consequence of $0 = 0' \rightarrow \bot$. If the n in $\mathcal{M}_1^n$ is large, the number of required axioms will be impressive, but obviously finite.

This approach too seems to involve difficulties. If the relevant implication is not extremely poor, one will have as a theorem $\forall x \forall y(x = y \rightarrow f(x) = f(y))$ for every one argument function f. So, in particular, one will have $\forall x \forall y(x = y \rightarrow x' = y')$ as a theorem. But then $0' = 0' \rightarrow 0'' = 0''$ is a theorem. As $\neg\, 0'' = 0''$ is a theorem of $\mathbf{PA}_1^2$ and $\rightarrow$ is contraposable, $\neg\, 0' = 0'$ would be a theorem of $\mathbf{PA}_1^2$. But this is wrong: $\neg\, 0' = 0'$ is false in $\mathcal{M}_1^2$ and so should not be a theorem of $\mathbf{PA}_1^2$. Of course, the difficulty will not occur if the relevant implication is weaker, for example is the one from Priest (2006, § 18.3). One wonders, however, whether this implication will be sufficient to formalise the whole body of our knowledge, empirical and mathematical. Indeed, Priest is a monologist. So he opposes using different logics in different contexts.

The presence of an enthymematic implication does not repair the situation. Indeed, while one might prefer to replace the relevant implication in $\neg\, 1 = 1 \rightarrow \bot$ by an enthymematic one, there is no reason to perform the same replacement in $\forall x \forall y(x = y \rightarrow f(x) = f(y))$ in case this is a theorem. However, the presence of a non-contraposable relevant implication would remove this specific difficulty, might very well be justifiable,[11] and seems to provide a sufficiently strong statement $\forall x \forall y(x = y \rightarrow f(x) = f(y))$.

More serious difficulties are lurking around the bend. First, the relevant implication is *ad hoc* in the present context—it occurs nowhere else in the inconsistent arithmetic, just like the classical negation from two paragraphs ago. Next, I cannot see any sense in which $\neg\, 0' = 0'$ can be said to *relevantly imply* every statement of the language. Adding the implicative axioms comes to a technical trick. It does the job, but can only be justified by the argument that it provides a warrant that is as good as the one the classical logician invokes by recurring to classical implication (which connects classical inconsistency to triviality)—but see below.

Another difficulty is related to the fact that everything is true in the trivial model, in the present context $\mathcal{M}_1^0$. So, even if it can be avoided that $\mathcal{M}_1^1$ is a model of $\mathbf{A}_1^2$, this theory still has both $\mathcal{M}_1^2$ and $\mathcal{M}_1^0$ as models, and so does not identify $\mathcal{M}_1^2$ in a unique way—please compare with the $\mathbf{LP}^m$-axiomatisation which *does* rule out the trivial model $\mathcal{M}_1^0$.

Incidentally, the classical logician seems to do better in this respect *on her understanding*. She can claim that adding $\check{\neg}\, 0' = 0''$ identifies $\mathcal{M}_1^2$ in a unique way. On the classical logician's understanding, there is no trivial model because the truth values, say t and f, are distinct, v_M is a function, and $v_M(\check{\neg} A) = t$ iff $v_M(A) = f$. So there are no models in which $v_M(\check{\neg} A) = t = v_M(A)$. Of course Graham Priest has argued that the classicist's understanding makes no sense, a point not discussed here.

---

[11]The most obvious justification for contraposition is consistency. So I always wondered why so many relevant logicians want their implications to be contraposable.

7.6.5 Describing the Models

Until now, I phrased the difficulty as one of axiomatising the formulas true in the models $\mathscr{M}_p^n$, while I took those models at face value. However, a similar difficulty affects the *description* of the models. Whether one considers the description I gave above or the description in Priest (2006), the model $\mathscr{M}_1^1$ actually agrees with the description of the model $\mathscr{M}_1^2$. The domain counts three different objects, $\mathfrak{f}$, $\mathfrak{m}$, and $\mathfrak{a}$. Of these, $\mathfrak{m}$ and $\mathfrak{a}$ are not only different but also identical and the successor function holds between them. Note, incidentally, that "$\mathfrak{m}$" and "$\mathfrak{a}$" are not the elements of the domain, but the names of these elements; just as the drawing is not the successor graph, but a representation of it. That the *characters* "$\mathfrak{m}$" and "$\mathfrak{a}$" are not identical, but different, and different only for that matter, does not prevent them from naming the same entity.[12] By a similar reasoning, the model $\mathscr{M}_1^0$ agrees with the description of $\mathscr{M}_1^2$.

So the description of $\mathscr{M}_1^2$ does not identify this model as we understood it, *unless* we presuppose that the description is as consistent as possible, viz. is presented in terms of $\mathbf{LP}^m$. Unlike **LPm**, $\mathbf{LP}^m$ will select the right description and will select the right models of the description—these are not the models described *by* the description.

7.7 Concluding Comment

Rather than commenting on the promise made in the introduction, I shall comment on a consequence of the preceding section.

In Mortensen (2008), Chris Mortensen writes that, according to inconsistency-adaptive logics, "only consistent conclusions are deduced *pro tem*" and continues "In the opinion of this (opinionated) writer, consistentising strategies are useful for the context of discovery, but fail to do justice to *a priori* reasoning from inconsistent premises, where one should be acknowledging the full role of all the premises without dodging the inconsistencies in them." These claims are actually false,[13] but the reason to quote them lies elsewhere, viz. in the presupposed status of *a priori* reasoning. If there is any truth in the previous section, one needs "consistentising

[12]One shouldn't make too much of the "different only" phrase. In Priest's view it may be true together with "the characters are the same", for otherwise "This sentence is false and only false." would produce triviality.

[13]The first quoted claim is obviously false: *all* formulas derivable by the lower limit logic are adaptively derivable, whether consistent or inconsistent. However, some *further* consequences are adaptively derivable by taking as many *other* inconsistencies to be false as the premises permit. So inconsistency-adaptive logics do acknowledge the full role of all the premises and do not dodge any inconsistencies in them. They presuppose that inconsistencies are false *unless and until* proven otherwise, from the premises that is.

strategies" in order to describe the models $\mathcal{M}_p^n$ and this is apparently required before any *a priori* reasoning about them can even start. Inconsistency-adaptive logics were always presented as *instruments* (or methods), which may be more or less suited to a specific context, and not as candidates for "the true logic" or "the standard of deduction" or "the canon of *a priori* reasoning". Nevertheless, the situation depicted in this section seems to present a further argument, apart from many others, to mistrust a strict separation between sensible reasoning instruments and *a priori* reasoning. It also suggests that, while it is easy to explain the paraconsistent viewpoint by relying on classical results, such as the supposedly consistent standard model of arithmetic, it might be more difficult for the monologist dialetheist to offer her teachings from scratch. That Graham Priest has been persistently working in that direction, including the development of a dialetheisticly sound set theory, deserves the admiration and sympathy of every logician, even of those who (like me) consider the standard of reasoning as context dependent.

Acknowledgements Research for this paper was supported by subventions from Ghent University and from the Fund for Scientific Research—Flanders. I am indebted to Graham Priest for comments on a former draft of this paper.

References

Batens, D. 2000. Towards the unification of inconsistency handling mechanisms. *Logic and Logical Philosophy* 8: 5–31.

Batens, D. 2001. A general characterization of adaptive logics. *Logique et Analyse* 173–175: 45–68.

Batens, D. 2005. A procedural criterion for final derivability in inconsistency-adaptive logics. *Journal of Applied Logic* 3: 221–250.

Batens, D. 2007. A universal logic approach to adaptive logics. *Logica Universalis* 1: 221–242.

Batens, D. 2009. Adaptive $\mathbf{C}_n$ logics. In *The many sides of logic*. Studies in logic, vol. 21, ed. W.A. Carnielli, M.E. Coniglio, and I.M. Loffredo D'Ottaviano, 27–45. London: College Publications.

Batens, D. 2009a. Towards a dialogic interpretation of dynamic proofs. In *Dialogues, logics and other strange things. Essays in honour of shahid rahman*, C. Dégremont, L. Keiff, and H. Rückert, 27–51. London: College Publications.

Batens, D. 201+. *Adaptive logics and dynamic proofs. Mastering the dynamics of reasoning*, forthcoming.

Batens, D., and K. De Clercq. 2004. A rich paraconsistent extension of full positive logic. *Logique et Analyse* 185–188: 227–257.

Batens, D., K. De Clercq, P. Verdée, and J. Meheus. 2009a. Yes fellows, most human reasoning is complex. *Synthese* 166: 113–131.

Batens, D., C. Straßer, and P. Verdée. 2009b. On the transparency of defeasible logics: Equivalent premise sets, equivalence of their extensions, and maximality of the lower limit. *Logique et Analyse* 207: 281–304.

Benferhat, S., D. Dubois, and H. Prade. 1997. Some syntactic approaches to the handling of inconsistent knowledge bases: A comparative study. Part 1: The flat case. *Studia Logica* 58: 17–45.

Carnielli, W.A., M.E. Coniglio, and J. Marcos. 2007. Logics of formal inconsistency. In *Handbook of philosophical logic*, vol. 14, ed. D. Gabbay and F. Guenthner, 1–93. Dordrecht/London: Springer.

Carnielli, W.A., and J. Marcos. 1999. Limits for paraconsistent calculi. *Notre Dame Journal of Formal Logic* 40: 375–390.

da Costa, N.C.A. 1974. On the theory of inconsistent formal systems. *Notre Dame Journal of Formal Logic* 15: 497–510.

Horsten, L., and P. Welch. 2007. The undecidability of propositional adaptive logic. *Synthese* 158: 41–60.

Kleene, S.C. 1952. *Introduction to metamathematics*. Amsterdam: North-Holland.

Makinson, D. 2005. *Bridges from classical to nonmonotonic logic*. Texts in computing, vol. 5. London: King's College Publications.

Mortensen, C. 2008. Fourth world congress on paraconsistency. *The Reasoner* 2(8): 8.

Paris, J.B., and N. Pathamanathan. 2006. A note on Priest's finite inconsistent arithmetics. *Journal of Philosophical Logic* 35: 529–537.

Paris, J.B., and A. Sirokofskich. 2008. On *LP*-models of arithmetic. *Journal of Symbolic Logic* 73: 212–226.

Priest, G. 1991. Minimally inconsistent **LP**. *Studia Logica* 50: 321–331.

Priest, G. 1994. Is arithmetic consistent? *Mind* 103: 337–349.

Priest, G. 1997. Inconsistent models of arithmetic. Part I: Finite models. *Journal of Philosophical Logic* 26: 223–235.

Priest, G. 2000. Inconsistent models of arithmetic. Part II: The general case. *Journal of Symbolic Logic* 65: 1519–1529.

Priest, G. 2006. *In contradiction: A study of the transconsistent*, Second expanded edn. Oxford: Oxford University Press.

Rescher, N., and R. Manor. 1970. On inference from inconsistent premises. *Theory and Decision* 1: 179–217.

Schütte, K. 1960. *Beweistheorie*. Berlin: Springer.

Smullyan, R.M. 1968. *First order logic*. New York: Dover.

Verdée, P. 201+. A proof procedure for adaptive logics. *Logic Journal of the IGPL*. In print.

Chapter 8
Consequence as Preservation: Some Refinements

Bryson Brown

8.1 Introduction

Preservationism was first developed by P.K. Schotch and R.E. Jennings in a series of papers going back to the late 1970s. (Jennings and Schotch 1984, 1981; Jennings et al. 1981, 1980; Jennings and Johnston 1983; Johnston 1978; MacLeod and Schotch 2000; Schotch and Jennings 1980b and Thorn 2000). The central theme of this program in philosophical logic is the suggestion that the standard semantic understanding of logical consequence as (guaranteed) preservation of truth is too narrow. In particular, if a set of sentences cannot all be true, any sentence is guaranteed to preserve what truth is contained in that set. The closure of such a premise set under a guaranteed truth-preservation consequence relation includes every sentence in the language. Thus the truth preservation view of consequence inevitably trivialises the consequences of unsatisfiable premise sets.

However, if there are other semantic or syntactic properties of such a premise set that are worth preserving, we can constrain the consequences that follow from those premises by insisting that the closure of a set under an acceptable consequence relation retain these valuable properties. This is a rather modest proposal; it allows us to motivate constraints on inference from unsatisfiable sets without proposing any revision in our views of what sets of sentences are satisfiable, and it imposes no particular view of what properties of sets of sentences are worth preserving.

Level was Schotch and Jennings' first suggestion for a preservable property that some inconsistent sets have; the level of a set of sentences Γ can be defined intuitively as the least cardinal n such that Γ can be divided amongst n classically consistent sets. On this definition, both $\{p \vee \neg p\}$ and $\{p, q\}$ have level 1, while $\{p, q, \neg p\}$ has level 2, $\{p \wedge q, p \wedge \neg q, \neg p \wedge \neg q\}$ has level 3 and $\{p \wedge \neg p\}$ (like all sets that include a contradiction) has no level at all, since it cannot be divided into

B. Brown (✉)
Department of Philosophy, University of Lethbridge, Lethbridge Alberta, Canada
e-mail: brown@uleth.ca

K. Tanaka et al. (eds.), *Paraconsistency: Logic and Applications*, Logic, Epistemology, and the Unity of Science 26, DOI 10.1007/978-94-007-4438-7_8,

n consistent sets for any cardinal n. Note also that some sets, such as the set of all literals in a language with an infinite collection of atoms, have level ω.

This first, rough definition can be refined in two ways. First, to ensure that every set of sentences has a level, we assign ∞ as the level of sets that cannot be divided into consistent sets. More subtly, in order to recognise the special consistency of the empty set and its consequences (the tautologies)—perhaps most vividly illustrated by the fact that unions of such sets with consistent sets are always consistent—we assign such sets the special level 0 as in Schotch and Jennings (1980a, 1989).

The classical consequence relation preserves levels 0 and 1. But Schotch and Jennings' forcing relation preserves all levels. The most obvious inferential restriction that forcing imposes (in comparison with the classical consequence relation) is the rejection of conjunction introduction, and its replacement with a sequence of level-dependent rules for aggregation: where n is the level of our premise set, the set is closed under singleton consequence and the rule

$$2/n+1: \quad \frac{\{\alpha_1 \ldots \alpha_{n+1}\}}{\bigvee_{1 \leq i \neq j \leq n+1}(\alpha_i \wedge \alpha_j)}$$ [1]

Preservationism is a broad tent; there are many other interesting properties worth preserving. The usual syntactic standard of acceptability is consistency; applying this standard, we get a variation on the usual account of the consequence relation:

$\Gamma \vdash \alpha \Leftrightarrow \alpha$ is a consistent extension of every consistent extension of Γ.

Similarly, the usual semantic standard of acceptability is satisfiability; applying this standard, we get:

$\Gamma \models \alpha \Leftrightarrow \alpha$ is a satisfiable extension of every satisfiable extension of Γ.

It's easy to see that, given classical accounts of consistency and satisfiability, these definitions will give us the familiar classical consequence relation—but these definitions focus our attention on the way in which closing a premise set under the logical consequence relation *preserves* something important: the consistent (satisifiable) extensions of our premises *remain* consistent (satisfiable) extensions when we extend the set by adding its logical consequences to it. We might say that extending a set in this way *begs no questions* that aren't already begged in accepting the set.

In Brown (1999), I proposed a general way of thinking about consequence relations as preserving desirable features of premise sets: given a standard of *acceptability* for extensions of sets of sentences, we say that:

α follows from Γ iff α is an acceptable extension of every acceptable extension of Γ.[2]

---

[1]It's worth pointing out here that the 2/n+1 rule can be replaced by any operation that forms the disjunction of conjunctions of the edges of an n-uncolourable hypergraph on a collection of input sentences. See the appendix of Brown and Schotch (1999) for details.

[2]We speak of extensions in two senses here: strictly speaking, a set Σ is an extension of a set Γ if and only if Σ is a superset of Γ, $\Sigma \supset \Gamma$. But we also sometimes describe a sentence α as an extension of a set Γ. What we mean in that case is the set that results from adding α to Γ, $\Gamma \cup \{\alpha\}$.

Given this general account of consequence relations, we can define new consequence relations by identifying other properties that can be used as standards of acceptability in the new consequence relation. For example, preserving a measure of the ambiguity required to produce a consistent image of a premise set leads us to a semantics for Priest's system LP (Priest 1979; Brown 1999) while preserving the level as defined above gives us the set-sentence ($\Gamma [\vdash \alpha$) version of Schotch and Jennings' *forcing* relation (Schotch and Jennings 1980a, 1989).

But approaching things this way is somewhat inelegant. Treating consequence relations as relations between two different types of entities, with premise sets on the left and single sentences on the right, disguises some important symmetries. In particular, we will be interested here in symmetries relating conjunction to disjunction, and assertion to denial.

In one direction, we can see how to reduce the usual set-sentence presentation of the classical consequence relation to a relation between sentences by pointing out that $\Gamma \vdash \alpha$ holds if and only if $\{\gamma\} \vdash \alpha$ holds, where γ is a conjunction of members of Γ: in classical logic (and in any other compact, *fully aggregative* logic that has a conjunction operator) closure under conjunction combines premises into single sentences from one or another of which any conclusion following from Γ must follow. However, bringing a connective into our account of consequence in this way is limiting: the resulting account will be restricted to languages that have such a conjunction, or to which one can be conservatively added. So it's more general, and more illuminating, to type-raise conclusions from sentences to sets of sentences, and treat the consequence relation as holding between premise sets and conclusion sets.

In classical multiple-conclusion logic we say that $\Gamma \vdash \Delta$ if the assertion of all members of Γ commits us to the assertion of at least one member of Δ. Assuming that denial and assertion are incompatible, i.e. that we cannot simultaneously deny and assert a sentence, this is equivalent to saying that $\Gamma \vdash \Delta$ if and only if the denial of all members of Δ commits us to the denial of at least one member of Γ. So the familiar logic of assertion runs from left to right, while the logic of denial runs from right to left.

In the rest of this paper we will consider two paraconsistent set-sentence consequence relations, examine the role of preservation in them, and develop corresponding set–set consequence relations. In the conclusion, reflections on these examples will lead to insights on the role of preservation in consequence relations.

8.2 Ambiguity, LP and FDE

The central idea of the logics we will discuss in this section is quite simple. Historically, a standard way of responding to apparent inconsistencies has been to say, instead, that there is an ambiguity in the sentences that appear to conflict. For example, when we say both that this wine is dry and that this wine (being wet) is not dry, we seem to be contradicting ourselves. But we know perfectly well that there is an ambiguity here: the use of 'dry' in 'this wine is dry' means (at least normally) something like 'not sweet' or 'having low sugar content', while the use of 'dry' in 'this wine is not dry' means something like 'is not a liquid' or 'contains very little water'. The apparent contradiction is *merely* apparent.

We can formulate a general approach to dealing with inconsistent premise sets along these lines: any inconsistent set of sentences can be rendered consistent by treating some of the atoms appearing in the set as ambiguous. We can make that ambiguity explicit by replacing instances of each of those atoms with one of two new atoms, creating a disambiguated image of the original set. However, without the informal guidance we have in the example above, we don't really know which such disambiguations are the right ones (if any are). So we will take all the different ways of disambiguating atoms of the original set into consideration. Some of these will produce consistent images of our premises. In general some will do so by treating more atoms as ambiguous, while others make do with less. Our logic will focus on *least* sets of atoms that allow the projection of a consistent image of our premises, i.e. sets of atoms that are enough to do the job, and for which no proper subset of those atoms is enough. Here we present the main ideas in simple form; more formal definitions of these ideas can be found in A.1.

We begin by defining how to project a consistent image of a set of sentences Γ using a set of sentence letters A:

Γ' is a consistent image of Γ *based on* A iff

1. A is a set of sentence letters.
2. Γ' is consistent.
3. Γ' results from the substitution, for each occurrence of each member a of A in Γ, of one of a pair of new sentence letters, a_f and a_t.

Acceptability is the central notion in the general preservationist proposal for consequence described above. The intuitive idea is that a set Δ is *acceptable*, given a commitment to a premise set Γ, if and only if:

1. Δ includes Γ.
2. Extending Γ to Δ does not *make things worse*, by our standards of acceptability.

In our first ambiguity logic, we will say Δ is ambiguity-acceptable relative to Γ iff there is some way of projecting a consistent image of Δ that does not require a larger base of sentence letters than the minimal projection bases for Γ. This means that Δ may resolve uncertainties about how we will project a consistent image that Γ leaves open, by eliminating *some* of the minimal projection bases that work for Γ. For example, we can project consistent images of the set $\Sigma = \{p, q, \neg(p \wedge q)\}$ based on either $\{p\}$ or $\{q\}$. So $\{p, q, \neg(p \wedge q), \neg p\}$ will be acceptable relative to Σ (as will $\{p, q, \neg(p \wedge q), \neg q\}$): consistent images of both these sets can be projected using one or the other of the least sets sufficient for projecting a consistent image of Σ. However $\{p, q, \neg(p \wedge q), \neg p, \neg q\}$ is not acceptable: both p and q must be treated ambiguously to create a consistent image of this set, so we cannot project a consistent image of this set based on either $\{p\}$ or $\{q\}$, the least sets that were sufficient to project consistent images of Σ. Subtler ways of making things worse retain some of the original ambiguity-set's members, but require that we *extend* others. We rule all these undesirable extensions out by requiring the extended set's least projection bases to be a *subset* of those of the original set.

To arrive at our new consequence relation we simply declare, following the pattern presented in our introduction, that α follows from Γ in our ambiguity-logic (which we write $\Gamma \vdash_{\text{amb}} \alpha$), if and only if α is an ambiguity-acceptable extension of every ambiguity-acceptable extension of Γ. In Brown (1999) I show that this consequence relation is identical to the consequence relation of Priest's logic of paradox (LP).

However, the LP consequence relation is inelegant from the point of view of this paper. Its set-sentence formulation leads to a number of ugly asymmetries. In particular, it treats absurdities on the left very differently from how it treats their duals, the theorems, on the right: In LP, classical contradictions on the left don't trivialise, but classical tautologies on the right do, that is, any such tautology follows from every premise set. These asymmetries persist even when we express LP in a multiple-conclusion form: multiple-conclusion LP blocks the trivialisation of inconsistent premise sets (in general, not every conclusion set follows from these) but it still trivialises (from right to left) all conclusion sets that cannot be *consistently denied*. These *right-trivial* sets are the sets whose closure under disjunction includes a tautology, the same sets that are right-trivial in classical logic.

First degree entailment (FDE) is a well-known logic closely related to LP that provides a symmetrical treatment of inconsistency on the left and its dual on the right. In Brown (2001) I present an ambiguity-based account capturing the consequence relation of FDE; the presentation here is a modified version of the original, emphasizing the idea of acceptability and the role of extensions of premise and conclusion sets.

Extending this use of ambiguity to provide a preservationist account of FDE requires careful development of the symmetries of the consequence relation. A direct approach to re-imposing the classical left-right symmetries on our ambiguity semantics for LP is to dualise the property preserved from left to right, and require that this dual property be preserved from right to left. Having used ambiguity to project consistent images of the premise set, we now use ambiguity in order to project *consistently deniable* images of the conclusion set.

In this sentence-set consequence relation, we say the set Δ follows from a sentence γ, or $\gamma \vdash_{\text{amb}^*} \Delta$ if and only if γ is an acceptable extension of every acceptable extension of Δ, considered as a set we are committed to denying. But acceptability of an extension is now defined in terms of preservation of the right-ambiguity set of Δ, the set of least sets of atoms whose ambiguous treatment allows us to project a consistently deniable image of Δ: the right-ambiguity set of an acceptable extension of Δ must be a subset of the right-ambiguity set of Δ.

We can combine these two asymmetrical consequence relations to construct a symmetrical one by treating sets on the left as closed under conjunction and sets on the right as closed under disjunction, and demanding that both these consequence relations apply:

$$\Gamma \vdash_{\text{Sym}} \Delta \Leftrightarrow \exists \delta \in \text{Cl}(\Delta, \vee) : \Gamma \vdash_{\text{Amb}} \delta \,\&\, \exists \gamma \in \text{Cl}(\Gamma, \wedge) : \gamma \vdash_{\text{Amb}^*} \Delta.$$

Alternatively (linking both to a purely sentential consequence relation, so that the symmetrical set–set relation arises from type-raising a symmetrical sentence-sentence relation), we can put it this way instead:

$$\Gamma \vdash_{\text{Sym}} \Delta \Leftrightarrow \exists \delta \in \text{Cl}(\Delta, \vee), \exists \gamma \in \text{Cl}(\Gamma, \wedge) : \gamma \vdash_{\text{Sym}} \delta.$$

Where $\gamma \vdash_{\text{Sym}} \delta \Leftrightarrow \{\gamma\} \vdash_{\text{Amb}} \delta \,\&\, \gamma \vdash_{\text{Amb*}} \{\delta\}$.

But the result of this manoeuvre is a logic sometimes called K*,[3] not the more elegant FDE. FDE and K* agree except when classically trivial sets lie on both the left and the right. In those cases the classical triviality of the set on the other side ensures that the property we're preserving is indeed preserved. So the logic just described trivialises when classically trivial sets appear on both the left and the right. Since our aim here is to use ambiguity to prevent trivialisation of the consequence relation as far as we can, we need to be a bit subtler about how we arrive at our symmetrical, set–set consequence relation.

The way forward here is to simultaneously consider consistent images of premise sets and consistently deniable images of conclusion sets, while requiring that the sets of sentence letters used to project these images be *disjoint*[4]:

> $\Gamma \vdash_{\text{FDE}} \Delta$ iff every such consistent image of Γ can be consistently extended by some member of each compatible non-trivial image of Δ (i.e. each non-trivial image of Δ based on a disjoint set of sentence letters),

or equivalently:

> $\Gamma \vdash_{\text{FDE}} \Delta$ iff every such non-trivial image of Δ can be extended by some element of each compatible non-contradictory image of the premise set while preserving its *consistent deniability*.

The new point I want to make here, however, is better seen in the light of another way of expressing this relation—one that focuses less on what feature of our premise and conclusion sets is preserved from left to right and right to left, and more on what we want to preserve regarding the *consequence relation* itself. The point lurking behind the characterisations of $\Gamma \vdash_{\text{FDE}} \Delta$ above is that what we are preserving is the classical consequence relation itself ($\vdash$), under a range of minimally ambiguous, consistent and consistently deniable images of our premises and conclusions, respectively:

> $\Gamma \vdash_{\text{FDE}} \Delta$ iff every image of the premise and conclusion sets, $I(\Gamma), I^*(\Delta)$ obtained by treating disjoint sets of sentence letters drawn from $\text{Amb}(\Gamma)$ and $\text{Amb}^*(\Delta)$ as ambiguous is such that $I(\Gamma) \vdash I^*(\Delta)$.

[3] Conversation with Graham Priest.

[4] In effect, ambiguity allows us to capture the results of using 'both' and 'neither' as (respectively) designated and non-designated fixed points for negation, while insisting that the two sets of ambiguously-treated letters be disjoint ensures that we never treat the same sentence letter in both these ways.

This suggests another interesting strategy for producing new consequence relations from old. We can say that the new consequence relation holds when and only when the old relation holds in all of a range of cases anchored to (or centered on) the original premise and conclusion sets. This strategy can reduce or eliminate trivialisation by ensuring that the *range* of cases considered includes some non-trivial ones, even when the instance forming our 'anchor' is trivial.

8.3 Forcing

In this section of the paper, we apply these ideas to Schotch and Jennings' weakly aggregative *forcing relation*, to arrive at a set–set version of the relation. Completeness for the resulting consequence relation is proved in an appendix.

In the more sophisticated multiple-conclusion version of the forcing relation in Schotch and Jennings (1989), we block right to left as well as left to right aggregative trivialisation. This requires giving up full aggregation on the right as well as the left: both the dual pair of classical rules,

$$\frac{\Gamma [\vdash \Delta, \alpha, \Gamma [\vdash \Delta, \beta}{\Gamma [\vdash \Delta, \alpha \wedge \beta}$$

$$\frac{\Gamma, \alpha [\vdash \Delta, \Gamma, \beta [\vdash \Delta}{\Gamma, \alpha \vee \beta [\vdash \Delta}$$

will fail in our new consequence relation. Following the pattern already applied to our ambiguity logic above, this logic requires a dualisation of level, ℓ', that applies to conclusion sets.

We begin by defining a generalisation of non-triviality for conclusion sets:

$$\text{NonTriv}(\Delta, \xi) \Leftrightarrow \exists A : \oslash, a_1, \ldots, a_i, 1 \leq i \leq \xi, \forall \delta \in \Delta, \exists a_i : \delta \vdash a_i \;\&\; \forall i, \oslash \nvdash a_i .$$

So just as before we will divide a set's content amongst a family of sets, but this time we require that the sets each *follow from* some member of our set, and that the sets all be consistently deniable.

Next we define a measure of Δ's degree of triviality based on this generalisation:

$\ell'(\Delta)$ = the least ξ such that NonTriv(Δ, ξ), else ∞.

The intermediate consequence relation corresponding to K* above results when we produce a symmetrical consequence relation by requiring preservation of both ℓ and ℓ' from left to right and right to left respectively. This consequence relation, like K*, holds trivially whenever classically trivial sets appear on both left and right: The inconsistency of Γ ensures that Δ can be extended by *some member* of Γ while preserving $\ell'(\Delta)$, and the inconsistency of denying Δ ensures that Γ can be extended by some member of Δ while preserving $\ell(\Gamma)$. Once again, focusing just on preserving desirable properties of the premise and conclusion sets when extending them leads us to a nicely symmetrical consequence relation, but one that doesn't capture the full potential of the logical tools we're using. And once again, we can improve on this by applying the division of Γ and Δ simultaneously:

> **Set-Set**: $\Gamma [\vdash \Delta \Leftrightarrow$ For every $\ell(\Gamma)$ covering family of Γ, A, and every $\ell'(\Delta)$ covering family of Δ, B, there is some pair $< a, b >, a \in A$ and $b \in B$, such that $a \vdash b$.

Aggregation of the premise and conclusion sets in our rules for the forcing relation is governed by a simple pigeon hole principle: given $\ell(\Gamma) + 1$ sentences contained in Γ, at least two of the sentences must follow from the same member of the $\ell(\Gamma)$-membered covering family of sets. So *some* conjunction amongst these sentences of Γ will follow from some member of every covering family. So our consequence relation will (as rule 3 below indicates) allow us to infer the disjunction of all the pairwise conjunctions amongst these sentences. In general, this disjunction of pairwise conjunctions will not follow from any of the sentences in Γ, so what we have here is a form of aggregation, though weaker than the familiar rule of conjunction-introduction. Similarly, given $\ell'(\Delta) + 1$ elements of Δ, at least two of the elements must be such that they imply the same member of our right-covering family of sets. This implies that the *disjunction* of those two elements will imply that member of the covering family, and there will be such a pair of elements for every covering family. So the conjunction of such disjunctions will imply some member of every right-covering family for Δ. In general, such a conjunction of disjunctions will not imply any of the original sentences drawn from Δ. So we have a weakening form of aggregation on the right, though not as powerful a weakening as closure under disjunction, the form that aggregation on the right takes in the classical case.

The upshot is that our **Set–Set** condition holds only if the aggregation the level of Γ allows us to apply on the left allows us to produce a formula a, and the aggregation that the right-level of Δ allows us to apply on the right allows us to produce a formula b, such that b follows from a. The equivalence of this characterisation and the official Set–Set condition above is not obvious; in fact, it is the main lemma in a completeness proof for the following rules for multiple-conclusion forcing:

1. Pres $\vdash$: $\dfrac{\Gamma [\vdash \alpha, \alpha \vdash \beta, \beta [\vdash \Delta}{\Gamma [\vdash \Delta}$
2. Ref: $\dfrac{\alpha \in \Gamma}{\Gamma [\vdash \alpha}, \quad \dfrac{\beta \in \Delta}{\beta [\vdash \Delta}$
3. 2/n+1(L): $\dfrac{\Gamma [\vdash \Delta, \alpha_1 \ldots \Gamma [\vdash \Delta, \alpha_{n+1}}{\Gamma [\vdash \Delta, \bigvee_{1 \leq i \neq j \leq n+1} (\alpha_i \wedge \alpha_j)}$ where $n = \ell(\Gamma)$
4. 2/n+1(R): $\dfrac{\Gamma, \alpha_1 [\vdash \Delta, \ldots, \Gamma, \alpha_{n+1} [\vdash \Delta}{\Gamma, \bigwedge_{1 \leq i \neq j \leq n+1} (\alpha_i \vee \alpha_j) [\vdash \Delta}$ where $n = \ell'(\Delta)$
5. Trans: $\dfrac{\Gamma, \alpha [\vdash \Delta, \Gamma [\vdash \alpha, \Delta}{\Gamma [\vdash \Delta}$

The proof of completeness for these rules, which generalises the proof of completeness for single conclusion forcing, appears in an appendix (see Apostoli and Brown 1995 for a version of the proof for single-conclusion forcing, applied to weakly aggregative modal logics). The new proof shows that the aggregation rules (3 and 4 above) suffice to capture all the aggregation that results from the division of the premise and conclusion set's contents amongst $\ell(\Gamma)$ and $\ell'(\Delta)$ cells, respectively. The rules, it emerges, are enough to provide a pair of single, bridging sentences to which we can apply rule 1 to complete our derivation, whenever $\Gamma[\vdash \Delta$.

The main point that I want to make about this logic here is that, like the version of FDE developed above, it can be described as preserving the classical consequence relation throughout a range of related premise and conclusion sets. Thus far, we say that $\Gamma[\vdash \Delta$ holds if and only if the classical consequence relation holds between some pair of cells in every $\ell(\Gamma)$, $\ell'(\Delta)$ covering of Γ and Δ's content. Since these levels are defined as the minimum cardinality required for a family of sets to cover Γ and Δ with only nontrivial members, this requirement will be non-trivial so long as some non-trivial divisions of Γ and Δ's content exist, i.e. so long as Γ and Δ have no individually trivial members. However, we can capture this preservation of $\vdash$ in yet another way, which emphasises the parallel between our ambiguity-based treatment of FDE and forcing. Instead of dividing Γ's content amongst the members of some family of sets indexed to $\ell(\Gamma)$, we can achieve the same effect by means of ambiguity, so long as we require that no ambiguity arise within a single sentence.

When $\ell(\Gamma) = n$, we replace the sentence letters of Γ with n sets of new sentence letters, and produce images of Γ that replace the sentence letters in each $\gamma \in \Gamma$ with sentence letters drawn from one of these sets. Supposing that no numerical subscripts appear in the sentence letters of Γ, we can replace the sentence letters of each sentence in Γ with the same letters combined with subscripts drawn from one of $1, \ldots, n$. Then we can say that $\Gamma[\vdash \delta$ if and only if every such image of Γ has *some* such image of δ (i.e. an image of δ produced by replacing its sentence letters with letters combined with one of our subscripts) as a classical consequence. Similarly, we can say that $\gamma[\vdash \Delta$ if and only if every such image of Δ is such that some such image of γ (i.e. an image of γ produced by replacing its sentence letters in the same way) is a premise from which Δ follows classically. Finally, we can invoke the singleton bridge principle, and say that $\Gamma[\vdash \Delta$ if and only if for some sentences $< \alpha, \beta >$, $\Gamma[\vdash \alpha$, $\beta[\vdash \Delta$, and $\alpha \vdash \beta$.

8.4 Final Reflections on Ambiguity and Aggregation

The existence of this ambiguity-based approach to forcing, together with the broader, strategic parallel between FDE and forcing as ways to ameliorate the trivialisation of classical logic suggests these logics are more closely related than their proponents have thought: ambiguity and aggregation are closely linked. Allowing for ambiguity weakens aggregation by dividing up the content of what is said, whether in a single sentence or between one sentence and another. FDE results from limiting the amount of ambiguity allowed according to the ambiguity sets of Γ and Δ, while insisting that when the ambiguity of a sentence letter is invoked to produce consistent images of Γ, we don't simultaneously use ambiguity of the same sentence letter to produce a consistently deniable image of Δ. Forcing results from limiting ambiguity in two different ways, first requiring that no ambiguity arise within a sentence, and second, allowing no more ambiguity than is then required to divide the content of Γ to the point of consistency, and to divide the content of Δ to the point of consistent deniability.

Weakening consequence relations when they give us undesirable results is an important job in philosophical logic. Preserving a base consequence relation under a range of images of the premise and conclusion sets is one general route to such weakenings, a route I suspect has more interesting results in store.

A.1 Appendix A: Formal Definitions for Ambiguity Logic

First, we define Amb(Γ) as the set of least sets each of which is the base of some projection of a consistent image of Γ:

$$\text{Amb}(\Gamma) = \{A | \exists\, \Gamma' : \text{ConIm}(\Gamma', \Gamma, A) \wedge \forall A', A' \subset A, \neg\exists\, \Gamma'' : \text{ConIm}(\Gamma'', \Gamma, A')\}$$

With this in hand, we can give a formal definition of the preservation relation:

$$\Delta \text{ is an Amb}(\Gamma)\text{-preserving extension of } \Gamma \Leftrightarrow \text{Amb}(\Gamma \cup \Delta) \subseteq \text{Amb}(\Gamma).$$

To indicate that this preservation relation determines the acceptability predicate for our logic, we define:

$$\text{Accept}(\Delta, \Gamma) \Leftrightarrow \Delta \text{ is an Amb}(\Gamma) - \text{preserving extension of } \Gamma,$$

That is,

$$\text{Accept}(\Delta, \Gamma) \Leftrightarrow \text{Amb}(\Gamma \cup \Delta) \subseteq \text{Amb}(\Gamma).$$

Combining this with our general reading of consequence relations as preserving the acceptability of all acceptable extensions, leads us to a new consequence relation:

$$\Gamma \vdash_{\text{Amb}} \alpha \Leftrightarrow \forall \Delta : \text{Accept}(\Delta, \Gamma) \rightarrow \text{Accept}(\Delta \cup \{\alpha\}, \Gamma)$$

Second, our sentence-set, right to left consequence relation preserves an ambiguity measure of consistent deniability from right to left. We begin by defining $\text{Amb}^*(\Delta)$:

$$\text{Amb}^*(\Delta) = \{A | \exists\, \Delta' : \text{ConDenIm}(\Delta', \Delta, A) \wedge$$
$$\forall A', A' \subset A, \neg\exists\, \Delta'' : \text{ConDenIm}(\Delta'', \Delta, A')\}$$

Where ConDenIm(Δ', Δ, A) says that by treating the set of atoms A as ambiguous, we can project Δ', a consistently deniable image of Δ. Then we define acceptable extensions of Δ:

$$\Gamma \text{ is an Amb}^*(\Delta)\text{–preserving extension of } \Delta \Leftrightarrow$$
$$\Delta \subseteq \Gamma \text{and Amb}^*(\Delta \cup \Gamma) \subseteq \text{Amb}^*(\Delta)$$

We write this $\text{Accept}^*(\Gamma, \Delta)$, and write the condition for our right-to-left, deniability-preserving consequence relation as:

$$\gamma \vdash_{\text{Amb}^*} \Delta \Leftrightarrow \forall \Delta : \text{Accept}^*(\Gamma, \Delta) \rightarrow \text{Accept}^*(\Gamma \cup \{\gamma\}, \Delta)$$

B.1 Appendix B: A Proof of Completeness for Set–Set Forcing, with Equivalence of Two Representation Theorems

This appendix presents a completeness proof for set–set forcing that I developed based on the singleton-bridge approach, a view of forcing that focuses separately on how the limited aggregation allowed by forcing enables us to combine the contents of our premises and our conclusions. A proof that the singleton-bridge approach is equivalent to a standard characterisation of forcing in terms of partitions of premise and conclusion sets, with a simple sort of preservation of the classical '$\vdash$' across the partitions, amounts, in the end, to a proof of completeness for our rules.

B.1.1 Definitions and Motivations

B.1.1.1 Preliminaries: Formal Definitions of Level and Set-Sentence Forcing

We begin with a more formal account of Schotch and Jennings' forcing relation, described in the introduction. First, we give a more general, definition of the level of a set of sentences: let A be a family of sets that *covers* Γ's content in the sense that, for each member of Γ, γ, A includes a set a such that $a \vdash \gamma$. Dividing Γ's content amongst such families of sets can allow us to "cover" all of Γ's members in this way even when Γ is inconsistent. We begin by defining a generalisation of consistency using the cardinality of such coverings.

$$\text{Con}(\Gamma, \xi) \Leftrightarrow \exists A : \oslash, a_1, \ldots, a_i, 1 \leq i \leq \xi, \forall \gamma \in \Gamma, \exists i : a_i | \gamma \,\&\, \forall i, a_i \nvdash \oslash$$

Next we define level, a measure of Γ's inconsistency based on this generalisation:

$$\ell(\Gamma) = \text{the least } \xi \text{ such that } \text{Con}(\Gamma, \xi), \text{ else } \infty.$$

Finally, we define a consequence relation that preserves ℓ just as the classical $\vdash$ preserves consistency:

$$\Gamma [\!\vdash \alpha \text{ iff, for } \xi = \ell(\Gamma), \text{ for all } A : a_1 \ldots a_i, 1 \leq i \leq \xi, \exists a_j \in A | a_j \vdash \alpha.$$

For sets of level 2 or more, forcing gives up adjunction (i.e. $\Gamma [\!\vdash \alpha, \Gamma [\!\vdash \beta / \Gamma [\!\vdash \alpha \wedge \beta$ fails). But so long as $\ell(\Gamma)$ is finite, we still obtain consequences that are not classical consequences of individual members of Γ. The forcing relation can be straightforwardly captured by means of the following rules:

1. Pres ⊢: $\dfrac{\Gamma [\!\vdash \alpha, \alpha \vdash \beta}{\Gamma [\!\vdash \beta}$
2. Ref: $\dfrac{\alpha \in \Gamma}{\Gamma [\!\vdash \alpha}$
3. 2/n+1: $\dfrac{\Gamma [\!\vdash \alpha_1, \ldots \Gamma [\!\vdash \alpha_{n+1}}{\Gamma [\!\vdash \bigvee_{1 \leq i \neq j \leq n+1} (\alpha_i \wedge \alpha_j),} \qquad n = \ell(\Gamma)$

This consequence relation preserves ℓ in the same way that classical logic preserves consistency—as a result, it avoids trivialisation for any set that has a well-defined level (i.e. any level other than ∞).

B.1.1.2 Towards a Set–Set Consequence Relation: Defining Con on the Right

The key here is to require that, when we partition (or cover, in general) according to the level, we get a consistent (consistently deniable) partition (covering), i.e. one that has no trivial elements. The level is the minimum cardinality of such a consistent partition. We define a new consistency predicate for the right side of the turnstile, following our definition of Con(Γ, ξ):

$$\text{Con}^*(\Delta, \xi) \Leftrightarrow \exists A : \oslash, a_1, \ldots, a_i, 1 \leq i \leq \xi, \forall \delta \in \Delta, \exists i : \delta \vdash a_i \,\&\, \forall i : \oslash \nvdash a_i$$

B.1.1.3 Defining Left- and Right-Levels of Incoherence for a Set–Set Consequence Relation

$$\ell_L(\Gamma) = \begin{cases} \xi : \text{Con}(\Gamma, \xi), & \text{if such a } \xi \text{ exists,} \\ \infty, \text{otherwise.} \end{cases}$$

$$\ell_R(\Delta) = \begin{cases} \xi : \text{Con}^*(\Delta, \xi), & \text{if such a } \xi \text{ exists,} \\ \infty, \text{otherwise.} \end{cases}$$

In what follows we may omit the subscripted L or R when which is meant is clear from the context.

B.1.1.4 Characterizing [⊢

Given these definitions, we can consider how best to define a condition for $\Gamma [\!\vdash \Delta$. The idea is that, in some sense, the existence of a classical consequence relation between Γ and Δ should be preserved in all the $\ell_L(\Gamma)$ and $\ell_R(\Delta)$ coverings of our premises and conclusions. For simplicity's sake, we will focus on partitions of Γ

and Δ, which can always meet the conditions Con and Con* if any covering can. We define $\Pi(\Gamma)$ as the set of all $\ell_L(\Gamma)$ partitions of Γ (and $\Pi(\Delta)$ for all the $\ell_R(\Delta)$ partitions of Δ); we use π for elements of $\Pi(\Gamma)$, and p for elements of these in turn. Then, starting with the singleton cases, we say:

1. $\Gamma[\vdash \alpha$ iff $\forall \pi \in \Pi(\Gamma), \exists p \in \pi : p \vdash \alpha$

2. $\alpha[\vdash \Delta$ iff $\forall \pi \in \Pi(\Delta), \exists p \in \pi : \alpha \vdash p$

Comments:

The first says that a consequence "survives" the partitioning of Γ if and only if it appears as a consequence of every partition, i.e. as a consequence of some cell of every partition. The second applies the same treatment to surviving as a "prover": a prover of some set survives the partitioning of that set if and only if it proves some cell of every partition.

Both fit well with the definition of levels. Since the level is the least number of cells such that a partition with no trivial cells exists, neither our consequences nor our provers can be trivial unless Γ or Δ (respectively) contains a singleton trivial sentence. Outside those cases, our consequences must result from some non-trivial premises, and our provers must prove some non-trivial consequence.

B.1.1.5 Defining [⊢ for the General Case

Now we need to consider the general, $\Gamma[\vdash \Delta$ case. One way of doing it is to require that whatever aggregation of Γ and of Δ that survives all the partitions in $\Pi(\Gamma)$ and $\Pi(\Delta)$ *guarantee* the preservation of any consequences that we will endorse. That is, we require that every pair of partitions of Γ and of Δ (according to their levels) π_Γ, π_Δ, be such that $\pi_\Gamma \uparrow\!\vdash \pi_\Delta$, that is, such that for some $p \in \pi_\Gamma$ and some $p' \in \pi_\Delta$, $p \vdash p'$. Then, given a set of rules for [⊢, we can state the:

Representation theorem for Forcing 1:

R1: $\Gamma[\vdash \Delta \Leftrightarrow \forall \pi \in \Pi(\Gamma), \pi' \in \Pi(\Delta), \pi \uparrow\!\vdash \pi'$.

But we can also capture the aggregation of Γ and Δ by means of the singular cases we've already settled the completeness of. That is, we can do the same job[1] by invoking our rules for the singular case:

Representation theorem for Forcing 2:

R2: $\Gamma[\vdash \Delta \Leftrightarrow \exists a : \Gamma[\vdash a \,\&\, a[\vdash \Delta$.

Proving that these are equivalent brings us very close to a completeness proof for the first representation theorem, so our strategy will be to establish the equivalence of R1 and R2 as a lemma, and then go on to prove R1.

[1]As a comment by Pol Nicholson suggested.

B.1.2 The Main Proofs

We begin by noting something important about certain singular cases—there are two rules that are sound for these singular cases but that fail as soon as we allow sets with cardinality 2 or more on both sides:

$$\vee[\vdash: \quad \frac{\Gamma, \alpha [\vdash \gamma, \Gamma, \beta [\vdash \gamma}{\Gamma, \alpha \vee \beta [\vdash \gamma} \qquad\qquad [\vdash \wedge: \quad \frac{\gamma [\vdash \Delta, \alpha, \gamma [\vdash \Delta, \beta}{\gamma [\vdash \Delta, \alpha \wedge \beta}$$

These rules play a crucial role in the completeness proof for the forcing relations—I think of them (in the context of set–set forcing) as extensions of pres $\vdash$: Pres $\vdash$ tells us that forcing respects the consequence relations between singletons on both sides of $[\vdash$ that classical logic provides. These special, singular versions of $\vee[\vdash$ and $[\vdash \wedge$ tell us that we will treat aggregation classically from right to left so long as we have a singleton on the right, and we will treat aggregation classically from left to right so long as we have a singleton on the left. And so, of course, we should: we are starting (in both cases) from a state of complete aggregation on the left or right.

Finally, we will need to bring in *fixed-level* forcing to deal with some variations on our initial premise and conclusion sets. So $[\vdash$ here will sometimes be replaced by ${}^{n}[\vdash^{m}$, where n and m are our initial premise and conclusion set levels. The rules for ${}^{n}[\vdash^{m}$ are precisely those for forcing, with the exception that the level-sensitive rules are set to the fixed numbers n and m, rather than varying in form depending on the levels of the premise and conclusion sets. The key points to keep in mind with respect to fixed level forcing are:

F1: For $n = \ell(\Gamma)$ and $m = \ell(\Delta)$, $\Gamma\, {}^{n}[\vdash^{m} \Delta \Leftrightarrow \Gamma [\vdash \Delta$.
F2: Fixed level forcing is strictly monotonic. (Only the effects of level increases prevent $[\vdash$ from being fully monotonic; but fixed level forcing ignores such increases—at the cost of trivializing ${}^{n}[\vdash^{m}$ for sets with levels higher than n, m. This monotonicity comes in handy in the proofs that follow.)

With this background in hand, the equivalence of R1 and R2 can be proved quite straightforwardly.

Lemma. $\forall \pi \in \Pi(\Gamma), \pi' \in \Pi(\Delta)\ \pi \uparrow\!\vdash \pi' \Leftrightarrow \exists \alpha : \Gamma [\vdash \alpha \,\&\, \alpha [\vdash \Delta$.

The proof is by induction on the cardinalities of Γ and Δ, paralleling the original completeness proof for $\Gamma [\vdash \alpha$; see Apostoli and Brown (1995). Given this proof, together with the completeness of $\Gamma [\vdash \alpha$ (and the perfectly dual completeness of $\alpha [\vdash \Delta$), we have a proof of completeness for a simple approach to the rules for $\Gamma [\vdash \Delta$.

$\Rightarrow$

1. Base: Consider the case where we have cardinality of $\Gamma \leq \ell(\Gamma)$, and cardinality of $\Delta \leq \ell(\Delta)$. Here the only way for $\forall \pi \in \Pi(\Gamma), \pi' \in \Pi(\Delta) \pi \uparrow\!\vdash \pi'$ to hold is by virtue of a singleton implication linking some elements of $\Gamma, \Delta : \exists \gamma \in \Gamma, \delta \in \Delta : \gamma \vdash \delta$. But then by pres $\vdash$, it follows that $\exists \alpha : \Gamma [\vdash \alpha$ and $\alpha [\vdash \Delta$.

2. Induction hypothesis: Suppose that this holds for cardinalities of Γ, Δ up to n, m.
3. Assume that Γ has level n', Δ has level m', so that $\Gamma [\!\vdash \Delta \Leftrightarrow \Gamma\, {}^{n'}[\!\vdash^{m'} \Delta$.

First induction step: Let Γ's cardinality be $n+1$ and assume that $\forall \pi \in \Pi(\Gamma), \pi' \in \Pi(\Delta) \pi \uparrow\!\vdash \pi'$.

We assume we are dealing with a case where $\mathrm{Card}(\Gamma) > \ell(\Gamma)$. Select an $\ell(\Gamma)+1$ tuple from $\Gamma, \{\gamma_1, \ldots \gamma_{n+1}\}$. Now consider the variations on Γ that we obtain by removing two elements from this $n+1$-tuple and replacing them with the conjunction of these two elements. In the context of fixed-level forcing, what we are doing is, in effect, simply restricting our attention to the set of partitions of Γ that keep γ_i and γ_j in the same cell. We already know from our induction assumption that every one of these partitions includes a cell proving some cell of each partition of Δ.

Let Γ^{ij} be the variation that does this with the ith and jth elements from our n-tuple. Then by our induction hypothesis and our assumption,

$$\text{If} \, \forall \pi \in \Pi(\Gamma), \pi' \in \Pi(\Delta) \pi \uparrow\!\vdash \pi', \exists \alpha : \Gamma^{ij\, n'}[\!\vdash^{m'} \alpha \& \alpha [\!\vdash \Delta 1 \leq i \neq j \leq n+1,$$

$$\forall \pi \in \Pi(\Gamma), \pi' \in \Pi(\Delta)\, \pi \uparrow\!\vdash \pi'$$

$$\text{So} \, \exists \alpha : \Gamma^{ij\, n'}[\!\vdash^{m'} \alpha \,\&\, \alpha [\!\vdash \Delta, 1 \leq i \neq j \leq n+1.$$

Now, by the monotonicity of ${}^{n}[\!\vdash^{m}$, it follows that $\Gamma, \wedge(\gamma_i, \gamma_j)\, {}^{n'}[\!\vdash^{m'} \alpha$. Since α is singular we can apply a series of $\vee [\!\vdash$ steps:

$$\frac{\Gamma, \alpha [\!\vdash \gamma, \Gamma, \beta [\!\vdash \gamma}{\Gamma, \alpha \vee \beta [\!\vdash \gamma}$$

to obtain

$$\Gamma, \bigvee_{1 \leq i \neq j \leq n+1} (\wedge(\gamma_i, \gamma_j))\, {}^{n'}[\!\vdash^{m'} \alpha$$

But $\Gamma\, {}^{n'}[\!\vdash^{m'} \bigvee_{1 \leq i \neq j \leq n+1} (\wedge(\gamma_i, \gamma_j))$ by our rule for aggregation from left to right. By the monotonicity of ${}^{n'}[\!\vdash^{m'}$, it follows that $\Gamma\, {}^{n'}[\!\vdash^{m'} \bigvee_{1 \leq i \neq j \leq n+1} (\wedge(\gamma_i, \gamma_j)), \Delta$. And then by trans:

$$\frac{\Gamma, \alpha [\!\vdash \Delta \,\&\, \Gamma [\!\vdash \alpha, \Delta}{\Gamma [\!\vdash \Delta}$$

we get $\Gamma\, {}^{n'}[\!\vdash^{m'} \alpha$.

But n', m' are the levels of Γ and Δ respectively. So $\Gamma\, {}^{n'}[\!\vdash^{m'} \alpha \Leftrightarrow \Gamma [\!\vdash \alpha$. Hence $\Gamma [\!\vdash \alpha$, and $\exists \alpha : \Gamma [\!\vdash \alpha \,\&\, \alpha [\!\vdash \Delta$ as required.

Note that it also follows that $\Gamma [\!\vdash \Delta$: Since $\Gamma [\!\vdash \alpha$ and $\alpha [\!\vdash \Delta, \alpha$ is a level-preserving extension of both Γ and Δ (just place α in whatever cells prove/are proved by α to get a consistent/consistently deniable partitioning of Γ, α and Δ, α respectively). So Mon* assures us that from $\Gamma [\!\vdash \alpha$ we get $\Gamma [\!\vdash \alpha, \Delta$ and from $\alpha [\!\vdash \Delta$ we get $\Gamma, \alpha [\!\vdash \Delta$. But then by trans, $\Gamma [\!\vdash \Delta$.

Second induction step: Let Δ's cardinality be $m+1$ and assume that $\forall \pi \in \Pi(\Gamma), \pi' \in \Pi(\Delta), \pi \uparrow\!\vdash \pi'$.

Select a $\ell(\Delta) + 1$-tuple from $\Delta, \delta_1, \dots \delta_{n+1}$. Now consider all the variations on Δ that we obtain by removing two elements from this $n + 1$-tuple and replacing them with the disjunction of these two elements. (Once again, in effect, we are restricting our attention to the partitions of Δ that keep δ_i and δ_j together.) Let Δ^{ij} be the variation that does this with δ_i and δ_j. Then our induction hypothesis and our assumption give us:

1. If $\forall \pi \in \Pi(\Gamma), \pi' \in \Pi(\Delta) \pi \uparrow\vdash \pi', \exists \alpha : \Gamma [\vdash \alpha \,\&\, \alpha\, {}^{n'}[\vdash^{m'} \Delta_{ij}$
2. $\forall \pi \in \Pi(\Gamma), \pi' \in \Pi(\Delta) \pi \uparrow\vdash \pi'$

Hence,

$$\exists \alpha : \Gamma [\vdash \alpha \,\&\, \alpha [\vdash \Delta_{ij}, 1 \leq i \neq j \leq n+1$$

Now, by the monotonicity of ${}^{n}[\vdash^{m}$,

$$\alpha'_n [\vdash m' \Delta, \vee(\delta_i, \delta_j), 1 \leq i \neq j \leq n+1$$

And since α is singular, a series of steps applying our special rule for singular aggregation from left to right ($[\vdash \wedge$) gives us:

$$\alpha\, {}^{n'}[\vdash^{m'} \Delta, \bigwedge (\vee(\delta_i, \delta_j))$$

By our general rule for aggregation from left to right, $\bigwedge_{1 \leq i \neq j \leq n+1} (\vee(\delta_i, \delta_j)) [\vdash \Delta$. And by the monotonicity of ${}^{n}[\vdash^{m}$ it follows that $\alpha, \bigwedge_{1 \leq i \neq j \leq n+1} (\vee(\delta_i, \delta_j))\, {}^{n'}[\vdash^{m'} \Delta$. So by trans

$$\frac{\Gamma, \alpha [\vdash \Delta \,\&\, \Gamma [\vdash \alpha, \Delta}{\Gamma [\vdash \Delta}$$

we get $\alpha\, {}^{n'}[\vdash^{m'} \Delta$.

But n', m' are the levels of Γ and Δ respectively. So $\alpha\, {}^{n'}[\vdash^{m'} \Delta \Leftrightarrow \alpha [\vdash \Delta$. Hence $\alpha [\vdash \Delta$, and $\exists \alpha : \Gamma [\vdash \alpha \,\&\, \alpha [\vdash \Delta$ as required.

Note again that it also follows that $\Gamma [\vdash \Delta$, by the same argument as above.

$\Leftarrow$

This direction is straightforward. Suppose $\exists \alpha : \Gamma [\vdash \alpha$ and $\alpha [\vdash \Delta . \exists \alpha : \Gamma [\vdash \alpha$ and $\alpha [\vdash \Delta$ holds only if every partition of Γ includes a cell proving α and every partition of Δ includes a cell proved by α. And by classical trans, any cell in $\pi \in \Pi(\Gamma)$ that proves α proves any cell $\pi' \in \Pi(\Delta)$ proved by α. □

Theorem. $\forall \pi \in \Pi(\Gamma), \pi' \in \Pi(\Delta) \pi \uparrow\vdash \pi' \Leftrightarrow \Gamma [\vdash \Delta$.

The notes at the end of each direction of the equivalence proof give us the hard direction ($\Rightarrow$); soundness follows from pigeon-hole considerations and well-known facts about classical singleton consequences.

References

Apostoli, P., and B. Brown. 1995. A solution to the completeness problem for weakly aggregative modal logics. *Journal of Symbolic Logic* 60(3): 832–842.

Brown, B. 1999. Yes, Virginia, there really are paraconsistent logics. *Journal of Philosophical Logic* 28: 489–500.

Brown, B. 2001. LP, FDE and ambiguity. In *IC-AI 2001 volume II: Proceedings of the 2001 meetings of the international conference on artificial intelligence*, ed. H. Arabnia. CSREA publications. Las Vegas, Nevada.

Brown, B., and P.K. Schotch. 1999. Logic and aggregation. *Journal of Philosophical Logic* 28: 265–287.

Jennings, R.E., and P.K. Schotch. 1981. Some remarks on (weakly) weak modal logics. *Notre Dame Journal of Formal Logic* 22: 309–314.

Jennings, R.E., and D.K. Johnston. 1983. Paradox-tolerant logic. *Logique et Analyse* 26: 291–308.

Jennings, R.E., and P.K. Schotch. 1984. The preservation of coherence. *Studia Logica* 43: 89–106.

Jennings, R.E., P.K. Schotch, and D.K. Johnston. 1980. Universal first order definability in modal logic. *Zeitschrift fur Mathematische Logik und Grundlagen der Mathematik* 26: 327–330.

Jennings, R.E., P.K. Schotch, and D.K. Johnston. 1981. The n-adic first order undefinability of the Geach formula. *Notre Dame Journal of Formal Logic* 22: 375–378.

Johnston, D.K. 1978. *A generalized relational semantics for modal logic*. M.A. thesis, Department of Philosophy, Simon Fraser University, Barnaby.

MacLeod, M.C., and P.K. Schotch. 2000. Remarks on the modal logic of Henry Bradford Smith. *The Journal of Philosophical Logic* 29: 603–615.

Priest, G. 1979. The logic of paradox. *Journal of Philosophical Logic* 8: 219–241.

Schotch, P.K. 2000. Skepticism and epistemic logic. *Studia Logica* 65: 187–198.

Schotch, P.K., and R.E. Jennings. 1980a. Inference and necessity. *Journal of Philosophical Logic* 9: 327–340.

Schotch, P.K., and R.E. Jennings. 1980b. Modal logic and the theory of modal aggregation. *Philosophia* 9: 265–278.

Schotch, P.K., and R.E. Jennings. 1989. On detonating. In *Paraconsistent logic*, G. Priest, R. Routley, and J. Norman, 306–327. Munich: Philosophia Verlag.

Thorn, P. 2000. Paraconsistency and the image of science. In *Logical Consequences*, ed. J. Woods and B. Brown. London: Hermes.

Chapter 9
On Modal Logics Defining Jaśkowski's D_2-Consequence

Marek Nasieniewski and Andrzej Pietruszczak

9.1 Introduction

The basic features of notions of deductive and deductive discussive systems used by Jaśkowski are as follows (see Jaśkowski 1948, p. 61 and Jaśkowski 1999, pp. 37–38).

- By theses of a deductive system Jaśkowski meant all expressions asserted within it, i.e. axioms and theorems deduced from them or proved in a specific way for a given system.
- A deductive system is based on a certain logic iff the set of its theses is closed under *modus ponens* rule with respect to theorems of the logic.
- A deductive system is overcomplete iff the set of its theses is equal to the set of all meaningful expressions of the language.
- A deductive system is inconsistent iff among its theses there are two theses such that one of them is the negation of the other.
- Usually, theses of a deductive system are formally expressed theorems of some consistent theory.
- If there is no assumption that theses of a deductive system express opinions which do not contradict each other, then such a system is called *discussive*.

Jaśkowski's aim was to formulate a logic, which when applied to inconsistent systems would not generally entail overcompleteness.

Jaśkowski gave an example of the way in which theses of discussive systems can be generated by referring to a discussion. Decisive for such a choice was the fact that during a discussion inconsistent voices can appear, however, we are not inclined to deduce every thesis from them.

M. Nasieniewski (✉) • A. Pietruszczak
Department of Logic, Nicolaus Copernicus University, Toruń, Poland
e-mail: mnasien@umk.pl; pietrusz@umk.pl

K. Tanaka et al. (eds.), *Paraconsistency: Logic and Applications*, Logic, Epistemology, and the Unity of Science 26, DOI 10.1007/978-94-007-4438-7_9,

One can treat voices appearing explicitly in the discussion as preceded by the following restriction: "according to the opinion of one of the participants of the discussion", which formally one can express by preceding the given statement with: 'it is possible that'. If we take a position of an external observer (i.e. someone that does not take part in a discussion) all voices appearing in a discussion are *only possible*. It is so, since a person who is not involved in the discussion has every right to treat particular voices in disbelief or to dissociate from discussants' statements. For the same reason, also conclusions following from explicitly expressed statements in a discussion are *only possible*. Conclusions one can treat as implicitly included into a discussion, since a given discussion consists not only of voices explicitly expressed, but also statements concluded from them. Summarizing, explicit voices, as well as their conclusions, are treated as theses of a discussive system.

Since the above pattern requires use of a modal language, one has to choose some specific modal logic. Jaśkowski himself chose the logic **S5**.

It is obvious that one needs to consider the language of full sentential logic, since otherwise one would have to treat all sentences as atomic ones, and it would not be possible to analyze logical deducibility based on the meaning of logical sentential constants.

In the present paper, 'p' and 'q' are propositional letters, used to built formulae (both discussive and modal). Capital Latin letters 'A', 'B' and 'C' (with or without subscripts) are metavariables for formulae, a Greek letter 'Π' is a metavariable for sets of formulae, while small Latin latter 'a' is a metavariable for propositional letters. Besides, following Jaśkowski's custom, Gothic letters are used to denote instances of concrete sentences of the natural language.

Jaśkowski observed that while formulating a discussive system one can not treat the implication '$\rightarrow$' as a material one, since sets of theses of discussive systems would not be closed under the *modus ponens* rule:

$$\frac{\begin{array}{c}\mathfrak{P} \rightarrow \mathfrak{Q}\\ \mathfrak{P}\end{array}}{\mathfrak{Q}}$$

[...] out of the two theses one of which is

$$\mathfrak{P} \rightarrow \mathfrak{Q},$$

and thus states: "it is possible that if $\mathfrak{P}$, then $\mathfrak{Q}$", and the other is

$$\mathfrak{P},$$

and thus states: "it is possible that $\mathfrak{P}$", it does not follow that "it is possible that $\mathfrak{Q}$", so that the thesis

$$\mathfrak{Q},$$

does not follow intuitively, as the rule of *modus ponens* requires. (Jaśkowski 1999, p. 43, see also Jaśkowski 1948, p. 66)

Jaśkowski meant that the formula:

$$\Diamond(p \to q) \to (\Diamond p \to \Diamond q)$$

is not a thesis of **S5**. Thus, not for all sentences $\mathfrak{P}$ and $\mathfrak{Q}$, the following sentence

$$\Diamond(\mathfrak{P} \to \mathfrak{Q}) \to (\Diamond\mathfrak{P} \to \Diamond\mathfrak{Q})$$

is a substitution of a logical thesis.

As an appropriate implication to be used in the formulation of a discussive logic Jaśkowski chose a discussive one. We will denote it by '$\to^{\mathrm{d}}$'. In the formal language Jaśkowski defined a formula

$$p \to^{\mathrm{d}} q$$

by

$$\Diamond p \to q\,.$$

Jaśkowski intuitively understood it in the following way: "if anyone states that p, then q" (Jaśkowski 1999, p. 44, see also Jaśkowski 1948, p. 67).

In the same fragment, Jaśkowski pointed to the fact that:

> In every discussive system two theses, one of the form:
>
> $$\mathfrak{P} \to^{\mathrm{d}} \mathfrak{Q}\,,$$
>
> and the other of the form:
>
> $$\mathfrak{P}\,,$$
>
> entail the thesis
>
> $$\mathfrak{Q}\,,$$
>
> and that on the strength of the theorem
>
> $$\Diamond\,(\Diamond p \to q) \to (\Diamond p \to \Diamond q). \tag{M$_2$1}$$

Thus, such an understanding of the implication ensures that sets of theses of deductive systems are closed under the *modus ponens* rule.

A discussive equivalence (notation: '$p \leftrightarrow^{\mathrm{d}} q$') Jaśkowski defined as:

$$(\Diamond p \to q) \wedge (\Diamond q \to \Diamond p).$$

In Jaśkowski (1948) (see also Jaśkowski (1969)), three classical connectives are used: negation ('$\neg$'), disjunction ('$\vee$') and conjunction ('$\wedge$'). Moreover, a discussive conjunction '$\wedge^{\mathrm{d}}$', was introduced in Jaśkowski (1949). Any sentence of the form '$p \wedge^{\mathrm{d}} q$' expresses a statement: "p and it is possible that q", i.e. formally: '$p \wedge \Diamond q$'. Notice that in Jaśkowski (1949) the classical conjunction was not dropped from the language of discussive systems.

> Dwuwartościowy dyskusyjny rachunek zdań oznaczony jako D_2 można wzbogacić definiując koniunkcję dyskusyjną *Kd*. [*In English*: The two-valued discussive propositional calculus denoted as D_2 can be enriched with a definition of the discussive conjunction $\wedge^d$]. (Jaśkowski 1949, p. 171, Jaśkowski 1999a, p. 57)

The question arises: *what is the natural interpretation of the classical conjunction in the context of discussive systems?* It seems that the classical conjunction can be used to "glue" particular statements of a given participant of the discussion. For example, if a given participant expresses two statements $\mathfrak{P}$ and $\mathfrak{Q}$ then she/he asserts $\ulcorner\mathfrak{P} \wedge \mathfrak{Q}\urcorner$, i.e. taking the external point of view we have in the modal language $\ulcorner\Diamond(\mathfrak{P}^\bullet \wedge \mathfrak{Q}^\bullet)\urcorner$, where $(-)^\bullet$ is the appropriate translation of discussive connectives which can appear within $\mathfrak{P}$ and $\mathfrak{Q}$. On the other hand discussive conjunction is usually meant as a tool adequate to express the status of a given discussion from the point of view of a given participant of the discussion. Thus, if we have assertions $\mathfrak{P}$ and $\mathfrak{Q}$ made by two participants, then the appearance of these two statements—taking the point of view of the first participant—can be expressed as follows: $\ulcorner\mathfrak{P} \wedge^d \mathfrak{Q}\urcorner$. From the external point of view such a statement becomes $\ulcorner\Diamond(\mathfrak{P}^\bullet \wedge \Diamond\mathfrak{Q}^\bullet)\urcorner$, which in the logic **S5** is equivalent to $\ulcorner\Diamond\mathfrak{P}^\bullet \wedge \Diamond\mathfrak{Q}^\bullet\urcorner$. We obtain the same formula if we start with the consideration of the point of view of the second participant. Indeed, we have the discussive record of the discussion from the point of view of the second participant: $\ulcorner\mathfrak{Q} \wedge^d \mathfrak{P}\urcorner$, while the external point of view of this statement becomes: $\ulcorner\Diamond(\mathfrak{Q}^\bullet \wedge \Diamond\mathfrak{P}^\bullet)\urcorner$, equivalently on the basis of **S5** we have $\ulcorner\Diamond\mathfrak{Q}^\bullet \wedge \Diamond\mathfrak{P}^\bullet\urcorner$.

Of course we are not interested only in the ≪external description≫ of a given discussion, but also whether $\mathfrak{Q}$ discussively follows from given statements $\mathfrak{P}_1, \ldots, \mathfrak{P}_n$ of n participants ($n > 0$). Using modal translations and the usual understanding of deduction in modal logics we inquire whether the following statements (equivalent by the positive logic):

(a) $(\Diamond\mathfrak{P}_1^\bullet \wedge \cdots \wedge \Diamond\mathfrak{P}_n^\bullet) \rightarrow \Diamond\mathfrak{Q}^\bullet$,
(b) $\Diamond\mathfrak{P}_1^\bullet \rightarrow (\ldots \rightarrow (\Diamond\mathfrak{P}_n^\bullet \rightarrow \Diamond\mathfrak{Q}^\bullet)\ldots)$

are valid in **S5**.[1] Equivalently we can look into the problem of validity of the following sentences in the discussive logic:

$(\mathrm{a})^d$ $(\mathfrak{P}_1 \wedge^d \cdots \wedge^d \mathfrak{P}_n) \rightarrow^d \mathfrak{Q}$,
$(\mathrm{b})^d$ $\mathfrak{P}_1 \rightarrow^d (\ldots \rightarrow^d (\mathfrak{P}_n \rightarrow^d \mathfrak{Q})\ldots)$.[2]

In both cases $(\mathrm{a})^d$ and $(\mathrm{b})^d$—using the logic **S5**—we obtain the equivalent translations of sentences into the modal language. We have to remember that in the case of validity in the discussive logic the translation obtained has to be preceded by '$\Diamond$',

[1]For $n = 0$ we inquire whether the sentence $\mathfrak{Q}$ is valid in the discussive logic, i.e. whether the modal sentence $\Diamond\mathfrak{Q}^\bullet$ is valid in **S5**.

[2]Notice that for $n = 1$ and any $m > 0$ a sentence $\ulcorner(\mathfrak{p}_1 \wedge \cdots \wedge \mathfrak{p}_m) \rightarrow^d \mathfrak{Q}\urcorner$ has a form $(\mathrm{a})^d$ as well as a form $(\mathrm{b})^d$, for $\mathfrak{P}_1 := \ulcorner\mathfrak{p}_1 \wedge \cdots \wedge \mathfrak{p}_m\urcorner$. Thus, it can be treated as expressing the external point of view where only one participant is considered.

since from the point of view of an external observer the sentences $(a)^d$ and $(b)^d$ are *only possible*. Thus indeed (a) and (b) are the modal counterparts of $(a)^d$ and $(b)^d$, respectively.

As it is known the formula (a), resp. (b), is valid in **S5** iff there is a finite sequence beginning with sentences $\ulcorner\Diamond\mathfrak{P}_1^\bullet\urcorner, \ldots, \ulcorner\Diamond\mathfrak{P}_n^\bullet\urcorner$, and ending with $\ulcorner\Diamond\mathfrak{Q}^\bullet\urcorner$, where the other elements (as well as $\ulcorner\Diamond\mathfrak{Q}^\bullet\urcorner$) are either theses of **S5** and/or are sentences obtained from some sentences preceding in the sequence obtained by *modus ponens*.

The main aim of our paper is to find the smallest normal logic and the smallest regular logic which could be used instead of **S5**. For these logics it is not enough to have the same theses beginning with '$\Diamond$' as **S5**; since we consider here the discussive deducibility relation, thus these logics have to include also ($M_2$1).

Remark 9.1. In the case of a sentence of the form (a), resp. (b), for $n = 0$ we only try to find out whether a wanted logic has the same thesis beginning with '$\Diamond$'. This problem has already been solved in the case of normal and regular classes of logics (Furmanowski 1975; Perzanowski 1975; Nasieniewski and Pietruszczak 2008). □

Nowadays in the considerations concerning the logic $\mathbf{D_2}$ the classical conjunction is usually omitted. It is justified by the functional completeness obtained by classical connectives of '$\neg$' and '$\vee$'. Thus, we also do not include the classical conjunction in the discussive language.

9.2 Basic Notions

Let For^d be the set of all formulae of the discussive language with constants: '$\neg$', '$\vee$', '$\wedge^d$', '$\rightarrow^d$', and '$\leftrightarrow^d$'. Let $\mathrm{For_m}$ be the set of all modal formulae.[3] *Jaśkowski's transformation* is the function $-^\bullet$ from For^d into $\mathrm{For_m}$ such that:

1. $(a)^\bullet = a$, for any propositional letter a,
2. and for any $A, B \in \mathrm{For}^d$:

   (a) $(\neg A)^\bullet = \ulcorner\neg A^\bullet\urcorner$,
   (b) $(A \vee B)^\bullet = \ulcorner A^\bullet \vee B^\bullet\urcorner$,
   (c) $(A \wedge^d B)^\bullet = \ulcorner A^\bullet \wedge \Diamond B^\bullet\urcorner$,
   (d) $(A \rightarrow^d B)^\bullet = \ulcorner\Diamond A^\bullet \rightarrow B^\bullet\urcorner$
   (e) $(A \leftrightarrow^d B)^\bullet = \ulcorner(\Diamond A^\bullet \rightarrow B^\bullet) \wedge \Diamond(\Diamond B^\bullet \rightarrow A^\bullet)\urcorner$.[4]

Assume that voices in a discussion are written formally by schemes: $A_1, \ldots, A_n$. We consider a possible conclusion B. Since formulae $A_1, \ldots, A_n$ and B may contain

[3] In Appendix we recall some chosen basic facts and notions concerning modal logic.

[4] If the classical conjunction were considered, one would have to add the following condition: $(A \wedge B)^\bullet = \ulcorner A^\bullet \wedge B^\bullet\urcorner$.

logical constants thus, instead of $\Diamond A_1, \ldots, \Diamond A_n$ and $\Diamond B$ we have to consider their discussive versions: $\Diamond A_1^\bullet, \ldots, \Diamond A_n^\bullet$ and $\Diamond B^\bullet$. Taking into account examples given by Jaśkowski we see that he used the following definition of a discussive relation: *B follows discussively from* $A_1, \ldots, A_n$ iff the following formula

$$\Diamond A_1^\bullet \to (\ldots \to (\Diamond A_n^\bullet \to \Diamond B^\bullet)\ldots)$$

belongs to **S5**.[5]

To conclude, discussive deductive systems are to be based on a certain logic connected with the following consequence relation for formulae from $\mathrm{For}^{\mathrm{d}}$.

Definition 9.1 *For any* $\Pi \subseteq \mathrm{For}^{\mathrm{d}}$ *and* $B \in \mathrm{For}^{\mathrm{d}}$: $\Pi \vdash_{\mathbf{D_2}} B$ *iff for some* $n \geqslant 0$ *and for some* $A_1, \ldots, A_n \in \Pi$ *we have*

$$\ulcorner \Diamond A_1^\bullet \to (\ldots \to (\Diamond A_n^\bullet \to \Diamond B^\bullet)\ldots)\urcorner \in \mathbf{S5}\,.$$

In other words,

$$\Pi \vdash_{\mathbf{D_2}} B \qquad \textit{iff} \qquad \{\Diamond A^\bullet : A \in \Pi\} \vdash_{\mathbf{S5}} \Diamond B^\bullet,$$

where $\vdash_{\mathbf{S5}}$ *is the pure modus-ponens-style inference relation based on* **S5** *(see Definition 9.A.1 and Fact 9.A.1 in Appendix).*

Jaśkowski used notation '$\mathbf{D_2}$' referring to a logic, i.e. a certain set of formulae.

Definition 9.2 $\mathbf{D_2} := \{\, A \in \mathrm{For}^{\mathrm{d}} : \ulcorner \Diamond A^{\bullet} \urcorner \in \mathbf{S5}\,\}$.

Thus, on the basis of $\mathbf{D_2}$ one can characterize the consequence relation for discussive systems in the following way:

Fact 9.1 *For any* $n \geqslant 0$, $A_1, \ldots, A_n$, $B \in \mathrm{For}^{\mathrm{d}}$:

$$\begin{aligned}
A_1, \ldots, A_n \vdash_{\mathbf{D_2}} B \quad & \textit{iff} \quad \ulcorner(\Diamond A_1^\bullet \wedge \cdots \wedge \Diamond A_n^\bullet) \to \Diamond B^{\bullet}\urcorner \in \mathbf{S5} \\
& \textit{iff} \quad \ulcorner \Diamond A_1^\bullet \to (\ldots \to (\Diamond A_n^\bullet \to \Diamond B^\bullet)\ldots)\urcorner \in \mathbf{S5} \\
& \textit{iff} \quad \ulcorner \Diamond (A_1 \to^{\mathrm{d}} (\ldots \to^{\mathrm{d}} (A_n \to^{\mathrm{d}} B)\ldots))^{\bullet}\urcorner \in \mathbf{S5} \\
& \textit{iff} \quad \ulcorner A_1 \to^{\mathrm{d}} (\ldots \to^{\mathrm{d}} (A_n \to^{\mathrm{d}} B)\ldots)\urcorner \in \mathbf{D_2} \\
& \textit{iff} \quad \ulcorner (A_1 \wedge^{\mathrm{d}} \cdots \wedge^{\mathrm{d}} A_n) \to^{\mathrm{d}} B\urcorner \in \mathbf{D_2}\,.
\end{aligned}$$

---

[5] In da Costa and Doria (1995) a similar relation was used, yet not for $\mathrm{For}^{\mathrm{d}}$, but for a modal language enriched with some discussive connectives. However, in this modal language the discussive conjunction was defined as follows: $\ulcorner(A \wedge^{\mathrm{d}} B) \leftrightarrow (\Diamond A \wedge B)\urcorner$. But, as it was proved in Ciuciura (2005), for a new transformation $-^*$ such that $(A \wedge^{\mathrm{d}} B)^* = \ulcorner \Diamond A^* \wedge B^* \urcorner$, we obtain another discussive logic $\mathbf{D}_2^*$ which differs from $\mathbf{D_2}$.

Proof. By **PL**, ($5^{\Diamond}!$), ($R^{\Diamond\Box}$), and definitions of the relation $\vdash_{\mathbf{D_2}}$, the function $-^{\bullet}$, and the logic $\mathbf{D_2}$.

Notice that, by the above fact, we can express the relation $\vdash_{\mathbf{D_2}}$ as the pure *modus-ponens*-style inference relation based on $\mathbf{D_2}$.

Fact 9.2 *For any* $\Pi \subseteq \mathrm{For^d}$ *and* $B \in \mathrm{For^d}$:

$\Pi \vdash_{\mathbf{D_2}} B$ *iff* *there exists a sequence* $A_1, \ldots, A_n = B$ *in which for any* $i \leqslant n$, *either* $A_i \in \Pi \cup \mathbf{D_2}$ *or there are* $j, k < i$ *such that* $A_k = \ulcorner A_j \to^{\mathrm{d}} A_i \urcorner$.

Proof. Because ($M_2 1$) belongs to **S5**, so $\mathbf{D_2}$ is closed under *modus ponens* for '$\to^{\mathrm{d}}$', i.e., for any $A, B \in \mathrm{For^d}$, if $A, \ulcorner A \to^{\mathrm{d}} B \urcorner \in \mathbf{D_2}$, then $B \in \mathbf{D_2}$. Moreover, $\mathbf{D_2}$ contains for any $A, B, C \in \mathrm{For^d}$ the following formulae:

$$A \to^{\mathrm{d}} (B \to^{\mathrm{d}} A)$$

$$(A \to^{\mathrm{d}} (B \to^{\mathrm{d}} C)) \to^{\mathrm{d}} ((A \to^{\mathrm{d}} B) \to^{\mathrm{d}} (A \to^{\mathrm{d}} C))$$

So the condition from the fact is equivalent to the following condition: for some $n \geqslant 0$ and for some $A_1, \ldots, A_n \in \Pi$ we have $\ulcorner A_1 \to^{\mathrm{d}} (\ldots \to^{\mathrm{d}} (A_n \to^{\mathrm{d}} B)\ldots)\urcorner \in \mathbf{D_2}$.[6] Thus, by Fact 9.1, it is equivalent to $\Pi \vdash_{\mathbf{D_2}} B$.

9.3 Other Logics Defining $\mathbf{D_2}$

Definition 9.3 *Let* $\boldsymbol{L}$ *be any modal logic.*

(i) We say that $\boldsymbol{L}$ *defines* $\mathbf{D_2}$ iff $\mathbf{D_2} = \{ A \in \mathrm{For^d} : \ulcorner \Diamond A^{\bullet} \urcorner \in \boldsymbol{L} \}$.
(ii) Let $\mathbf{S5}_{\diamond}$ be the set of all modal logics which have the same theses beginning with '$\Diamond$' as **S5**, i.e., $\boldsymbol{L} \in \mathbf{S5}_{\diamond}$ iff $\forall_{A \in \mathrm{For_m}} (\ulcorner \Diamond A \urcorner \in \boldsymbol{L} \iff \ulcorner \Diamond A \urcorner \in \mathbf{S5})$.

Fact 9.3 *Nasieniewski and Pietruszczak (2008). For any classical modal logic* $\boldsymbol{L}$: $\boldsymbol{L}$ *defines* $\mathbf{D_2}$ *iff* $\boldsymbol{L} \in \mathbf{S5}_{\diamond}$.

In Furmanowski (1975), it was shown that **S4** and **S5** have the same members beginning with '$\Diamond$'—thus, one can use weaker modal logics to define $\mathbf{D_2}$. In Perzanowski (1975), the smallest normal modal logic (denoted by $\mathbf{S5^M}$) possessing this property was indicated.

[6] So notice that for the logic $\mathbf{D_2}$ we have an analogous fact to Fact 9.A.1.

In Perzanowski (1975) $\mathbf{S5^M}$ was defined as the smallest normal logic containing $\ulcorner\Diamond\top\urcorner$,[7]

$$\Diamond\Box(\Diamond\Box p \to \Box p) \tag{ML5}$$

$$\Diamond\Box(\Box p \to p) \tag{MLT}$$

and closed under the following rule:

$$\text{if } \ulcorner\Diamond\Diamond A\urcorner \in \mathbf{S5^M} \text{ then } \ulcorner\Diamond A\urcorner \in \mathbf{S5^M}. \tag{RM^2_1}$$

Let $\mathsf{NS5}_\Diamond$ and $\mathsf{RS5}_\Diamond$ be respectively the sets of all normal and regular logics from $\mathsf{S5}_\Diamond$.

Fact 9.4 *Perzanowski (1975).* $\mathbf{S5^M}$ *is the smallest logic in* $\mathsf{NS5}_\Diamond$.

Notice that one can drop two out of the three axioms of the original formulation of $\mathbf{S5^M}$ (see also Fact 9.8ii).

Fact 9.5 *Nasieniewski and Pietruszczak (2008).* $\mathbf{S5^M}$ *is the smallest normal logic which contains* (MLT) *and is closed under* (RM^2_1).

Besides, it was proved in Błaszczuk and Dziobiak (1977) that one can define the logic $\mathbf{S5^M}$ without the rule (RM^2_1), using instead—as an additional axiom—the following formula ("semi-4"):

$$\Box p \to \Diamond\Box\Box p \tag{4_s}$$

Fact 9.6 *Błaszczuk and Dziobiak (1977).* $\mathbf{S5^M}$ *is the smallest normal logic containing* (4_s) *and* (MLT), *i.e.* $\mathbf{S5^M} = \mathbf{K4_s(MLT)}$.[8]

Additionally, in Nasieniewski (2002) another axiomatisation of the logic $\mathbf{S5^M}$ without the rule (RM^2_1) was given.

Fact 9.7 *Nasieniewski (2002).* $\mathbf{S5^M}$ *is the smallest normal logic which contains* (4_s) *and the converse of* (5)

$$\Box p \to \Diamond\Box p \tag{5_c}$$

i.e. $\mathbf{S5^M} = \mathbf{K4_s5_c}$.

In Nasieniewski and Pietruszczak (2008) a regular version of the logic $\mathbf{S5^M}$ was considered. It was proved that while defining the logic $\mathbf{D_2}$ one can use a weaker modal logic than $\mathbf{S5^M}$.

---

[7] As it is well known, in all regular logics (and so in normal ones) the formula $\ulcorner\Diamond\top\urcorner$ is equivalent to the formula (D) (see Lemma 9.A.7). The smallest normal logic containing (D) (equivalently $\ulcorner\Diamond\top\urcorner$) is denoted by '**KD**' or simply by '**D**'. We have, $\mathbf{D} \subsetneq \mathbf{S5^M}$.

[8] For an explanation of the *Lemmon code* $\mathbf{KX}_1 \ldots \mathbf{X}_n$ or $\mathbf{CX}_1 \ldots \mathbf{X}_n$ see page 160.

Definition 9.4 *Let* $\mathbf{rS5^M}$ *denote the smallest regular logic which contains* (MLT) *and is closed under the rule* (RM^2_1).

Fact 9.8 *Nasieniewski and Pietruszczak (2008).*

(i) The logic $\mathbf{rS5^M}$ is not normal. In other words, $\mathbf{rS5^M}$ has no thesis of the form $\ulcorner\Box B\urcorner$. Thus, $\mathbf{rS5^M} \subsetneq \mathbf{S5^M}$.
(ii) (D), (ML5) $\in \mathbf{rS5^M}$.
(iii) $\mathbf{rS5^M}$ is the smallest logic in $\mathsf{RS5}_\diamond$; so $\mathbf{rS5^M}$ is the smallest regular logic defining $\mathbf{D_2}$.

From Fact 9.8(iii) we obtain:

Corollary 9.1. *For any modal logic L: if $\mathbf{rS5^M} \subseteq L \subseteq \mathbf{S5}$, then $L \in \mathsf{S5}_\diamond$.*

In Nasieniewski and Pietruszczak (2009) three axiomatisations of $\mathbf{rS5^M}$ where given: two of them were formulated without (RM^2_1) rule, while one was using (RM^2_1). Axiomatisations of $\mathbf{rS5^M}$ correspond to axiomatisations of the logic $\mathbf{S5^M}$. These results have been summarized below.

Fact 9.9 *Nasieniewski and Pietruszczak (2009).*
$\mathbf{rS5^M}$ *is the smallest regular logic which*:

(i) Contains (MLT) and (4_s), i.e. $\mathbf{rS5^M} = \mathbf{C4_s(MLT)}$;
(ii) Contains (5_c) and (4_s), i.e. $\mathbf{rS5^M} = \mathbf{C4_s5_c}$;
(iii) Contains (5_c) and is closed under (RM^2_1).

Besides, we have the upward analogue of the result from Fact 9.8(iii).

Fact 9.10 *Nasieniewski and Pietruszczak (2008).*
If L is a regular logic defining $\mathbf{D_2}$, then $L \subseteq \mathbf{S5}$.[9]

9.4 KD45 in the Formulation of D$_2$-Consequence

It appears that the consequence relation $\vdash_{\mathbf{D_2}}$ is closely related to the normal logic $\mathbf{KD45}$ (= $\mathbf{K5!} = \mathbf{K55_c}$; see Lemma 9.A.8(v)). To start an investigation of this relationship, we will prove the following lemma.

Lemma 9.1.

(i) (4_s) $\in \mathbf{CD4} \subsetneq \mathbf{KD4}$.
(ii) (4), (5) $\notin \mathbf{K4_s5_c} = \mathbf{S5^M}$.
(iii) $\mathbf{S5^M} \subsetneq \mathbf{KD4} \subsetneq \mathbf{KD45}$.

Proof. (i) By (4), (US) and **PL**, the formula '$\Box p \to \Box\Box\Box p$' belongs to **C4**. Moreover, by (D), (US) and **PL**, we obtain that (4_s) $\in$ **CD4**.

[9] It was proved in Błaszczuk and Dziobiak (1975) that if $L \in \mathsf{NS5}_\diamond$, then $L \subseteq \mathbf{S5}$.

(ii) By "the corresponding Hintikka condition" from Segerberg (1971), Theorem 6.5 (see also Błaszczuk and Dziobiak 1977; Nasieniewski 2002) we know that normal logics defined by (5_c), and (4_s) are determined by frames $\langle W, R\rangle$ fulfilling, respectively, the following conditions:

$$\forall_u \exists_x \big(u R x \;\&\; \forall_v (x R v \Longrightarrow u R v)\big) \tag{h5$_c$}$$

$$\forall_u \exists_x \big(u R x \;\&\; \forall_v (x R^2 v \Longrightarrow u R v)\big) \tag{h4$_s$}$$

We can indicate a model whose frame fulfils this conditions in which (4) and (5) are falsified. Thus, (5), (4) $\notin$ $\mathbf{K4_s5_c}$. By Fact 9.7, $\mathbf{K4_s5_c} = \mathbf{S5^M}$.

(iii) By (i), (ii) and Lemma 9.A.8(iii) we have $\mathbf{S5^M} \subsetneq \mathbf{KD4} = \mathbf{K45_c} \subsetneq \mathbf{KD45}$.

Since $\mathbf{S5^M} \subseteq \mathbf{KD45} \subseteq \mathbf{S5}$, so from Fact 9.3 and Corollary 9.1 we obtain:

Corollary 9.2. $\mathbf{KD45} \in \mathsf{NS5}_\Diamond$ *and* **KD45** *defines* $\mathbf{D_2}$.

We can define a discussive consequence on the basis of any modal logic $\boldsymbol{L}$.

Definition 9.5 *For any* $\Pi \subseteq \mathrm{For^d}$ *and* $B \in \mathrm{For^d}$: $\Pi \vdash_{\mathbf{D}_L} B$ *iff for some* $n \geqslant 0$ *and for some* $A_1, \ldots, A_n \in \Pi$ *we have* $\ulcorner \Diamond A_1^\bullet \to (\ldots \to \Diamond A_n^\bullet \to \Diamond B^\bullet)\ldots)\urcorner \in \boldsymbol{L}$. *In other words,*

$$\Pi \vdash_{\mathbf{D}_L} B \qquad \textit{iff} \qquad \{\Diamond A^\bullet : A \in X\} \vdash_L \Diamond B^\bullet,$$

where $\vdash_L$ *is the pure modus-ponens-style inference relation based on* $\mathbf{L}$ *(see Definition 9.A.1 and Fact 9.A.1).*

If $\Pi = \{A_1, \ldots, A_n\}$, *then we will use notation:* $A_1, \ldots, A_n \vdash_{\mathbf{D}_L} A$.

By $(\mathrm{R}^{\Diamond\Box})$ and (5 !) we obtain

Lemma 9.2. *Let* $\boldsymbol{L}$ *be any normal logic such that* $\mathbf{KD45} \subseteq \boldsymbol{L}$. *Then for any* $A_1, \ldots, A_n, B \in \mathrm{For_m}$:

$$\ulcorner \Diamond(\Diamond A_1 \to (\ldots \to (\Diamond A_n \to B)\ldots))\urcorner \in \boldsymbol{L} \quad \textit{iff}$$
$$\ulcorner \Diamond A_1 \to (\ldots \to (\Diamond A_n \to \Diamond B)\ldots)\urcorner \in \boldsymbol{L}\,.$$

Corollary 9.3. *For any* $A_1, \ldots, A_n, B \in \mathrm{For^d}$:

$$\ulcorner A_1 \to^{\mathrm{d}} (\ldots \to^{\mathrm{d}} (A_n \to^{\mathrm{d}} B)\ldots)\urcorner \in \mathbf{D_2} \quad \textit{iff}$$
$$\ulcorner \Diamond(A_1 \to^{\mathrm{d}} (\ldots \to^{\mathrm{d}} (A_n \to^{\mathrm{d}} B)\ldots))^{\bullet}\urcorner \in \mathbf{KD45} \quad \textit{iff}$$
$$\ulcorner \Diamond A_1^\bullet \to (\ldots \to (\Diamond A_n^\bullet \to \Diamond B^\bullet)\ldots)\urcorner \in \mathbf{KD45}\,.$$

By definitions, Corollaries 9.2 and 9.3, and Fact 9.1 we obtain

Theorem 9.1. $\vdash_{\mathbf{D_2}} = \vdash_{\mathbf{D_{KD45}}}$.

Proof. For any $A_1, \ldots, A_n, B \in \mathrm{For}^{\mathrm{d}}$ we obtain

$$\begin{aligned} A_1, \ldots, A_n \vdash_{\mathbf{D_{KD45}}} B \quad &\text{iff} \quad \ulcorner \Diamond A_1^{\bullet} \to (\ldots \to (\Diamond A_n^{\bullet} \to \Diamond B^{\bullet})\ldots)\urcorner \in \mathbf{KD45} \\ &\text{iff} \quad \ulcorner \Diamond(A_1 \to^{\mathrm{d}} (\ldots \to^{\mathrm{d}} (A_n \to^{\mathrm{d}} B)\ldots))^{\bullet}\urcorner \in \mathbf{KD45} \\ &\text{iff} \quad \ulcorner \Diamond(A_1 \to^{\mathrm{d}} (\ldots \to^{\mathrm{d}} (A_n \to^{\mathrm{d}} B)\ldots))^{\bullet}\urcorner \in \mathbf{S5} \\ &\text{iff} \quad A_1, \ldots, A_n \vdash_{\mathbf{D_2}} B \end{aligned}$$

Thus, for any $\Pi \subseteq \mathrm{For}^{\mathrm{d}}$ and $B \in \mathrm{For}^{\mathrm{d}}$: $\Pi \vdash_{\mathbf{D_2}} B$ iff $\Pi \vdash_{\mathbf{D_{KD45}}} B$.

In what follows, we prove that **KD45** is the smallest, while **S5** is the largest among normal logics which define the same consequence relation $\vdash_{\mathbf{D_2}}$. But neither $\mathbf{S5^M}$ nor **S4** is appropriate for this purpose.

Fact 9.11 $\vdash_{\mathbf{D_{S5M}}} \subsetneq \vdash_{\mathbf{D_{S4}}} \subsetneq \vdash_{\mathbf{D_2}}$. □

The inclusions "$\subseteq$" are obvious. For "$\subsetneq$" we can use either the following examples or the next fact.

Example 9.1. (i) $(p \vee \neg p) \wedge^{\mathrm{d}} p \vdash_{\mathbf{D_{S4}}} p$, while $(p \vee \neg p) \wedge^{\mathrm{d}} p \nvdash_{\mathbf{D_{S5M}}} p$. Indeed, $(p \vee \neg p) \wedge^{\mathrm{d}} p \vdash_{\mathbf{D_{S5M}}} p$ iff '$\Diamond((p \vee \neg p) \wedge \Diamond p) \to \Diamond p$' belongs to $\mathbf{S5^M}$ iff $(4^{\Diamond}) \in \mathbf{S5^M}$ iff $(4) \in \mathbf{S5^M}$. But $(4) \notin \mathbf{S5^M}$, by Lemma 9.1(ii).

(ii) $p, q \vdash_{\mathbf{D_2}} p \wedge^{\mathrm{d}} q$, while $p, q \nvdash_{\mathbf{D_{S4}}} p \wedge^{\mathrm{d}} q$.

(iii) $(p \vee \neg p) \wedge^{\mathrm{d}} p, q \vdash_{\mathbf{D_2}} p \wedge^{\mathrm{d}} q$, while $(p \vee \neg p) \wedge^{\mathrm{d}} p, q \nvdash_{\mathbf{D_{S4}}} p \wedge^{\mathrm{d}} q$. □

Fact 9.12

(i) *Let* $\boldsymbol{L}$ *be any regular logic such that* $\vdash_{\mathbf{D_2}} \subseteq \vdash_{\mathbf{D}_L}$. *Then* $\boldsymbol{L}$ *contains* (D), (4), *and* $\ulcorner \Box\top \to (5)\urcorner$, *so* **CD45(1)** $\subseteq \boldsymbol{L}$.[10]

(ii) *Let* $\boldsymbol{L}$ *be any normal logic such that* $\vdash_{\mathbf{D_2}} \subseteq \vdash_{\mathbf{D}_L}$. *Then* $\boldsymbol{L}$ *contains* (D), (4), *and* (5), *so* **KD45** $\subseteq \boldsymbol{L}$.

Proof. (i) For (D): Since $\emptyset \vdash_{\mathbf{D_2}} p \vee \neg p$, so—by the assumption—also $\emptyset \vdash_{\mathbf{D}_L} p \vee \neg p$. Hence '$\Diamond(p \vee \neg p)$' $\in \boldsymbol{L}$, by the definition of $\vdash_{\mathbf{D}_L}$. By Lemmas 9.A.5 and 9.A.7 we have that (D) $\in \boldsymbol{L}$.

For $\ulcorner \Box\top \to (5)\urcorner$: Since $p \to^{\mathrm{d}} \neg(p \vee \neg p), p \vdash_{\mathbf{D_2}} \neg(p \vee \neg p)$, so—by the assumption—also $p \to^{\mathrm{d}} \neg(p \vee \neg p), p \vdash_{\mathbf{D}_L} \neg(p \vee \neg p)$. Therefore, by the definition of $\vdash_{\mathbf{D}_L}$, we get that '$\Diamond[\Diamond p \to \neg(p \vee \neg p)] \to [\Diamond p \to \Diamond\neg(p \vee \neg p)]$' belongs to $\boldsymbol{L}$. Thus, by **PL**, '$\neg\Diamond(\Diamond p \to \neg\top) \vee (\Diamond p \to \Diamond\neg\top)$' belongs to $\boldsymbol{L}$. Thus, by $(\mathrm{R}^{\Diamond\Box})$ and **PL**, also '$\neg(\Box\Diamond p \to \Diamond\neg\top) \vee (\Diamond p \to \Diamond\neg\top)$', '$(\Box\Diamond p \wedge \neg\Diamond\neg\top) \vee \neg\Diamond p \vee \Diamond\neg\top$', '$(\Box\Diamond p \vee \neg\Diamond p \vee$

---

[10] The name '**CD45(1)**' is used in the sense of Segerberg (1971), vol. II. Notice that **CD45** = **KD45**.

$\Diamond \neg \top) \wedge (\neg \Diamond \neg \top \vee \neg \Diamond p \vee \Diamond \neg \top)$', and '$\Box \Diamond p \vee \neg \Diamond p \vee \Diamond \neg \top$' belong to $\boldsymbol{L}$. Thus, '$\Diamond p \wedge \Box \top \to \Box \Diamond p$' and $\ulcorner \Box \top \to (5^\Diamond) \urcorner$ belong to $\boldsymbol{L}$. Hence, by the standard duality result, $\ulcorner \Box \top \to (5) \urcorner \in \boldsymbol{L}$ as well.

For (4): Since $p \wedge^{\mathrm{d}} q \vdash_{\mathbf{D_2}} q$, so '$\Diamond(p \wedge \Diamond q) \to \Diamond q$' and '$\Diamond(\top \wedge \Diamond q) \to \Diamond q$' belong to $\boldsymbol{L}$. However '$\Diamond \Diamond q \to \Diamond(\top \wedge \Diamond q)$' is a thesis of all regular logics. Thus, by transitivity, we obtain that $(4^\Diamond) \in \boldsymbol{L}$; so also $(4) \in \boldsymbol{L}$.

(ii) Since $\boldsymbol{L}$ is normal, so $\boldsymbol{L}$ is regular and $\ulcorner \Box \top \urcorner \in \boldsymbol{L}$.

Let $\mathbf{Cn}_\Diamond\mathbf{S5}$ be the set of modal logics which satisfy the following condition: for any logic $\boldsymbol{L}$

$$\boldsymbol{L} \in \mathbf{Cn}_\Diamond\mathbf{S5} \overset{\mathrm{df}}{\Longleftrightarrow} \text{for any } \Pi \subseteq \mathrm{For_m} \text{ and } B \in \mathrm{For_m},$$

$$\Diamond \Pi \vdash_{\boldsymbol{L}} \Diamond B \text{ iff } \Diamond \Pi \vdash_{\mathbf{S5}} \Diamond B\,.$$

Let $\mathbf{NCn}_\Diamond\mathbf{S5}$ be the set of all normal logics from $\mathbf{Cn}_\Diamond\mathbf{S5}$. By definitions, Lemma 9.2, and Corollary 9.2 we obtain

Fact 9.13 $\mathbf{KD45} \in \mathbf{NCn}_\Diamond\mathbf{S5}$.

Lemma 9.3. (5_{c}) *and* (5) *belong to all logics from* $\mathbf{NCn}_\Diamond\mathbf{S5}$. *Thus, every logic from* $\mathbf{NCn}_\Diamond\mathbf{S5}$ *includes* **KD45**.

Proof. Firstly, '$\Diamond(\Diamond p \to p)$' and '$(\Diamond p \wedge \Diamond \neg \Diamond p) \to \Diamond \neg \top$' are theses of **S5**; so they are also theses of all logics from $\mathbf{NCn}_\Diamond\mathbf{S5}$. Secondly, these formulae are equivalent, respectively, to $(5_{\mathrm{c}}^\Diamond)$ and $(5^\Diamond)$, on the basis of any normal modal logic. Thus, $(5_{\mathrm{c}}^\Diamond)$ and $(5^\Diamond)$ belong to all logics from $\mathbf{NCn}_\Diamond\mathbf{S5}$. So every logic from $\mathbf{NCn}_\Diamond\mathbf{S5}$ includes $\mathbf{K55_c}$ (= **KD45**).

By Fact 9.13 and Lemma 9.3 we obtain:

Theorem 9.2. **KD45** *is the smallest element in* $\mathbf{NCn}_\Diamond\mathbf{S5}$.

Below we introduce a transformation from $\mathrm{For_m}$ to $\mathrm{For^d}$. It allows us to prove that if any normal logic defines the $\mathbf{D_2}$-consequence, it has to be located between **KD45** and **S5**.

Definition 9.6 *Let* $-^\circ$ *be the function from* $\mathrm{For_m}$ *into* $\mathrm{For^d}$ *such that:*

1. $(a)^\circ = a$, *for any propositional letter* a,
2. *And for any* $A, B \in \mathrm{For_m}$:
   (a) $(\neg A)^\circ = \ulcorner \neg A^\circ \urcorner$,
   (b) $(A \vee B)^\circ = \ulcorner A^\circ \vee B^\circ \urcorner$,
   (c) $(A \wedge B)^\circ = \ulcorner \neg(\neg A^\circ \vee \neg B^\circ) \urcorner$,
   (d) $(A \to B)^\circ = \ulcorner \neg A^\circ \vee B^\circ \urcorner$,
   (e) $(A \leftrightarrow B)^\circ = \ulcorner \neg(\neg(\neg A^\circ \vee B^\circ) \vee \neg(\neg B^\circ \vee A^\circ)) \urcorner$,
   (f) $(\Diamond A)^\circ = \ulcorner (p \vee \neg p) \wedge^{\mathrm{d}} A^\circ \urcorner$,
   (g) $(\Box A)^\circ = \ulcorner \neg A^\circ \to^{\mathrm{d}} \neg(p \vee \neg p) \urcorner$.

Lemma 9.4. *For any* $A \in \mathrm{For_m}$: $\ulcorner A \leftrightarrow A^{\circ\bullet} \urcorner$ *is a thesis of all classical logics.*

Lemma 9.5. *For any classical modal logic* $\boldsymbol{L}$

$$\vdash_{\mathbf{D}_L} = \vdash_{\mathbf{D}_2} \quad \textit{iff} \quad L \in \mathsf{Cn}_{\diamond}\mathsf{S5}\,.$$

Proof. "$\Rightarrow$" $\diamond A_1, \ldots, \diamond A_n \vdash_L \diamond B$ iff $\ulcorner \diamond A_1 \to (\ldots \to (\diamond A_n \to \diamond B)\ldots)\urcorner \in$ $\boldsymbol{L}$ iff, by Lemma 9.4, **PL**, and (REP), $\ulcorner \diamond A_1^{\circ\bullet} \to (\ldots \to (\diamond A_n^{\circ\bullet} \to \diamond B^{\circ\bullet})\ldots)\urcorner \in$ $\boldsymbol{L}$ iff $A_1^{\circ}, \ldots, A_n^{\circ} \vdash_{\mathbf{D}_L} B^{\circ}$ iff $A_1^{\circ}, \ldots, A_n^{\circ} \vdash_{\mathbf{D}_2} B^{\circ}$ iff $\ulcorner \diamond A_1^{\circ\bullet} \to (\ldots \to (\diamond A_n^{\circ\bullet} \to$ $\diamond B^{\circ\bullet})\ldots)\urcorner \in \mathbf{S5}$ iff, by Lemma 9.4, **PL**, and (REP), $\diamond A_1, \ldots, \diamond A_n \vdash_{\mathbf{S5}} \diamond B$.

"$\Leftarrow$" Obvious.

Finally, we get the following

Theorem 9.3. *For any normal modal logic* $\boldsymbol{L}$:

$$\vdash_{\mathbf{D}_L} = \vdash_{\mathbf{D}_2} \quad \textit{iff} \quad \mathbf{KD45} \subseteq L \subseteq \mathbf{S5}\,.$$

Proof. "$\Rightarrow$" For **KD45** $\subseteq \boldsymbol{L}$ see Fact 9.12(ii).

For any $A \in \mathrm{For_m}$ we have: $\emptyset \vdash_{\mathbf{D}_2} A^{\circ}$ iff $\emptyset \vdash_{\mathbf{D}_L} A^{\circ}$. So by Definitions 9.1 and 9.5 we have: $\ulcorner \diamond A^{\circ\bullet}\urcorner \in \mathbf{S5}$ iff $\ulcorner \diamond A^{\circ\bullet}\urcorner \in L$. Thus, by Lemma 9.4, **PL**, and (REP), we obtain that: $\ulcorner \diamond A\urcorner \in \mathbf{S5}$ iff $\ulcorner \diamond A\urcorner \in L$. Thus, $L \in \mathsf{NS5}_{\diamond}$. Therefore $L \subseteq \mathbf{S5}$, by Facts 9.3 and 9.10.

"$\Leftarrow$" By Corollary 9.2 and Fact 9.3, $L \in \mathsf{NS5}_{\diamond}$. Thus, $L \in \mathsf{NCn}_{\diamond}\mathsf{S5}$, by Lemma 9.2. Hence $\vdash_{\mathbf{D}_2} = \vdash_{\mathbf{D}_L}$, by Lemma 9.5.

9.5 The Smallest Regular Modal Logic Defining D2-Consequence

We will show that consequence relation $\vdash_{\mathbf{D}_2}$ is also closely connected with the regular logic **CD45(1)**.

Definition 9.7 *Let* **CD45(1)** *be the smallest regular logic which contains* (D), (4), *and* (5(1)), *i.e.* $\ulcorner \Box\top \to (5)\urcorner$.

Remark 9.2. In the notation of Segerberg a regular logic **CN¹D(1)4(1)5(1)** corresponds, by the definition, to the normal logic **KD45**. Yet in **C2** the formulae (D), (4) and (5_c) are respectively equivalent to (D(1)), (4(1)) and (5_c(1)), i.e., $\ulcorner \Box\top \to (\Box p \to \diamond p)\urcorner$, $\ulcorner \Box\top \to (\Box p \to \Box\Box p)\urcorner$ and $\ulcorner \Box\top \to (\Box p \to$ $\diamond\Box p)\urcorner$ (see Segerberg 1971, p. 208). Moreover, the formula (N^1), i.e. $\ulcorner \Box\top \to$ $\Box\Box\top\urcorner$ (see Segerberg 1971, p. 198), is an instance of (4). Thus, **CD45(1)** $=$ **CN¹D(1)4(1)5(1)**. Hence, by Lemma 9.A.9, i.e. Corollary 2.4 from Segerberg (1971), vol. II, we obtain:

$$\mathbf{CD45(1)} = \mathbf{CF^1} \cap \mathbf{KD45}\,,$$

$$\mathbf{CN^15_c5(1)} = \mathbf{CF^1} \cap \mathbf{K55_c}\,,$$

where $\mathbf{CF^1}$ is the *falsum* logic. □

By the above remark and the equality **KD45** $=$ **K55$_c$** we obtain[11]:

Fact 9.14 **CD45(1)** $=$ **CN^{1}5$_c$5(1)**.

Fact 9.15 *The logic* **CD45(1)** *is not normal. In other words,* **CD45(1)** *has no thesis of the form* $\ulcorner \Box B \urcorner$.

Proof. It is enough to use a model from Fact 3.1 of Nasieniewski and Pietruszczak (2008): Let v be a valuation from $\mathrm{For_m}$ into $\{0, 1\}$ which preserves classical truth conditions for classical connectives and let $v(\Box A) = 0$ and $v(\Diamond A) = 1$, for any $A \in \mathrm{For_m}$. Notice that for any thesis of **CD45(1)** we have $v(A) = 1$, while, for example, $v(\Box\top) = 0$.

Fact 9.16 **rS5^{M}** $\subsetneq$ **CD4** $\subsetneq$ **CD45(1)** $\subsetneq$ **KD45** $\subsetneq$ **S5**.

Proof. Notice that, by Lemma 9.9, **rS5^{M}** $=$ **C4$_s$5$_c$**. Moreover, $(5^{\Diamond}_{c})$, $(4_s) \in$ **CD4** $=$ **C45$_c$**, respectively by Lemmas 9.A.8(ii) and 9.1(i). Thus, **rS5^{M}** $\subseteq$ **CD4**. This inclusion is proper, since **rS5^{M}** $\subsetneq$ **S5^{M}** $\subsetneq$ **KD4** and $(4) \notin$ **S5^{M}** (see Lemma 9.1).

Besides, we have **CD4** $\subseteq$ **KD4**. But $(5) \notin$ **KD4**, so also $(5(1)) \notin$ **KD4**, since in all normal logics we have the thesis '$(5) \leftrightarrow (5(1))$'. Hence $(5(1)) \notin$ **CD4**. Moreover, **CD45(1)** $\subseteq$ **KD45**. This inclusion is proper by Fact 9.15.

Lemma 9.6. *The formulae* (†) *and for any* $n \geqslant 2$

$$\Diamond p_1 \to (\Diamond p_2 \to \ldots (\Diamond p_n \to (\Diamond(p_1 \wedge (\Diamond p_2 \wedge \ldots (\Diamond p_{n-1} \wedge \Diamond p_n) \ldots)))))$$

and for any $n \geqslant 1$

$$\Diamond (\Diamond p_1 \to (\Diamond p_2 \to \ldots (\Diamond p_n \to q) \ldots)) \to$$
$$\to (\Diamond p_1 \to (\Diamond p_2 \to \ldots (\Diamond p_n \to \Diamond q) \ldots))$$

are theses of **CN^{1}5(1)** $\subseteq$ **CD45(1)**.

Proof. By Lemma 9.A.8(vi), (†) $\in$ **K5**. Obviously (†) $\in$ **CF^1**. So, we use Lemma 9.A.9. The proof in the case of remaining formulae is analogous. It is by induction on n.

Let **RCn$_\Diamond$S5** be the set of all regular logics from **Cn$_\Diamond$S5**. We have:

Lemma 9.7. **CD45(1)** $\in$ **RCn$_\Diamond$S5**.

---

[11] We have also a proof of the following fact without the use of Lemma 9.A.9. Firstly, by Lemma 9.A.8(ii), $(5_c) \in$ **CD4**; so **CN^{1}5$_c$5(1)** $\subseteq$ **CD45(1)**.

Secondly, $5^{\Diamond}(1)$ belongs to **C5$_c$5(1)**, so by (US) we have: '$\Box\top \to (\Diamond\Box p \to \Box\Diamond\Box p)$'. Moreover, by $5(1)$, (RM), (K) and **PL**, we obtain: '$\Box\Box\top \to (\Box\Diamond\Box p \to \Box\Box p)$'. So, by **PL**, we receive: '$(\Box\Box\top \wedge \Box\top) \to (\Diamond\Box p \to \Box\Box p)$'. Hence, by (5_c) and **PL**, we get '$(\Box\Box\top \wedge \Box\top) \to (\Box p \to \Box\Box p)$'. Hence, by (N^1), **PL** and (RM), we have that $(4) \in$ **CN^{1}5$_c$5(1)**. Thus, **CD45(1)** $\subseteq$ **CN^{1}5$_c$5(1)**, since by Lemma 9.A.8(i), $(\mathrm{D}) \in$ **C5$_c$**.

Proof. For any $A_1, \ldots, A_n, B \in \mathrm{For_m}$ by Lemma 9.2 and Fact 9.16, and Fact 9.8(iii): $\ulcorner \Diamond A_1 \to (\ldots \to (\Diamond A_n \to \Diamond B)\ldots)\urcorner \in \mathbf{S5}$ iff $\ulcorner \Diamond(\Diamond A_1 \to (\ldots \to (\Diamond A_n \to B)\ldots))\urcorner \in \mathbf{S5}$ iff $\ulcorner \Diamond(\Diamond A_1 \to (\ldots \to (\Diamond A_n \to B)\ldots))\urcorner \in \mathbf{CD45(1)}$.

By Lemma 9.6, it follows from the last statement that $\ulcorner \Diamond A_1 \to (\ldots \to (\Diamond A_n \to \Diamond B)\ldots)\urcorner \in \mathbf{CD45(1)}$. The reverse implication is obvious.

By the above lemma we have directly:

Corollary 9.4. $\vdash_{\mathbf{D_2}} = \vdash_{\mathbf{D_{CD45(1)}}}$.

By Fact 9.12(i) and Definition 9.7 we obtain:

Lemma 9.8. *For any regular logic* $\boldsymbol{L}$ *such that* $\vdash_{\mathbf{D_2}} = \vdash_{\mathbf{D}_{\boldsymbol{L}}}$ *it is the case that* $\mathbf{CD45(1)} \subseteq \boldsymbol{L}$.

By Lemmas 9.7, 9.5, and 9.8 we conclude that

Corollary 9.5. $\mathbf{CD45(1)}$ *is the smallest element in* $\mathbf{RCn_\Diamond S5}$.

We have of course also a regular version of Theorem 9.3:

Lemma 9.9. $\mathbf{S5}$ *is the biggest element in* $\mathbf{RCn_\Diamond S5}$.

Proof. Let us assume that $\boldsymbol{L} \in \mathbf{RCn_\Diamond S5}$ and $A \in \boldsymbol{L}$. By the classical logic we have $(p \vee \neg p) \to A \in \boldsymbol{L}$ and by monotonicity $\Diamond\Box(p \vee \neg p) \to \Diamond\Box A \in \boldsymbol{L}$ i.e, $\Diamond\Box(p \vee \neg p) \vdash_{\boldsymbol{L}} \Diamond\Box A$. Thus, by the assumption $\Diamond\Box(p \vee \neg p) \to \Diamond\Box A \in \mathbf{S5}$ and by (MP) we obtain that $\Diamond\Box A \in \mathbf{S5}$, so using the standard reduction of modalities we obtain that $A \in \mathbf{S5}$.

We have a lemma that is analogous to Lemma 9.8:

Lemma 9.10. *For any regular logic* $\boldsymbol{L}$ *such that* $\vdash_{\mathbf{D_2}} = \vdash_{\mathbf{D}_{\boldsymbol{L}}}$ *it is the case that* $\boldsymbol{L} \subseteq \mathbf{S5}$.

Proof. Assume that $A \in \boldsymbol{L}$. By Lemma 9.4 we have also $A^{\circ\bullet} \in \boldsymbol{L}$.

Since $\Diamond(\Diamond \neg(p \vee \neg p) \to \neg(p \vee \neg p)) \in \mathbf{S5}$ thus, $\neg(p \vee \neg p) \to^{\mathrm{d}} \neg(p \vee \neg p) \in \mathbf{D_2}$ and by the assumption also $\neg(p \vee \neg p) \to^{\mathrm{d}} \neg(p \vee \neg p) \in \mathbf{D}_{\boldsymbol{L}}$. By the definition of $\mathbf{D}_{\boldsymbol{L}}$ it means that $\Diamond(\Diamond \neg(p \vee \neg p) \to \neg(p \vee \neg p)) \in \boldsymbol{L}$. But for every regular modal logic the last statement is equivalent to: $\Diamond\Box(p \vee \neg p) \in \boldsymbol{L}$. It follows from Lemma 9.A.6 that $\Diamond\Box A^{\circ\bullet} \in \boldsymbol{L}$. But again for every regular modal logic this condition is equivalent to $\Diamond(\Diamond \neg A^{\circ\bullet} \to \neg(p \vee \neg p)) \in \boldsymbol{L}$, which means that $\neg A^{\circ} \to^{\mathrm{d}} \neg(p \vee \neg p) \in \mathbf{D}_{\boldsymbol{L}}$, so $\neg A^{\circ} \to^{\mathrm{d}} \neg(p \vee \neg p) \in \mathbf{D_2}$. Therefore, $\Diamond(\neg A^{\circ} \to^{\mathrm{d}} \neg(p \vee \neg p))^{\bullet} \in \mathbf{S5}$, equivalently $\Diamond\Box A^{\circ\bullet} \in \mathbf{S5}$. From this follows that $A^{\circ\bullet} \in \mathbf{S5}$ while by Lemma 9.4 we conclude that $A \in \mathbf{S5}$.

So taking together Lemmas 9.8 and 9.10 we receive:

Corollary 9.6. *For any regular logic* $\boldsymbol{L}$ *such that* $\vdash_{\mathbf{D}_{\boldsymbol{L}}} = \vdash_{\mathbf{D_2}}$ *we have* $\mathbf{CD45(1)} \subseteq \boldsymbol{L} \subseteq \mathbf{S5}$.

Lemma 9.11. *For any regular logic* $\boldsymbol{L}$ *such that* **CD45(1)** $\subseteq \boldsymbol{L} \subseteq$ **S5** *we have* $\boldsymbol{L} \in \mathbf{RCn}_{\Diamond}\mathbf{S5}$.

Proof. Assume that **CD45(1)** $\subseteq \boldsymbol{L} \subseteq$ **S5**. We have to prove that for any $A_1, \ldots, A_n$, $B \in \mathrm{For_m}$: $\ulcorner\Diamond(\Diamond A_1 \to (\ldots \to (\Diamond A_n \to B)\ldots))\urcorner \in$ **S5** iff $\ulcorner\Diamond(\Diamond A_1 \to (\ldots \to (\Diamond A_n \to B)\ldots))\urcorner \in \boldsymbol{L}$. Left-to-right implication follows from Lemma 9.7. The reverse implication is obvious.

From this lemma and Lemma 9.5 we obtain

Theorem 9.4. *For any regular logic* $\boldsymbol{L}$ *such that* **CD45(1)** $\subseteq \boldsymbol{L} \subseteq$ **S5** *we have* $\vdash_{\mathbf{D}_{\boldsymbol{L}}} = \vdash_{\mathbf{D}_2}$.

Finally, directly from Corollary 9.6 and Lemma 9.11 we get the following

Theorem 9.5. *For any regular modal logic* $\boldsymbol{L}$

$$\vdash_{\mathbf{D}_{\boldsymbol{L}}} = \vdash_{\mathbf{D}_2} \quad \textit{iff} \quad \mathbf{CD45(1)} \subseteq \boldsymbol{L} \subseteq \mathbf{S5}\,.$$

Appendix: Some Facts from Modal Logic

As in Chellas (1980) modal formulae are formed in a relational way from propositional letters: 'p', 'q', 'p_0', 'p_1', 'p_2', ...; truth-value operators: '$\neg$', '$\vee$', '$\wedge$', '$\to$', and '$\leftrightarrow$' (connectives of negation, disjunction, conjunction, material implication, and material equivalence, respectively); modal operators: the necessity sign '$\Box$' and the possibility sign '$\Diamond$'; and brackets. Let $\mathrm{For_m}$ be the set of modal formulae, and—as in Chellas (1980)—let **PL** be the set of modal formulae which are instances of classical tautologies. Let $\top :=$ '$p \to p$'.

As in Bull and Segerberg (1984) and Chellas and Segerberg (1996), a set $\boldsymbol{L}$ of modal formulae is a (*modal*) *logic* iff

- **PL** $\subseteq \boldsymbol{L}$,
- For any $C, A \in \mathrm{For_m}$: $\boldsymbol{L}$ contains the following formula

$$C[{}^{\neg\Box\neg A}/_{\Diamond A}] \leftrightarrow C\,, \qquad (\mathtt{rep}^{\Box})$$

  where $C[{}^{A}/_{B}]$ is any formula that results from C by replacing one or more occurrences of A, in C, by B, i.e. using ($\mathtt{rep}^{\Box}$) we are *replacing* in C one or more occurrences of '$\neg\Box\neg$' by '$\Diamond$'.[12]

[12]In Bull and Segerberg (1984) and Chellas and Segerberg (1996) the symbol '$\Diamond$' is only an abbreviation of '$\neg\Box\neg$'. In the present paper '$\Diamond$' is a primary symbol, thus, we have to admit an axiom of the form ($\mathtt{rep}^{\Box}$). Theses of this form are equivalent to the usage of '$\Diamond$' as the abbreviation of '$\neg\Box\neg$'.

- $\boldsymbol{L}$ is closed under the following three rules: *modus ponens* for '$\rightarrow$':

$$\text{if } A \text{ and } \ulcorner A \rightarrow B \urcorner \text{ are members of } \boldsymbol{L}, \text{ so is } B. \tag{MP}$$

  uniform substitution:

$$\text{if } A \in \boldsymbol{L} \text{ then } \mathrm{s}\, A \in \boldsymbol{L}, \tag{US}$$

  where $\mathrm{s}\, A$ is the result of uniform substitution of formulae for propositional letters in A.

Definition 9.A.1 *Let* $\boldsymbol{L}$ *be any modal logic. We define the consequence* $\vdash_{\boldsymbol{L}}$ *as follows. For any* $\Pi \subseteq \mathrm{For_m}$ *and* $B \in \mathrm{For_m}$*:* $\Pi \vdash_{\boldsymbol{L}} B$ *iff for some* $n \geqslant 0$ *and for some* $A_1, \ldots, A_n \in \Pi$ *we have* $\ulcorner A_1 \rightarrow (\ldots \rightarrow (A_n \rightarrow B)\ldots)\urcorner \in \boldsymbol{L}$.

Notice that $\Pi \vdash_{\boldsymbol{L}} B$ iff there is a derivation of B from $\boldsymbol{L} \cup \Pi$ with the help of *modus ponens* for '$\rightarrow$' as the only rule of inference, i.e., $\vdash_{\boldsymbol{L}}$ is the pure *modus-ponens*-style inference relation based on $\boldsymbol{L}$.

Fact 9.A.1 *Lemmon (1977).* $\Pi \vdash_{\boldsymbol{L}} B$ *iff there exists a sequence* $A_1, \ldots, A_n = B$ *in which for any* $i \leqslant n$*, either* $A_i \in \Pi$*, or* $A_i \in \boldsymbol{L}$*, or there are* $j, k < i$ *such that* $A_k = \ulcorner A_j \rightarrow A_i \urcorner$.

All members of the set $\boldsymbol{L}$ are called *theses* of the logic $\boldsymbol{L}$. By ($\mathtt{rep}^{\Box}$), every modal logic has the following thesis:

$$\Diamond\, p \leftrightarrow \neg \Box \neg\, p. \tag{$\mathtt{df}\,\Diamond$}$$

A modal logic $\boldsymbol{L}$ is *classical* (*congruent*) iff $\boldsymbol{L}$ is closed under the following rule for any $A, B \in \mathrm{For_m}$:

$$\text{if } \ulcorner A \leftrightarrow B \urcorner \in \boldsymbol{L} \text{ then } \ulcorner \Box A \leftrightarrow \Box B \urcorner \in \boldsymbol{L}. \tag{RE}$$

Every classical logic $\boldsymbol{L}$ is closed under the rule of replacement, i.e. for any $A, B, C \in \mathrm{For_m}$:

$$\text{if } \ulcorner A \leftrightarrow B \urcorner \in \boldsymbol{L} \text{ then } \ulcorner C \leftrightarrow C[{}^{A}/{}_{B}] \urcorner \in \boldsymbol{L}. \tag{REP}$$

It is known (cf. e.g. Chellas 1980) that while defining classical logics one uses ($\mathtt{df}\,\Diamond$) instead of ($\mathtt{rep}^{\Box}$), i.e. treats them (logics) as subsets of $\mathrm{For_m}$ which include **PL** and ($\mathtt{df}\,\Diamond$) and which are closed under rules (MP), (US) and (RE). We also have an analogous situation in the case of monotonic, regular, and normal modal logics defined further.

Every classical modal logic has the following thesis

$$\Box p \leftrightarrow \neg \Diamond \neg\, p \tag{$\mathtt{df}\,\Box$}$$

Lemma 9.A.1 *A classical modal logic contains, respectively, the following formulae*

$$\Box(p \to q) \to (\Box p \to \Box q) \tag{K}$$

$$\Box(p \wedge q) \leftrightarrow (\Box p \wedge \Box q) \tag{R}$$

$$\Box p \to p \tag{T}$$

$$\Box p \to \Box\Box p \tag{4}$$

$$\Diamond\Box p \to \Box p \tag{5}$$

$$\Box p \to \Diamond\Box p \tag{5$_\mathtt{c}$}$$

$$\Box p \leftrightarrow \Diamond\Box p \tag{5!}$$

if and only if it contains, respectively, their dual versions

$$\Box(p \to q) \to (\Diamond p \to \Diamond q) \tag{K$^\Diamond$}$$

$$\Diamond(p \vee q) \leftrightarrow (\Diamond p \vee \Diamond q) \tag{R$^\Diamond$}$$

$$p \to \Diamond p \tag{T$^\Diamond$}$$

$$\Diamond \Diamond p \to \Diamond p \tag{4$^\Diamond$}$$

$$\Diamond p \to \Box \Diamond p \tag{5$^\Diamond$}$$

$$\Box \Diamond p \to \Diamond p \tag{5$_\mathtt{c}^\Diamond$}$$

$$\Diamond p \leftrightarrow \Box \Diamond p \tag{5$^\Diamond$!}$$

Lemma 9.A.2 *For any classical modal logic* $\boldsymbol{L}$ *the following conditions are equivalent*:

(a) *For any* $\tau \in \mathbf{PL}$, $\ulcorner\Box\tau\urcorner \in \boldsymbol{L}$ *(resp.* $\ulcorner\Diamond\tau\urcorner \in \boldsymbol{L}$, $\ulcorner\Diamond\Box\tau\urcorner \in \boldsymbol{L}$*)*.
(b) $\ulcorner\Box\top\urcorner \in \boldsymbol{L}$ *(resp.* $\ulcorner\Diamond\top\urcorner \in \boldsymbol{L}$, $\ulcorner\Diamond\Box\top\urcorner \in \boldsymbol{L}$*)*.

Lemma 9.A.3 *Let* $\boldsymbol{L}$ *be any classical modal logic such that*

(a) *either* $\ulcorner\Box\top\urcorner \in \boldsymbol{L}$,
(b) *or* (5), $\ulcorner\Diamond B\urcorner \in \boldsymbol{L}$, *for some* $B \in \mathrm{For_m}$.[13]

Then $\boldsymbol{L}$ *is closed under the rule of necessitation*:

$$\textit{if } A \in \boldsymbol{L} \textit{ then } \ulcorner\Box A\urcorner \in \boldsymbol{L}. \tag{RN}$$

[13]Notice that (b) implies (a).

Lemma 9.A.4 *Chellas (1980). Let* $\boldsymbol{L}$ *be any classical modal logic such that* $(\mathtt{T}), (\mathtt{5}) \in \boldsymbol{L}$. *Then* $\boldsymbol{L}$ *has as its theses* $\ulcorner\Box\top\urcorner$, $\ulcorner\Diamond\top\urcorner$, $\ulcorner\Diamond\Box\top\urcorner$, $(\mathtt{4})$, *and*

$$\Box p \to \Diamond p \tag{D}$$

and $\boldsymbol{L}$ *is closed under* (RN) *and the following rules*:

$$\text{if } A \in \boldsymbol{L}, \text{ then } \ulcorner\Diamond A\urcorner \in \boldsymbol{L}, \tag{RP}$$

$$\text{if } A \in \boldsymbol{L}, \text{ then } \ulcorner\Diamond\Box A\urcorner \in \boldsymbol{L}. \tag{RPN}$$

A modal logic $\boldsymbol{L}$ is *monotonic* iff $\boldsymbol{L}$ is closed under the monotonicity rule, i.e. for any $A, B \in \mathrm{For_m}$:

$$\text{if } \ulcorner A \to B\urcorner \in \boldsymbol{L}, \text{ then } \ulcorner\Box A \to \Box B\urcorner \in \boldsymbol{L}, \tag{RM}$$

Every monotonic logic $\boldsymbol{L}$ is classical and it is closed under the dual form of (RM), i.e. for any $A, B \in \mathrm{For_m}$:

$$\text{if } \ulcorner A \to B\urcorner \in \boldsymbol{L}, \text{ then } \ulcorner\Diamond A \to \Diamond B\urcorner \in \boldsymbol{L}. \tag{RM$^\Diamond$}$$

Lemma 9.A.5 *For any monotonic logic* $\boldsymbol{L}$ *the following conditions are equivalent*:

(a) *For any* $\tau \in \mathbf{PL}$, $\ulcorner\Box\tau\urcorner \in \boldsymbol{L}$ (resp. $\ulcorner\Diamond\tau\urcorner \in \boldsymbol{L}$, $\ulcorner\Diamond\Box\tau\urcorner \in \boldsymbol{L}$).
(b) $\ulcorner\Box\top\urcorner \in \boldsymbol{L}$ *(resp.* $\ulcorner\Diamond\top\urcorner \in \boldsymbol{L}$, $\ulcorner\Diamond\Box\top\urcorner \in \boldsymbol{L}$).
(c) For some $B \in \mathrm{For_m}$, $\ulcorner\Box B\urcorner \in \boldsymbol{L}$ *(resp.* $\ulcorner\Diamond B\urcorner \in \boldsymbol{L}$, $\ulcorner\Diamond\Box B\urcorner \in \boldsymbol{L}$).

Lemma 9.A.6 *Let a monotonic logic* $\boldsymbol{L}$ *has a thesis of the form* $\ulcorner\Box B\urcorner$ *(resp.* $\ulcorner\Diamond B\urcorner$, $\ulcorner\Diamond\Box B\urcorner$). *Then* $\boldsymbol{L}$ *is closed under the rule* (RN) *(resp.* (RP), (RPN)).

A modal logic $\boldsymbol{L}$ is *regular* iff $\boldsymbol{L}$ is monotonic and $(\mathtt{K}) \in \boldsymbol{L}$. A logic $\boldsymbol{L}$ is regular iff $\boldsymbol{L}$ is closed under the *regularity rule*, i.e. for any $A, B, C \in \mathrm{For_m}$:

$$\text{if } \ulcorner A \wedge B \to C\urcorner \in \boldsymbol{L} \text{ then } \ulcorner\Box A \wedge \Box B \to \Box C\urcorner \in \boldsymbol{L}. \tag{RR}$$

Every regular modal logic has the following theses: $(\mathtt{K}^\Diamond)$, $(\mathtt{R})$, $(\mathtt{R}^\Diamond)$ and

$$\Diamond\,(p \to q) \leftrightarrow (\Box p \to \Diamond q) \tag{R$^{\Diamond\Box}$}$$

By $(\mathtt{R}^{\Diamond\Box})$ we obtain.

Lemma 9.A.7 *For any regular logic* $\boldsymbol{L}$: $\ulcorner\Diamond\top\urcorner \in \boldsymbol{L}$ *iff* $(\mathtt{D}) \in \boldsymbol{L}$.

A modal logic is *normal* iff it contains $(\mathtt{K})$ and is closed under (RN) iff it is regular and contains $\ulcorner\Box\top\urcorner$.

Let **K** (resp. **C2**) be the smallest normal (resp. regular) modal logic. Using names of formulae from Lemma 9.A.1, to simplify naming normal (resp. regular) logics we write the *Lemmon code* $\mathbf{KX}_1 \ldots \mathbf{X}_n$ (resp. $\mathbf{CX}_1 \ldots \mathbf{X}_n$) to denote the smallest normal (resp. regular) logic containing formulae $(\mathtt{X}_1)$, ..., $(\mathtt{X}_n)$ (see Bull and Segerberg 1984; Chellas 1980; Lemmon 1977). We standardly put $\mathbf{T} := \mathbf{KT}$, $\mathbf{S4} := \mathbf{KT4}$ and $\mathbf{S5} := \mathbf{KT5}$. As it is known, $\mathbf{T} \subsetneq \mathbf{S4} \subsetneq \mathbf{S5}$, $\mathbf{KD45} \subsetneq \mathbf{S5}$, $\mathbf{KD45} \nsubseteq \mathbf{S4}$ and $\mathbf{T} \nsubseteq \mathbf{KD45}$.

Lemma 9.A.8

(i) $(\mathtt{D}) \in \mathbf{C5_c} \subseteq \mathbf{K5_c}$; $(\mathtt{D}) \in \mathbf{KT}$.
(ii) $(\mathtt{5_c}) \in \mathbf{CD4} \subseteq \mathbf{KD4}$.
(iii) $\mathbf{KD4} = \mathbf{K45_c}$ and $\mathbf{CD4} = \mathbf{C45_c}$.
(iv) $(\mathtt{4}) \in \mathbf{K5!}$.
(v) $\mathbf{KD45} = \mathbf{K5!} = \mathbf{K55_c}$.
(vi) *In* $\mathbf{K}$ *the formula* $(\mathtt{5})$ *is equivalent to the following formula*

$$(\Diamond p \wedge \Diamond q) \to \Diamond(p \wedge \Diamond q) \tag{\dagger}$$

Proof. (i) '$\Diamond(p \to \Box p)$' belongs to $\mathbf{C5_c}$, by $(\mathtt{R}^{\Diamond\Box})$. So, we use Lemma 9.A.7.
(ii) By $(\mathtt{4})$, (US), $(\mathtt{D})$ and **PL** we obtain that $(\mathtt{5_c}) \in \mathbf{CD4}$.
(iii) By (i) and (ii).
For (iv) see Exercise 4.46 in Chellas (1980).
(v) By (i), (ii) and (iv).
For (vi) see Exercise 4.37 in Chellas (1980).

Notice that from Lemmas 9.A.3, 9.A.4, and 9.A.7 we obtain:

Corollary 9.A.1 $\mathbf{CD5} = \mathbf{KD5}$, $\mathbf{CD45} = \mathbf{KD45}$ *and* $\mathbf{CT5} = \mathbf{KT5} := \mathbf{S5}$.

Thus, while defining strictly regular logics one uses some additional formulae. We adopt a convention from Segerberg (1971), p. 206. For the formula $(\mathtt{X})$ and any $i \geqslant 0$ we put $(\mathtt{X(i)}) := \ulcorner \Box^i \top \to (\mathtt{X}) \urcorner$.

Lemma 9.A.9 *Segerberg (1971), vol. II, Corollary 2.4. For any* $i > 0$:

$$\mathbf{CN^iX_1(i)} \ldots \mathbf{X}_n\mathbf{(i)} = \mathbf{CF^i} \cap \mathbf{KX}_1 \ldots \mathbf{X}_n\,,$$

where

$$\Box^i \top \to \Box^{i+1} \top \tag{$\mathtt{N^i}$}$$

$$\Diamond^i \neg \top \tag{$\mathtt{F^i}$}$$

Of course, in any modal logic $\mathtt{N}^0$ is equivalent to $\ulcorner \Box \top \urcorner$; so $\mathbf{CN^0} = \mathbf{K}$.

References

Bull, R.A., and K. Segerberg. 1984. Basic modal logic. In *Handbook of philosophical logic*, vol. II, ed. D.M. Gabbay and F. Guenthner, 1–88. Dordrecht: Reidel.

Błaszczuk, J.J., and W. Dziobiak. 1975. Modal systems related to $S4_n$ of Sobociński. *Bulletin of the Section of Logic* 4: 103–108.

Błaszczuk, J.J., and W. Dziobiak. 1977. Modal logics connected with systems $S4_n$ of Sobociński. *Studia Logica* 36: 151–175.

Chellas, B.F. 1980. *Modal logic: An introduction*. Cambridge: Cambridge University Press.

Chellas, B.F., and K. Segerberg. 1996. Modal logics in the vicinty of S1. *Notre Dame Journal of Formal Logic* 37(1): 1–24.

Ciuciura, J. 2005. On the da Costa, Dubikajtis and Kotas' system of the discursive logic, D_2^*. *Logic and Logical Philosophy* 14(2): 235–252.

da Costa, N.C.A., and F.A. Doria. 1995. On Jaśkowski's discussive logics. *Studia Logica* 54(1): 33–60.

Furmanowski, T. 1975. Remarks on discussive propositional calculus. *Studia Logica* 34: 39–43.

Jaśkowski, S. 1948. Rachunek zdań dla systemów dedukcyjnych sprzecznych. *Studia Societatis Scientiarum Torunensis* Sect. A, I(5): 57–77.

Jaśkowski, S. 1949. O koniunkcji dyskusyjnej w rachunku zdań dla systemów dedukcyjnych sprzecznych. *Studia Societatis Scientiarum Torunensis* Sect. A, I(8): 171–172.

Jaśkowski, S. 1969. Propositional calculus for contradictory deductive systems. *Studia Logica* 24: 143–157; the first English version of Jaśkowski (1948).

Jaśkowski, S. 1999. A propositional calculus for inconsistent deductive systems. *Logic and Logical Philosophy* 7: 35–56; the second English version of Jaśkowski (1948).

Jaśkowski, S. 1999a. On the discussive conjunction in the propositional calculus for inconsistent deductive systems. *Logic and Logical Philosophy* 7: 57–59; the English version of Jaśkowski (1949).

Lemmon, E.J. (in collaboration with D. Scott). 1977. *"Lemmon notes": An introduction to modal logic*. No. 11 in the American philosophical quarterly monograph series, ed. K. Segerberg. Oxford: Basil Blackwell.

Nasieniewski, M. 2002. A comparison of two approaches to parainconsistency: Flemish and Polish. *Logic and Logical Philosophy* 9: 47–74.

Nasieniewski, M., and A. Pietruszczak. 2008. The weakest regular modal logic defining Jaśkowski's logic D_2. *Bulletin of the Section of Logic* 37(3–4): 197–210.

Nasieniewski, M., and A. Pietruszczak. 2009. New axiomatisztions of the weakest regular modal logic defining Jaśkowski's logic D_2. *Bulletin of the Section of Logic* 38(1–2): 45–50.

Perzanowski, J. 1975. On M-fragments and L-fragments of normal modal propositional logics. *Reports on Mathematical Logic* 5: 63–72.

Segerberg, K. 1971. *An essay in classical modal logic*, vol. I, II. Uppsala: Uppsala University.

Chapter 10
FDE: A Logic of Clutters

R.E. Jennings and Y. Chen

10.1 Introduction: The Centrality of Truth

Paraconsistent studies are in no ideological struggle with classical logic: in fact they offer detailed understandings of proper sublogics of PL that had not hitherto been systematically studied. Labels such as *paraconsistentism* and *classicalism* are misapplied. Within paraconsistent studies, what is somewhere referred to as *preservationism*, and seen as competing with what is called *dialetheism* might better be understood as a very general proposal for the study of inference, including inference codified by systems previously given dialetheic semantic analyses. Preservational studies have themselves admitted a multiplicity of methods. It is, for example, historically incorrect to characterise preservational logic as 'the non-aggregative approach to paraconsistency'. The preservational claims about $\wedge$–I consist only in the observation that in a classical setting, $\wedge$–I does not preserve certain well-defined measures. One of them is the *coherence level* of a set, Σ of sentences, that is, the least ζ for which there is a ζ–decomposition of Σ into consistent subsets (see Schotch and Jennings 1980). Another is the *incoherence dilution* of a set, Σ, of sentences, that is the least ζ for which Σ includes a ζ–member inconsistent subset. However, as we shall see in the sequel, measures of level and dilution can be defined that are indifferent to aggregation (see Jennings and Schotch 1984). The preservational veto on $\wedge$–I is to be contrasted with the dialetheic rejection of disjunctive syllogism, which fails to preserve truth, albeit in a highly specialized semantic setting.

No logical consideration requires us to choose between the preservational and dialetheic approaches. From a philosopher's point of view the dialetheic approach is the more radical. It requires us to contemplate true contradictions, and not contradictions that reflect the incapacity of a descriptive language to be both complete

R.E. Jennings (✉) • Y. Chen
Laboratory for Logic and Experimental Philosophy, Simon Fraser University, Burnaby, Canada
e-mail: jennings@sfu.ca

K. Tanaka et al. (eds.), *Paraconsistency: Logic and Applications*, Logic, Epistemology, and the Unity of Science 26, DOI 10.1007/978-94-007-4438-7_10,

and consistent, but contradictions that reflect the nature of the world. Again the simultaneous satisfaction of contradictory sentences requires a abandonment of the classical understanding of the connectives. From the same point of view, preservational methods seem more conservative, since they do not require us to accept true contradictions, nor do they impose re-interpretations of connectives. To repeat, there is no purely logical grounds for regarding one as correct, and the other as not. The one says, 'Suppose that there were true contradictions.' The other, 'Suppose a central processor were to receive contradictory data from distinct sources.' In each case, an obvious question arises: how are we to draw principled inferences in such imagined circumstances?

Nevertheless, there are profound *political* differences, potentially transforming the role of logic within philosophy. From this larger point of view, it is the dialetheic approach that emerges as the more conservative, and the preservational approach as the more radical. Each requires a broadening of the classical point of view. But which presents philosophers with the more profound, and one might ask, the more salutary adjustment? One could argue that it is the preservational approach, for the dialetheic enshrines the traditional philosophical centrality of truth in the constitution of logic; the latter would see it overthrown.

Now philosophers have no better understanding of truth than anyone else. Indeed the nature of truth has been from very early times a central preoccupation of their discipline. It is among the principal instances of the Socratic paradox. To put the matter plainly, in our acquisition of natural language we acquire a conversational facility with the vocabulary of *truth*. But since that acquisition does not require that a deeper understanding of alethic vocabulary be accessible, there is no reason to suppose that such a deeper understanding can be achieved. In speaking of truth, there is no need for us to know what we are talking about, nor is there a need for there to be some way for us to come to know what we are talking about. Certainly there is no reason to suppose that talking about truth, the only method that philosophy has so far applied, is itself a reliable method by which to come to know what truth is. Since, as we are often told, philosophy absorbs its own metatheory, all of these observations ought by now to be trite philosophy. So ought this: neither a god nor conversation vouchsafes to us the right vocabulary for understanding. It is the responsibility of the theorist is to choose the language of his theory.

The language of models appropriates the *vocabulary* of truth, but only as a reading for what is independently well defined mathematically. It cannot supply depth, only clarity. Logicians who study inferential preservation in general, do so because they follow Tarski in being wary of the language of truth.

> the languages (either the formalized languages or—what is more frequently the case—the portions of everyday language) which are used in scientific discourse do not have to be semantically closed. This is obvious in case of linguistic phenomena and, in particular, semantic notions do not enter in any way into the subject-matter of a science; for in such a case the language of this science does not have to be provided with any semantic terms at all... Semantically closed languages can be dispensed with even in those scientific discussions in which semantic notions are essentially involved. (Tarski 1944, pp. 341–375)

Instead such logicians study the capacities of inference systems to preserve mathematically well-defined properties of sets of sentences. Their aim is to investigate features of data sets that are preserved classically beneath the superficies of truth-preservation, and then to study systems that preserve those features, even in the absence of truth.

It is obvious that the semantic representation of a sentence is dictated by its composition. This is evident in the case of classical truth-set representations: the truth-set of the negation of α is the complement of the truth-set of α; that of a disjunction the union of the truth-sets of its disjuncts, and so on. But truth-sets also destroy compositional information. Thus the truth-set of $p \vee \neg p$ is identical to the truth-set of $q \vee \neg q$ despite the compositional differences between those sentences; similarly for their negations. An inference relation that respects the classical understanding of the connectives cannot both preserve *that* compositional information and take relations between truth-sets as its semantic currency.

The earliest preservational studies did retain classical truth-sets, and accordingly did not preserve all compositional information about classical theorems and absurdities. It did however preserve compositionally *accessible* information about the rest. As an example, 3-forcing would permit any inference of the form

$$\{(\alpha \vee \beta \vee \gamma), (\alpha \vee \beta \vee \delta), (\alpha \vee \gamma \vee \delta), (\beta \vee \gamma \vee \delta)\}/\alpha \vee \beta \vee \gamma \vee \delta.$$

because the set,

$$\{\{\alpha, \beta, \gamma\}, \{\alpha, \beta, \delta\}, \{\alpha, \gamma, \delta\}, \{\beta, \gamma, \delta\}\}$$

is 3-harmonic hypergraph[1] of which the premise set is a particular formulation. It is a rule of n-forcing that every n-harmonic hypergraph[2] on a set of sentences, formulated as a set of disjunctions of vertices of hyperedges, 3-forces the disjunction of its vertices. In the example, the wff is a direct formulation of a 3-harmonic hypergraph, but the disjunction is also n-forced by any set of propositionally equivalent sentences. Again, 2-forcing permits any inference of the following form

$$\alpha, \beta, \gamma/(\alpha \vee \beta) \wedge (\alpha \vee \gamma) \wedge (\beta \vee \gamma)$$

because the conclusion formulates a 2−uncolourable[3] hypergraph as a disjunction of conjunctions of the vertices of its hyperedges. And again, any wff classically equivalent to that conclusion would be 2-forced by that set of premises. In general

---

[1]A hypergraph on a set is a collection of collections of objects from that set whose elements are called the vertices of the hypergraph. The 3-harmonic hypergraph is a hypergraph on the premise set. Members of the hypergraph are called edges. A hypergraph is n-harmonic for any natural number n if $n + 1$ is the least number of edges whose intersection is non-empty.

[2]Every n−tuple of edges has a common vertex.

[3]A hypergraph H is n-uncolourable if the width of the narrowest colouring of the set which will not result in monochromatic cells is greater that n; else it is n−colourable. The least n such that H is n−colourable is called the *chromatic number* of the set.

any set, Σ of wffs n-forces any wff equivalent to the formulation of a n-uncolourable hypergraph on Σ. So the question as to whether a wff is n-forced by a given set of wffs can be reduced to the question as to whether a hypergraph formulation of it is (a) a hypergraph on the set of premises, and (b) n-uncolourable.

This is a significant observation, for every propositional wff is equivalent to a conjunction of disjunctions of literals (also to a disjunction of conjunctions of literals). Thus every sentence of propositional logic can be mapped to a hypergraph on its language. But given a model $\mathscr{M} = \langle \mathscr{U}, \mathscr{V} \rangle$ for a propositional language, every hypergraph on a set of sentences of the language can be mapped to a hypergraph on $\wp(U)$. That is every sentence of propositional language can be represented by a powerset hypergraph. Moreover, for every pair of wffs in non-identical sets of atoms, there is a propositional model in which they are represented by distinct hypergraphs. Such a representation may be described as *articular*, it can be understood as articulating the information content of the wff in the hyperedges, and ultimately in the vertices of a powerset hypergraph. The question as to what constitutes entailment between two wffs has no single answer. It will depend upon what information content we wish to preserve from entailing wff to entailed. In the remainder of this paper we introduce the general methodology of the articular idiom while presenting more specifically an articular analysis of first degree entailment (FDE), and providing one illustrative variant. Other systems are studied in more detail in Jennings and Chen (2010, 2011) and Chennings and Sahasrabudhe (2011).

An immediate semantic consequence of this discussion is asserted in the following:

Principle of Articulation: *Every propositional wff, α has a classical semantic representation as a clutter H_α on the power set of a set of states.*

Definition 10.1. A clutter $H = \{E_1, E_2, \ldots, E_n\}$ is a hypergraph such that if $\forall E_i, E_j \in H, E_i \not\subset E_j$.

A logician who has habitually said that for logicians, a proposition is a set, can, with equal propriety, say, 'A proposition is a clutter.' The difference is one of purpose: there is a purpose for which the latter is the better formulation, and the former the worse. From such a point of view the truth-set mentioned in the former is merely the intersection of the union of the edges of the clutter.

Accordingly, we can take as the semantic entailment of β by α that, for any set of states, clutter H_β of β, subsumes the clutter H_α of α. That is, every (hyper)edge of H_β extends some edge of H_α. Intuitively, we can say that α entails β iff every informational atom of β is classically entailed by some informational atom of α. The resulting first degree implication we set out here in a binary system, AL. AL is like preservationist systems in adopting a more discriminating account of what is to be preserved, and represents a redeployment of classical methods, rather than an introduction of semantically new connectives. Even in the simplified setting of propositional logic, the strategy yields a simple sublogic of *PL* that is both relevant and paraconsistent. It bears an apparently close family resemblance to the logic of paradox, *LP* in Priest (1989), but, as we shall see, it bears an even closer family resemblance to *FDE* with which it is identical.

10.2 The Systems of Articular Inference

The language L of articular inferences has

1. A denumerable set At of atoms $p_1, p_2, \ldots, p_i, \ldots$;
2. A set K of logical connectives $\{\neg, \vee, \wedge\}$;
3. A set Φ of well-formed formulae defined in the usual recursive manner.

In the following, α, β, γ and so on are arbitrary well-formed formulae and Γ, Σ, and so on are sets of well-formed formulae. The articular logics, AL that arise from the systems of articular inference we will introduce are *binary logics* in the sense of Goldblatt (1974), that is, sets of ordered pairs of the form $\langle\beta, \alpha\rangle$. We write $\beta \vdash \alpha$ if and only if $\langle\beta, \alpha\rangle \in AL$. For convenience, we write $\langle\Sigma, \alpha\rangle \in AL$ $(\Sigma \vdash \alpha)$ as an abbreviation of $\exists\beta_1, \beta_2, \ldots, \beta_n \in \Sigma$ such that $\langle\beta_1 \wedge \beta_2 \ldots \wedge \beta_n, \alpha\rangle \in AL$.

AL has two variants, AL_1 and AL_2.

10.3 Articular Semantics and AL_1

An articular model is a triple $\mathscr{M} = \langle \mathscr{U}, \mathbb{H}, \mathrm{H} \rangle$ where

1. $U \neq \varnothing$ is a set;
2. $\mathbb{H} \subseteq \wp\wp\wp(U)$;
3. H: $At \rightarrow \{H \mid H \in \mathbb{H}\ \&\ H$ is a clutter$\}$.

That is, to each p_i, H assigns a clutter on $\wp(U)$, denoted by $H(p_i)$.

The account of H(.), which extends H to Φ, requires some preliminary definitions.

Definition 10.2. If $A \subseteq \wp(U)$, then b is an intersector of A iff $\forall a \in A, b \cap a \neq \phi$.

Definition 10.3. If $A \subseteq \wp(U)$, then $\tau(A) = \{b \mid b$ is a minimal intersector of A$\}$.

Definition 10.4. If $A \subseteq \wp(U)$, then $\overline{[A]} = \{\overline{a} \mid a \in A\}$.[4]

Simplification Not all of the set-theoretic operations introduced below preserve simplicity. However, for present purposes, every hypergraph H[5] has an equivalent clutter, $\star H$, which is obtained by casting out super-edges.

Definition 10.5. Let $\mathbf{H}(U)$ be the set of hypergraphs on $\wp(U)$. Then $\forall H \in \mathbf{H}(U)$, $\star$H is $H - (E \in H \mid \exists E' \in H : E' \subset E)$.

---

[4]By $\overline{a}$ we mean the set theoretic complement of a.

[5]H so far has been used to denote clutters. However, when there is no danger of confusion, we refer to hypergraph by it as well.

Definition 10.6. Before entering into the definition of logical operators, we define some preliminary operations on clutters:

- $H \sqcup H' = \star\{\{a \cup b\} \mid a \in H,\ b \in H'\}$
- $H \sqcap H' = \star\{a \mid a \in H \text{ or } a \in H'\}$
- $\overline{H} = H\{\overline{[B_i]} \mid B_i \in \tau(H)\}$.

H(·) extends H to Φ as follows:

$$H_{P_i} = \mathrm{H}(P_i)$$
$$H_{\neg\alpha} = \overline{H}$$
$$H_{\alpha\vee\beta} = H_\alpha \sqcup H_\beta$$
$$H_{\alpha\wedge\beta} = H_\alpha \sqcap H_\beta.$$

Definition 10.7. $\forall H, H' \in \mathbf{H}(U), H \sqsubseteq H'$, ($H$ *is properly subsumed by* H') iff $\forall b \in H', \exists a \in H$ such that $a \subseteq b$ and $\forall a' \in H, \exists b' \in H'$ such that $a' \subseteq b'$.

Definition 10.8. $\forall \alpha, \beta, \alpha \vdash_1 \beta$ (α entails β) iff $\forall \mathcal{M} = \langle\, \mathcal{U},\ \mathbb{H}, \mathrm{H}\,\rangle, H_\alpha \sqsubseteq H_\beta$. Alternatively, we say that $\alpha \vdash \beta$ is valid. So, *mutatis mutandis*, for $\Gamma \vDash \alpha$.

Lemma 10.1. *$\langle\star[\boldsymbol{H}(U)],\ \sqsubseteq\rangle$ is a lattice.*

Proof. It is easily seen that $\sqsubseteq$ is a partial ordering and that $\forall \alpha, \beta \in \Phi$, $sup(H_\alpha, H_\beta) = H_{\alpha\vee\beta}$ and $\forall \alpha, \beta \in \Phi, inf(H_\alpha, H_\beta) = H_{\alpha\wedge\beta}$.

In representing the rules and axioms that $\vdash_1$[6] picks up, we use $\vdash$ to refer to $\vdash_1$. It is straightforward to verify that the $\vdash_1$ of AL_1 satisfies the following rules:

$$\text{Cut} \quad \Gamma,\ \alpha \vdash \beta,\ \Gamma \vdash \alpha \;/\; \Gamma \vdash \beta.$$

In addition, $\vdash$ satisfies the following rule governing conjunction:

$$\text{RC} \quad \Gamma \vdash \alpha,\ \Gamma \vdash \beta \;/\; \Gamma \vdash \alpha \wedge \beta;\ \text{(rightconjunctivity)}$$

AL_1 has the following eight binary axioms.

1. $\alpha \vdash \alpha$;
2. $\neg(\alpha \wedge \beta) \dashv\vdash \neg\alpha \vee \neg\beta$;
3. $\neg(\alpha \vee \beta) \dashv\vdash \neg\alpha \wedge \neg\beta$;
4. $\neg\neg\alpha \dashv\vdash \alpha$;
5. $\alpha \wedge \beta \vdash \beta \wedge \alpha$;
6. $\alpha \vee \beta \vdash \beta \vee \alpha$;
7. $\alpha \vee (\beta \wedge \gamma) \dashv\vdash (\alpha \vee \beta) \wedge (\alpha \vee \gamma)$;
8. $\alpha \wedge (\beta \vee \gamma) \vdash (\alpha \wedge \beta) \vee (\alpha \wedge \gamma)$.

[6] The subscript suggests that the binary axioms belong to AL_1.

10.4 AL_2

AL_2 can be obtained from AL_1 by relaxing the subsumption relation in the manner of the following definition

Definition 10.9. $\forall H, H' \in \mathbf{H}(U), H \sqsubseteq H'$, (*$H$ is subsumed by H'*) iff $\forall b \in H'$, $\exists a \in H$ such that $a \subseteq b$.

It can be verified that AL_2 has all the rules that AL_1 has, in addition to the following:

$$\text{Mon} \quad \alpha \vdash_2 \beta \;/\; \Gamma, \alpha \vdash_2 \beta;$$

$$\text{Ref} \quad \alpha \in \Gamma \;/\; \Gamma \vdash_2 \alpha.$$

and that therefore $\vdash_2$ is a consequence relation in the sense of Scott (1974). In addition to [RC], $\vdash_2$ also satisfies the following rule governing disjunction:

$$\text{LD} \quad \beta \vdash_2 \alpha, \ \gamma \vdash_2 \alpha \;/\; \beta \vee \gamma \vdash_2 \alpha; \text{(leftdisjunctivity)}$$

Apart from the nine binary axioms for AL_1, AL_2 has two other axioms.

1. $\alpha \wedge \beta \vdash_2 \beta$;
2. $\alpha \vdash_2 \beta \;/\; \neg\beta \vdash_2 \neg\alpha$.

It is easily checked that the rules [LD] and [RC] preserve validity, and that the listed binary axioms are valid.

10.5 Metatheory of *AL*

Metatheorem 10.1. *Both AL_1 and AL_2 are sound with respect to a class of articular models.*

The first axiom for AL_1 becomes a particular instance of [Mon], therefore can be omitted for AL_2.

The proof of completeness of AL requires some preliminary definitions.

Definition 10.10. Σ is a *complete set of literals* iff $\Sigma \subseteq At \cup \neg[At]$ and $\forall p_i \in At$, $p_i \in \Sigma \Leftrightarrow \neg p_i \notin \Sigma$.

Definition 10.11. Σ is an $AL-$ full theory if and only if it is the $\vdash$-closure of a complete set of literals.

Notice that the set of $AL-$ full theories is identical to the set of maximal $PL-$consistent sets. $|\alpha|$ (the proof set of α is the set of $AL-$ full theories that contain α).

The AL-canonical model $\mathscr{M}^*$ is the ordered pair $\langle U^*, \text{H}^*\rangle$ where

1. U^* is the set of $AL-$ full theories;
2. H^* is defined piecewise by: $\text{H}^*(p_i) = \{\{|p_i|\}\}$.

The burden of proof for the fundamental theorem lies in the generalization of H^* to every wff of AL, so that each wff is assigned a corresponding clutter on the power set of $AL-$ full theories.

Definition 10.12. $\forall\alpha \in \Phi$, $\mathscr{A}(\alpha)$ is the family of proof-sets of literals, disjoined in the conjuncts of the CNF of α. Thus, where $\text{CNF}(\alpha) = \bigwedge_{i=1}^{n} \Delta_i$ and $\Delta_i = \bigvee_{i=1}^{m} \delta_i$, then $\forall\alpha \in \Phi$, $H_\alpha^* = \{\{|\delta_i^1|, |\delta_i^2|, \ldots, |\delta_i^n|\} \mid \delta_i \in \Delta_i \ \&\ \Delta_i \in \mathscr{A}(\alpha)\}$.

It is easily shown that

Metatheorem 10.2. $H_\alpha^* =_{def.} \mathscr{A}(\alpha)$.

Definition 10.13. $\forall\alpha \in \Phi$, $\mathscr{L}(\alpha)$ (the language of α) is the set of atoms having occurrences in α.

Metatheorem 10.3. *If* $\mathscr{L}(\alpha) = \{p_1, p_2, \ldots, p_n\}$, *then* $\alpha \dashv\vdash \bigwedge_{i=1}^{n} \Delta_i$ *where* $\Delta_i = \bigvee_{i=1}^{m} \delta_i$, $\mathscr{L}(\bigwedge_{i=1}^{n} \Delta_i) = \mathscr{L}(\alpha)$, *and* δ_i *is either* p_j *or* $\neg p_j (1 \leq j \leq n)$.

Metatheorem 10.4. $\alpha \vDash \beta \Rightarrow \alpha \vdash \beta$

AL is complete with respect to the class of articular models.

Metatheorem 10.5. *AL is decidable.*

The demonstration, which we omit here, requires only standard classical filtration methods, since failure of AL-provability is a failure of a classical provability.

Metatheorem 10.6. *AL_2 is first degree entailment.*

In Anderson and Belnap (1961), the first degree entailment $A \rightarrow B$ is valid if and only if $\forall a \in \tau(\mathscr{A}(A))$, and $\forall b \in \mathscr{A}(\mathscr{B})$, $a \rightarrow b$ is explicitly tautological, i.e. a and b share a common variable. Such is the case in the canonical model for AL_2. In the ordinary models of AL_2, the semantic representation of first degree entailment is $\forall a \in \tau(H)$, and $\forall b \in H'$, $a \cap b \neq \varnothing$. It is easily demonstrated that this is equivalent with $H \sqsubseteq H'$.

10.6 Some Concluding Remarks About AL

The theorems of AL and the validities are binary; the notion of a tautology as a wff entailed by the empty set is inimical to the idea of articular inference, as perhaps it ought to be inimical to relevant inference more generally. Since the $\rightarrow$ of AL is defined as $\neg\alpha \vee \beta$, AL does not distinguish between *modus ponens* and *disjunctive syllogism*. Both fail in AL.

Although our semantic account of inference relies wholly upon hypergraphs, truth sets, though derivative, are not wholly absent. They are obtained by melting articulations into single sets.

$$[[\alpha]]^{\mathcal{M}} = \bigcap_{i=1}^{n}\{\bigcup f_i \mid f_i \in H_\alpha,\ 1 \leq i \leq n\}$$

Thus, if $x \in [[\alpha]]^{\mathcal{M}}$ & $\alpha \models \beta$, then $x \in [[\beta]]^{\mathcal{M}}$. Modus Ponens is a rule of inference of AL. That is, $\vdash$ preserves satisfaction.

Evidently the subsumption lattice imposes strict variable-sharing requirements upon articular inference. We would maintain that they are the intuitively correct requirements. At any rate, they are sufficient to distinguish inferentially both pairs of classical tautologies and pairs of classical inconsistencies that share no variables. Thus, for example,

$$p \vee \neg p \nvdash q \vee \neg q$$

$$p \wedge \neg p \nvdash q \wedge \neg q$$

And of course, AL-inference is non-explosive.

$$p \wedge \neg p \nvdash q$$

Not surprisingly, the inference relation $\vdash$ for AL is sensitive to the definition of the subsumption relation. As an example, we can adopt the stronger subsumption relation of Jennings and Nicholson (2007), which requires that every hyperedge of the subsumed clutter be a subedge of some edge of the subsuming clutter. Such an adoption will validate a rule of transposition for $\vdash$.

Acknowledgements The research presented in this paper is supported by SSHRC research grant # 410-2008-2330. We would like to express our thanks to Professor J. Michael Dunn for his stimulating correspondence and to Bryson Brown, Julian Sahasrabudhe and Andrew Withy for their valuable observations.

References

Anderson, A.R., and N.D. Belnap. 1961. Tautological entailments. *Philosophical Studies* 13: 9–24.

Chennings, and J. Sahasrabudhe. 2011. A new idiom for the study of entailment. *Logica Universalis* 5(1): 101–113.

Goldblatt, R. 1974. Semantic analysis of orthologic. *Journal of Philosophical Logic* 3: 19–35.

Jennings, R.E., and Y. Chen. 2010. Articular models for first degree paraconsistent systems. forthcoming.

Jennings, R.E., and Y. Chen. 2011. Necessity and materiality. forthcoming.

Jennings, R.E., and D. Nicholson. 2007. An axiomatization of family resemblance. *Journal of Applied Logic* 5: 577–585.

Jennings, R.E., and P.K. Schotch. 1984. The preservation of coherence. *Studia Logica* 43: 1–2.

Priest, G. 1989. The logic of paradox. *Journal of Philosophical Logic* 8: 219–241.

Schotch, P.K., and R.E. Jennings. 1980. Inference and necessity. *Journal of Philosophical Logic* 9: 327–340.

Scott, D. 1974. Completeness and axiomatizability in many-valued logic. In *Proceedings of symposia in pure mathematics*, 411–436. Providence: AMS.

Tarski, A. 1994. The semantical concept of truth and the foundations of semantics. *Philosophy and Phenomenological Research* 4: 341–375.

Chapter 11
A Paraconsistent and Substructural Conditional Logic

Francesco Paoli

11.1 Classifying Conditionals: Where Does the Demarcation Lie?

11.1.1 Entailments, Implications, Defeasible Conditionals

Although propositional logic is about the analysis of *all* logical connectives, we must undoubtedly recognise a *primus inter pares* in this class: the *conditional* connective "if... then". Since the ancient times reams of paper have been depleted, and rivers of ink have been spilt, in order to discuss the logical properties of conditionals—even crows on the roofs once did so, according to an oft-quoted passage by Callimachus. Here I'll beg those birds to move over and let me join them in croaking about which conditionals are sound and which are not.

Given the massive proportions of such a debate, it is to some extent surprising that there is comparably little agreement among the specialists on how to classify conditional sentences in natural languages like English. For the purpose of the present discussion, let us focus on what is in my opinion the most accurate taxonomy of conditionals sentences from a *logical* viewpoint. This taxonomy, or something closely resembling it, is to be found in several places in the literature (e.g. Routley et al. 1982; Mares 2004); conditionals are ranked in decreasing order according to the logical cogency of the connection between their antecedents and their consequents.

- At the top of the ladder we find *entailments*, where the degree of logical cogency is maximal: *necessarily*, if the antecedent holds true, then so does the consequent. For example,

F. Paoli (✉)
Department of Pedagogy, Psychology, Philosophy, University of Cagliari, Cagliari, Italy
e-mail: paoli@unica.it

K. Tanaka et al. (eds.), *Paraconsistency: Logic and Applications*, Logic, Epistemology, and the Unity of Science 26, DOI 10.1007/978-94-007-4438-7_11,

(1) If it rains and it is hot, then it rains.

Using a dichotomy suggested by Meyer (1986), entailment is the kind of notion that logical systems like **S4** or the relevant system **E** from Anderson and Belnap (1975) mean to *express*, while systems like classical propositional logic or the relevant system **R** only content themselves with *indicating*: in the theorems of **R** or of **CPC** one is invited to read only asserted (principal) occurrences of the conditional connective as entailments, while in the theorems of **S4** or of **E** we are justified in interpreting *any* occurrence of that connective as formalising an entailment.

- The next sentence is an example of an *implication*, or of a *sufficiency conditional:*

  (2) If I am in Melbourne, then I am in Australia.

  This is not an entailment, at least if we are persuaded that there may well be some possible world where Melbourne fails to be in Australia (which could be doubted by partisans of rigid designation). In any case, the obtaining of the antecedent is a *sufficient condition* for the obtaining of the consequent. Classical logicians and relevant logicians who adhere to **R** usually invite us to read non-principal occurrences of the conditional connective in the theorems of their own favourite systems as implications, not entailments.
- Finally, we have *defeasible conditionals* like

  (3) If this match is struck, it will light.

  Here the connection between antecedent and consequent is still looser. The obtaining of the former is not even a sufficient condition for the obtaining of the latter, in general: it is such only *under "normal" conditions* (for example, if the match at issue does not happen to be wet). According to Priest (2001), the real logical form of conditionals like (3) is "If A and C_A, then B", where the *ceteris paribus* clause C_A can be read as "other things being equal", or "if everything else relevant remains unchanged". More precisely, C_A captures an open-ended set of conditions and depends strongly on A, a feature which is notationally represented through the use of the subscript.

Now, it looks like most scholars have inadvertently followed a "division of labour" plan that led them to focus on one rung or another of the previously described ladder, losing sight of the whole. For example, one can find a copious literature about paradoxes of entailment or implication where defeasible conditionals hardly ever get a mention. On the other hand, conditional logicians (e.g., Nute 1984; Bennett 2003) traditionally disregard most of such debate in their analysis of "if... then" sentences in natural language. Only a few authors (e.g., Sanford 1989; Mares 2004; Humberstone 2011) seem to have undertaken the praiseworthy task of giving a unified account of the phenomenon. Before trying to join them and offering my view of the problem, however, I must dwell a little longer on the internal subdivision of the category of *ceteris paribus* conditionals.

11.1.2 Defeasible Conditionals: The Main Competing Theories

The classification of defeasible conditionals in English is one of the most controversial issues in the whole area of philosophical logic. Do English sentences having the grammatical form "If A, then B" share the same logical form as well, or else may such hypothetical clauses express different connectives according to circumstances? If the latter alternative is correct, where should the dividing lines be drawn?

The former option (i.e. the claim that the meaning of "if... then" is basically uniform) has enjoyed some popularity from time to time (Bryant 1981; Lowe 1995). The difference between such counterfactual sentences as the famous

(4) If kangaroos had no tails, they would topple over.

and non-counterfactual conditionals has been explained e.g. in epistemic terms, claiming that it does not depend on an ambiguity of "if... then", but merely on the speaker's subjective opinion about the truth value of the antecedent. However, it is well-known that a heavy burden of proof lies upon the supporters of such uniform (or *monist*, as they are also labelled) theories of conditionals. In fact, they owe us a plausible account of the contrast between (5) and (6) below by Adams (1970):

(5) If Oswald didn't kill Kennedy, someone else did.
(6) If Oswald hadn't killed Kennedy, someone else would have.

(5) and (6) seem to have different truth conditions: if we take on trust the Warren report—and its claim that Oswald killed Kennedy unassisted by any accomplice—(6) is false, while (5) is trivially true given only that Kennedy has been murdered by someone.

The traditional *dualist* view (Adams 1970; Lewis 1973; Jackson 1987), therefore, has it that conditional sentences whose antecedents are in the indicative mood—in plain words, *indicative* conditionals—actually express a different connective from *subjunctive* conditionals, whose protases are in the subjunctive mood. In particular, sentences like (6) can be rephrased by forming new conditionals whose verbs are indicative, and therefore fully susceptible of being assigned a truth value:

(7) If it had been the case that Oswald didn't kill Kennedy, it would have been the case that someone else did.

Hence, it is argued that on the level of "deep structure" (5) and (6) have exactly the same antecedent and the same consequent, and that the moods of the verbs in (6) are parts not of the component sentences, but rather of the conditional construction, viz. of a "subjunctive conditional" connective which is different from its indicative counterpart (Nute 1984).

Although supporters of the standard dualist view agree that indicative and subjunctive conditionals have distinct truth conditions, it is a matter of dispute what these conditions really amount to. According to Lewis (1973, 1976), for example, subjunctive conditionals have an intensional nature, while indicative conditionals are truth-functional; Stalnaker (1968, 1975), on the other side, believes that although both kinds of conditionals can be modelled by means of possible worlds semantics, in the case of indicative conditionals a decisive role is played by appropriate contextual presuppositions.

The strongest competitor of the previous approach is surely the classification in Dudman (1983, 1984, 1989), which gained increasing support during the 1980s and beyond. Roughly put, Dudman claims that the difference between "hadn't-would" (HW) conditionals like (6) and "doesn't-will" (DW) conditionals like

(8) If Oswald doesn't kill Kennedy, someone else will.

is one of *tense*, not of *mood*: sentences of the former type express at time t what sentences of the latter would have expressed at a particular time $t' < t$. As Bennett (1988) once put it, "Every hadn't-would was once a doesn't-will". Actually, Dudman contends that HW conditionals are only seemingly subjunctive: careful linguistic analysis reveals that the verbs contained therein are indicative—more precisely, the antecedent is in the past perfect tense, the consequent in the simple past tense. Both kinds of conditionals, in Dudman's opinion, correspond to *imaginative projections* highlighted, on the linguistic level, by a "forward tense shift": "The imagined course of events is appended to the course of previous actual history, and the use of 'Vs', 'Vd', or 'had Vd' locates at present, past or 'past past' the point at which history gives way to imagination. And since the satisfaction of the antecedent is always part of what is imagined, it is always later than this 'changeover point" (Smiley 1984, p. 249). On the other side, "didn't-did" (DD) conditionals like (5) express *condensed arguments* whose antecedents "signal the entertainment of a hypothesis from which a conclusion is deduced" (Smiley 1984, p. 248).

A similar distinction is drawn by Gibbard (1981): *epistemic* conditionals, whose assertion is guided by a subjective connection in the utterer's belief system and whose paradigmatic examples are DD conditionals, must be kept separate from *factual* conditionals, whose assertion is guided by an objective connection between states of affairs and whose paradigmatic examples are HW conditionals. The main difference between the accounts by Gibbard and Dudman is that for the former DW hypothetical clauses may fall into either category according to circumstances, while the latter (supported by Bennett 1988), as already remarked, claims that DD stay on the one side of the fence and DW and HW on the other.

The traditional account, disparagingly labelled in Bennett (1988) the "phlogiston theory of conditionals", experienced a resurgence over the last 15 years. Edgington (1995), Weatherson (2001),[1] and Bennett (1995) have advocated it against Dudman's attacks. In particular, examples have been provided both of epistemic DW conditionals and of factual DD conditionals, showing that the "objectivity point" (as Bennett calls it) cannot be used to defend Dudman's view of the matter. Later we shall examine in greater detail some of Bennett's allegations.

[1]Observe that Weatherson is defending a substantially different theory in his more recent Weatherson (2009).

11.2 Two Conditionals or Three?

11.2.1 The Ambiguity of Disjunction

As we have just seen, according to the received view—pleaded e.g. by Stalnaker and Lewis—all indicative conditionals must be assigned to the same class, whereas Dudman and Gibbard claim that some of them belong together with the subjunctives. I agree partly with the former and partly with the latter. They possess some degree of semantical uniformity,[2] in that they can be rephrased with no essential alteration of meaning as "Either not-A or B". Yet, they do not belong to the same class, in so far as the previous disjunction is inherently *ambiguous*. I will contend that there are at least three different kinds of indicative conditionals in English, and that what distinguishes them from one another are the operational properties of the disjunctions underlying each conditional—i.e. of the disjunctions in terms of which each conditional can be rephrased.

Let us examine disjunction first. Consider the following three sentences:

(9) Either $2 + 2 = 4$, or London is in Alaska.
(10) Either the butler did it, or the gardener did it.
(11) Either it will rain, or the match will be played.

Has the "either...or" construction the same meaning in each of these sentences? Since none of (9)–(11) expresses an *exclusive* disjunction, it could be believed that it has. However, in the tradition of relevant and substructural logics (see e.g. Read 1988; Restall 2000; Paoli 2002, 2005; Allo 2011), it has been argued at length that *inclusive* disjunction is, in turn, ambiguous. The arguments are altogether well-known, and I will not try to recapitulate them. Put quite roughly: in sentences like (9), no special relationship is presumed to hold between the disjuncts; such disjunctions are asserted simply on the ground of the acceptance of at least one of the disjuncts themselves. This kind of disjunction has been labelled *lattice-theoretical*, or *additive* (especially by linear logicians), or also *extensional* (especially by relevant logicians). Here, it will be denoted by means of the symbol $\sqcup$. It is an associative, commutative and idempotent disjunction: as to the last property, it is easily realised that $A \sqcup A$ is accepted in virtue of the acceptance of one of its disjuncts if and only if A itself is accepted.

On the other hand, suppose that (10) is uttered in a context where it is not known who committed a given crime, but the only suspects are the butler and the gardener (who, possibly, may have acted by common consent). (10) presupposes then a *connection* between the disjuncts: it is such a connection that produces the acceptance of the disjunction, not the previous acceptance of one or of the other disjunct. In substructural logics, this kind of disjunction has been termed *group-theoretical*,

[2]Of course we must rule out such pseudoconditionals as Austin's "There are biscuits on the sideboard if you want some". These sentences will not be considered further in this paper.

or *multiplicative* (especially by linear logicians), or also *intensional* (especially by relevant logicians). Here, it will be referred to by means of the symbol $\oplus$. In terms of its operational properties, it is an associative and commutative connective, but it is not idempotent. For example, I can accept the first disjunct of (10) without accepting

(12) Either the butler did it or the butler did it.

(with an intensional "or"), if I entertain as a genuine alternative the possibility that the culprit was the gardener.

So far, nothing new under the sun. Yet, I want to go a step farther and claim that even (10) and (11) cannot belong in the same lot—a difference which seems to have passed unnoticed in the debates on the ambiguity of logical constants. Like (10), (11) requires a connection between the disjuncts, but of a somewhat different kind. While the meaning of (10) is not disturbed by the inversion of the disjuncts—both alternatives are entertained together, and enjoy so to speak equal rights—this is not the case for (11), where a tacit *ceteris paribus clause* attached to the first disjunct seems to award it a privileged role in the sentence. In other words, (11) appears to mean something like: either it will rain, or something wholly unexpected other than the rain will prevent the match from being played, or the match will be played. Permutation of disjuncts would render the clause idle, thus affecting the meaning of the whole sentence. The disjunction in (11) (hereafter called *superintensional* and referred to by the symbol $\curlyvee$), therefore, lacks not only the property of idempotency, but also that of *commutativity*.[3] And, just to anticipate a bit, the noncommutativity of $\curlyvee$ is tightly related to the failure of contraposition for the corresponding conditional.

11.2.2 From Multiple Disjunctions to Multiple Conditionals

Virtually all authors who denied the equivalence between the (indicative) conditional and material implication have also denied the equivalence between "If A then B" and "Either not-A or B". There is something more in an English indicative conditional, so goes the received view, than there is in its disjunctive paraphrase: the latter, unlike the former, is a truth-functional sentence where the meaning connection between the antecedent and the consequent of the corresponding conditional is irremediably lost. As anticipated above, I disagree with this analysis. Once we acknowledge the ambiguity of disjunction, we can vindicate the correctness of the disjunctive paraphrase of conditionals without being committed to accepting the equivalence between indicative conditionals and material implications. To clarify this point, let us now rewrite (9)–(11) in conditional form:

[3] What about associativity? I do not have univocal intuitions either way. In my formal theory below, I will not assume that this property holds.

(13) If $2 + 2 \neq 4$, then London is in Alaska.
(14) If the butler didn't do it, then the gardener did.
(15) If it doesn't rain, then the match will be played.

Apparently, no gross change in their meanings has been produced. This should come as no surprise, since the equivalence of "Either A or B" and "If not-A, then B" for indicative conditionals has long been taken for granted before being challenged by some non-classical logics. In my opinion, indeed, such darts were not aimed at the proper target: it is not such an equivalence that must be dropped, it is the ambiguity of disjunction that ought to be duly acknowledged. Once this is done, it becomes reasonable to suppose that the features which distinguish the above kinds of disjunctions are mirrored by different features of the corresponding conditionals. In fact, we will now see that such typologies of sentences abide by different logical laws.

For a start, consider (13). Just like there is a sense in which (9) was assertible simply on the ground of the acceptance of its first disjunct, so there is a sense in which (13) is acceptable simply on the ground of the rejection of its antecedent, i.e. simply in virtue of the fact that its subordinate clause is ruled out: if $2 + 2 \neq 4$, then anything whatsoever can be the case.[4] In particular, therefore, it is implied by the negation of its antecedent.[5] As the debate about the paradoxes of the material conditionals has demonstrated, this is the case neither for (14) nor for (15): there is another sense of "if... then" in which the fact that if it wasn't the butler who did it then it was the gardener does not arise as a consequence of the butler's failing to do

[4]And, if the order of the disjuncts in the original disjunction had been reversed, the resulting conditional would instead have been acceptable simply in virtue of the fact that its main clause was taken for granted.

[5]Do these conditionals actually occur in everyday speech? People often use conditionals like "If Pete is a good barber, then I'm a monkey's uncle". Nonetheless, they are often dismissed as mere rhetorical devices. It is remarked in Anderson and Belnap (1975), p. 163: "It is of course sometimes said that the 'if-then' we use admits that false or contradictory propositions imply anything you like, and we are given the example 'If Hitler was a military genius, then I'm a monkey's uncle'. But it seems to us unsatisfactory to dignify as a principle of logic what is obviously no more than rhetorical figure of speech, and a facetious one at that". In my opinion, however, the decisive issue is not whether to dignify or not the *ex absurdo quodlibet* as a principle of logic: it is only that we must correctly determine for which logical constants it holds and for which ones it fails. "Monkey's uncle" conditionals are commonly used by ordinary English speakers and have an "if... then" grammatical form; moreover, people freely use the *ex absurdo quodlibet* in dealing with these sentences. If it is thought that they are not proper hypotheticals, independent reasons should be given to justify this claim. I believe that the linguistic data do not lead to dispose of them offhand as pseudoconditionals—but, as long as they are conveniently kept distinct from conditionals which do not abide by that law, there is little to worry about. This issue, by the way, marks a first difference between my perspective and the relevant one (more will emerge soon): since the implicational paradoxes plainly fail for the relevant conditional, Anderson and Belnap conclude that they are fallacious altogether. On the contrary, I think that they are false of other kinds of conditionals but true of this one.

it, while from the fact that it rains it does not follow that if it doesn't, the match will be played.[6] Thus, (13) cannot belong to the same category as (14) or (15), for the *ex absurdo quodlibet* holds for the former but not for the latter.

Next, consider the following variants of (15):

(16) If the match isn't played, it will rain.
(17) If it doesn't rain and no player turns up, the match will be played.
(18) If it doesn't rain, the match will be played. If it snows, it won't rain. Therefore, if it snows, the match will be played.

Failures of contraposition, transitivity, and monotonicity for some kinds of conditionals sentences are, as a matter of fact, widely recognized in the literature (see e.g. Sainsbury 1991). The need for a distinction between ordinary conditionals and conditionals for which these principles may fail has been widely recognized. For Lewis, Jackson and others (see Lewis 1976 or Jackson 1979), validation of such laws marks a watershed between indicative conditionals, which are truth-functional, and subjunctive conditionals, which lead us into the realms of intensionality and are captured by possible worlds semantics. However, the above examples show that a conditional need not be in the subjunctive mood to exhibit intensional features and fail to abide by the laws of transitivity, contraposition, monotonicity: it is easy to imagine situations where (15) is true while (16) and (17) are false and (18) has true premisses and a false conclusion. On the other hand, with (14) at least contraposition does not seem to cause trouble: its contrapositive sounds as a fair paraphrase of the original sentence, whose meaning it does not seem to affect. Thus, (15) cannot belong to the same category as (14), for contraposition (and, possibly, monotonicity and transitivity) holds for the latter but not for the former.

If failure of the above principles is not a distinctive property of subjunctive conditionals but is shared also by some indicatives, when exactly does it arise? An interesting suggestion, which we already hinted at when introducing our noncommutative disjunction, has been advanced for example by Priest (2001), who maintains that these logical principles break down when a hypothetical clause expresses a *ceteris paribus* enthymeme: what we are assenting to when we endorse (15), for example, is not the conditional itself, but rather something like

(19) If it doesn't rain then, other things being equal, the match will be played.

This duly accounts e.g. for the failure of contraposition: while the original conditional meant "If *A* and nothing relevant about *A* changes, then *B*", the meaning

[6] Advocates of the Gricean conversational account of paradoxes of material implication, like Lewis, will disagree: however, Read (1988) and others have convincingly shown that the recourse to implicature yields no advantage in the case of *nested* conditionals and wherever the conditional is not the principal connective of the sentence at issue. Therefore, it does not offer a viable solution to the paradoxes.

of the contraposed sentence is "If $\neg B$, and nothing relevant about $\neg B$ changes, then $\neg A$". It is evident how this phenomenon is linked to the noncommutativity of the corresponding disjunction.[7]

The previous observations lead to surmise that there are (at least) three kinds of indicative conditionals in English. In short, some conditionals—like (13) above—satisfy the *ex absurdo quodlibet*: we call them *extensional* or *squiggle* conditionals and use for them the symbol $\rightsquigarrow$. Among those which do not, some—like our (14)—satisfy contraposition, other ones—for example, (15)—provide patent violations of this schema and can be assimilated, at least under this respect, to subjunctive conditionals. We call the former *intensional* or *arrow* conditionals and the latter *superintensional* or *corner* conditionals, hereafter denoting them, respectively, by the symbols $\rightarrow$ and $>$.

It seems appropriate, now, to try and answer some questions which may have occurred to the reader:

1. In the preceding section we surveyed a number of taxonomies of conditionals. How does our classification relate to them?
2. What is the relationship between the above connectives and the material conditional (here referred to by the symbol $\supset$) of classical logic?
3. Does the relevant conditional of Anderson and Belnap (1975) coincide with any of the previous conditionals?

Let us face such issues one by one.

1. According to my proposal, some indicative conditionals share with subjunctive hypotheticals strongly intensional (I used, in fact, the term "superintensional") features, which determine the failure of such logical principles as transitivity, contraposition, monotonicity. This placement of the cut-off point signals an irreconcilable rift with the traditional theory, and brings my suggestion closer to the approaches by Priest, Dudman and Gibbard. I guess that Priest's distinction between ordinary and *ceteris paribus* conditionals, as well as Dudman's distinction between "condensed arguments" and "imaginative projections", or Gibbard's one between epistemically and factually based conditionals, are all approximately correct and hinge at least in part on similar intuitions. I only think that marking the boundaries of each class of conditionals by means of an appeal to the operational properties of the respective underlying disjunctions allows one to remain on a firm logical ground, while Dudman's grammatical criterion (DD on the one side, DW and HW on the other) or Gibbard's epistemic criterion do not.

   Furthermore, it seems to me that my approach to the issue yields an extra bonus. I believe that the appeal to grammatical or epistemic criteria in the classification of conditionals has befuddled the debate to the extent that it has

[7]In the framework of standard possible world semantics, in fact, $A > B$ would be true at w just in case $f_A(w) \subseteq [B]]$, while $\neg B > \neg A$ would be true at w just in case $f_{\neg B}(w) \subseteq [\neg A]$. Likewise, $A \curlyvee B$ would be true at w just in case $f_{\neg A}(w) \subseteq [B]$, while $B \curlyvee A$ would be true at w just in case $f_{\neg B}(w) \subseteq [A]$. Obviously, these conditions need not be equivalent.

prevented some authors from recognising that DD (or epistemic) conditionals *obey exactly the same logical laws* as sufficiency conditionals. If this observation is correct, there is no need to create a separate category for implications: they belong together with epistemic conditionals to the class of arrow conditionals. In this way, we can finally have hope to bridge the theories of entailment, of implication and of natural language conditionals, which have been kept artificially separate for such a long time, by aiming at a logical theory which *expresses* entailment and *indicates* both sufficiency and defeasible conditionals. In the final section of this paper I will try to flesh out in mathematical terms this basic informal intuition.

2. Although I am joining here the majority of relevant logicians (e.g. Anderson and Belnap 1975 or Read 1988) in drawing a sharp distinction between extensional and intensional connectives, mainstream relevant logicians also claim that the material conditional is no conditional connective, while I maintain that it is rather two conditional connectives in one: as argued more thoroughly in Paoli (2007), it is an ambiguous concept which has been paralogistically assigned the properties of both the squiggle and the arrow. Such a difference, to some extent, could be disregarded as merely verbal; more importantly, however, relevant logicians do not seem to distinguish sharply enough between the material conditional and the squiggle. In fact, even though modus ponens for the squiggle is not relevantly valid, if you replace $\supset$ by $\rightsquigarrow$ all the classical implicational tautologies[8] become theorems of the logic which is at the forefront of all relevant logical systems—Anderson's and Belnap's **R**. Hence, the horseshoe and the squiggle come to obey exactly the same laws, though not the same inference rules. This consequence of the presence of suitably strong contraction principles in **R** is, in my opinion, a further reason not to favour the adoption of such principles. If A is neither accepted nor rejected, in fact, we have no ground for accepting $A \rightsquigarrow A$, because we can neither reject its antecedent nor accept its consequent; nonetheless, $A \rightsquigarrow A$ should hold true if the squiggle obeyed the same laws as the material conditional. Hence, the material conditional cannot coincide with the squiggle; but it cannot coincide with the arrow either, because $\rightarrow$ does not satisfy the paradoxes of material implication while $\supset$ does. On the contrary, there is no connective fulfilling both the laws characterizing $\rightsquigarrow$ (such as the law of *a fortiori*, or the *ex absurdo quodlibet*) and those holding of $\rightarrow$ (such as the principles of identity, assertion and transitivity, or the rule of modus ponens). An endless series of paralogisms, of which C.I. Lewis's "independent proof" is the prototypical example, originated from this equivocation.[9]

[8]Caution: Anderson and Belnap use the horseshoe to denote both the classical conditional, which obeys modus ponens in any arbitrary theory, and the extensional conditional of their relevant logic, which does not. My chosen notation avoids any possible misunderstanding.

[9]See again Paoli (2007) for a more detailed discussion of this point.

3. It is instructive to notice that Anderson and Belnap, while insisting that the intensional disjunction means nothing else than "if not-A, then B", where "if... then" stands for the relevant conditional, are not completely clear about whether such a conditional should be read as an indicative or a subjunctive conditional. Our impression is that the relevant conditional is seen by Anderson and Belnap as an all-purpose logical concept which can be used to model both the arrow and the corner:

> The truth of A-or-B, with truth functional "or", is not a sufficient condition for the truth of "If it were not the case that A, then it would be the case that B". [...] On the other hand the intensional varieties of "or" which do support the disjunctive syllogism are such as to support corresponding (possibly counterfactual) subjunctive conditionals. When one says "That is either *Drosophila melanogaster* or *Drosophila virilis*, I'm not sure which" and on finding that it wasn't *Drosophila melanogaster* concludes that it was *Drosophila virilis*, no fallacy is being committed. But this is precisely because "or" in this context means "if it isn't the one, then it is the other". [...] But it should be equally clear that it is not simply the truth functional "or" either, from the fact that a speaker would naturally feel that if what he said was true, then if it hadn't been Drosophila virilis, it would have been *Drosophila melanogaster* (Anderson and Belnap 1975, p. 176).

The remark by Anderson and Belnap can be disputed (cp. Paoli 2007). Let us return to our early example of Oswald and Kennedy. When one says "Either Oswald or someone else killed Kennedy, I don't know who", and on finding that Oswald didn't do it concludes that someone else did, no fallacy is being committed. But, as we already noticed, a speaker would not naturally feel that if what he said was true, then if Oswald hadn't done it, someone else would have! What Anderson and Belnap seem to do, here, is blending into their concept of relevant conditional the properties of two different connectives (the arrow and the corner), both of intensional nature - although to a different degree.[10]

Summing up: my suggested taxonomy provides for at least three kinds of indicative conditionals: squiggles ($\rightsquigarrow$), arrows ($\rightarrow$) and corners ($>$), respectively definable in terms of an associative, commutative and idempotent disjunction ($\sqcup$), an associative and commutative, but non-idempotent disjunction ($\oplus$), and a possibly non-associative and surely non-commutative and non-idempotent disjunction ($\curlyvee$). Subjunctive conditionals can be assimilated to corner conditionals, except for the fact that—for grammatical reasons—it is unclear how to obtain therefrom syntactically adequate disjunctive paraphrases.[11] The next Table summarises the information just given.

[10] Anderson and Belnap are not the sole authors in the relevant tradition who seem committed to such an equivocation. Hunter (1993), for one, followed their lead. A happy exception is the paper Mares and Fuhrmann (1995), where the arrow is carefully distinguished from the corner, although the squiggle is assigned no special status.

[11] The last statement is not always correct, though. Consider the perfectly grammatical disjunctive paraphrase of a subjunctive conditional: "I had to jot that down or I would have forgotten it" (Dowing 1975, p. 86).

Symbol	Name	Disj. and Conj.	Properties of such	Type of connection
$\rightsquigarrow$	squiggle	$\sqcup, \sqcap$	assoc., comm., idemp.	no connection
$\rightarrow$	arrow	$\oplus, \otimes$	assoc., comm.	subjective
$>$	corner	$\curlyvee, \curlywedge$		objective

As far as I can see, even though such a picture is new on the whole, each single aspect of it has been anticipated in the debate over conditionals:

- Lewis (1973, 1976) and Jackson (1979), among others, realised the need for a distinction between corner conditionals (even though they mistakenly equated them with subjunctive conditionals) and some other kind of conditional, but neglected the divide between arrow and squiggle conditionals, identifying them with the hybrid concept of *material conditional.*
- Dually, Anderson and Belnap (1975) correctly acknowledged the need for a distinction between squiggle conditionals (even though they mistakenly had them obey the same laws as material conditionals) and some other kind of conditional, but overlooked the divide between arrow and corner conditionals, identifying them with the hybrid concept of a *relevant conditional.*

It seems to me that only by amending the faults in the partly correct intuitions of both sides one can get the right demarcation lines.

To the best of my knowledge, the sole author who advocated a three-level logic with three different kinds of conjunctions, disjunctions and conditionals, characterised by distinct operational properties—even though in view of different applications—was Casari (1997). His research was a major source of inspiration for the present paper, in ways that will become more and more evident in the subsequent pages.

11.2.3 An Evaluation of Some Arguments by Bennett

In a paper on the classification of conditionals, Bennett (1995), Jonathan Bennett discusses some features of conditional sentences in order to corroborate the traditional dualist view, to which he reverted on that occasion after having taken sides with Dudman's reforming proposal for a number of years. In this subsection I will examine some of his arguments, trying to assess them in the light of my own suggestion.

Bennett mainly discusses the placement of indicative DW conditionals; his chosen example is the sentence

(20) If Booth doesn't kill Lincoln, someone else will.

Bennett imagines the following situation:

> Suppose that a bit before the fatal time, one conspirator is sure that plans are in place for Booth to make the attempt and for someone else to take over in the event that he fails. This conspirator, Oscar, has objectively connecting grounds for accepting something

> which he expresses in the words "If Booth doesn't kill Lincoln, then someone else will". Another conspirator, Sam, has subjectively connecting grounds for accepting something that he expresses in the very same sentence [... e.g.] he hears someone being ordered to kill Lincoln; he thinks he was Booth, but he isn't sure; he is sure that whoever gets the order will carry it out (Bennett 1995, pp. 334–336).

The main point of disagreement between Bennett and me arises precisely over the following issue: Bennett takes the sentence believed by Sam and the sentence believed by Oscar to be tokens of the very same proposition. The grounds for asserting it may be different in each case, but this is thought to have little to do with the meaning of the sentence itself:

> Edgington issues this challenge: why not say simply that [it] expresses a single proposition—or means just one thing—which Oscar and Sam accept for different reasons? [...] Suppose that Oscar has his objectively connecting reasons and doesn't know Sam's reason. Sam says "I think that if Booth doesn't kill Lincoln, someone else will—don't you agree?". It would be excessively odd for Oscar to reply "It depends on what you mean" (Bennett 1995, p. 336).

The reason why the sentences uttered by Oscar and Sam might not express a single proposition was discussed above: epistemic conditionals imply their own contrapositives, while factual conditionals need not. Failure of contraposition may not affect (20) in particular; however, that something might go wrong with the *ceteris paribus* clause after the contraposition move is emphasised by the fact that, if someone says

(21) If no one else kills Lincoln, Booth will.

probably Oscar will not assent; rather, he might correct his interlocutor by bringing into play an appropriate backtracking version of the conditional: "No, but perhaps what you mean is that if no one else kills Lincoln, it is because Booth already did it". Vice versa, Sam is more likely to agree with (21) ("If it's no one else, it means that I got it right, it's going to be Booth"). This asymmetry should at least cast some doubts on Bennett's claim.[12]

The second argument is connected to Lewis's celebrated triviality result according to which no non-trivial conditional proposition is such that a person's confidence in it is proportional to the confidence that she accords to the consequent on the supposition that the antecedent holds true (i.e. no non-trivial conditional has what Bennett calls the *confidence property*). Bennett argues as follows: suppose that DW conditionals express factually based propositions (corner conditionals) on some occasions and subjectively based propositions (non-corner conditionals) on other occasions. In both cases, such conditionals patently have the confidence property. While non-corner conditionals, however, are amenable to be treated as

[12] Cp. the remark in Smiley (1984): "When a conditional conveys temporal succession [...] it becomes an understatement to say that contraposition fails. Either the contraposed conditional conveys a message unrelated to the original (compare 'If the surgeon didn't operate the patient would die' with 'If the patient didn't die the surgeon would operate') or it fails to convey a coherent message at all (try contraposing 'If the surgeon didn't operate tonight the patient would die tomorrow')".

either material conditionals or conditional assertions, and so do not contradict the theorem by Lewis (for material conditionals can after all be denied the confidence property, while conditional assertions do not express conditional propositions), such an escape is not available when corner conditionals are at issue. Thus, the triviality result causes a contradiction: the indicated instances of DW should both have and lack the confidence property.

I believe that the argument rests on a dubious premise—namely, that DW conditionals have the confidence property. In my opinion, on the contrary, none of $\rightsquigarrow, \rightarrow, >$ really has the property. That squiggles lack it should be fairly obvious. But the same can be repeated also for arrows and corners, because they require a connection (of subjective or objective kind) between the antecedent and the consequent. My confidence in the conditional

(22) If England beat France today, then the sun will rise tomorrow.

is virtually null whether the "if... then" is interpreted as an arrow or as a corner, but the probability I am willing to accord to "The sun will rise tomorrow" on the supposition that England beat France is as high as it can be. Thus, there is no contradiction in supposing that a DW conditional can express, according to circumstances, a squiggle, an arrow or a corner. At the very least, no such contradiction is entailed by Lewis' theorem.

Let us now examine a couple of the remaining arguments advanced by Bennett in defence of the traditional taxonomy.

(A) The opt-out property. A subjunctive conditional, according to Bennett, has the opt-out property: "It can properly be accepted by someone who would, if he became sure of its antecedent's truth, simply drop it, opt out, say that his conditional had presupposed something false and was therefore inoperative" (Bennett 1995, p. 341). On the other side, Bennett claims that indicative conditionals lack the property.

However, some counterexamples can be devised. Suppose that Oscar, who wants the death of Lincoln, hears that Booth is probably going to shoot him on that same day. Such a course of events would obviously suit Oscar, who could achieve his own treacherous end without exposing himself. Therefore, he will simply stand by and wait—if Booth has the nerve to kill the president, all the better; otherwise, he will do it personally. After a few months, Booth is convicted for the murder; Sam, who had come to know about Oscar's plot, remarks: "If Booth hadn't killed Lincoln, Oscar would have". Finally, suppose that Booth is fully discharged on appeal. Now that he knows that the antecedent of his previous conditional was true, does Sam have any reason to opt out? Not quite: the most reasonable thing to do would seem to presume that, after all, Oscar actually carried out his plan. The above HW conditional does not seem to have the opt-out property.

Now, let us switch tó our conditional (13) above, which is indicative and therefore, according to Bennett, should not allow the opt-out move. It is evident, however, that it does: squiggle conditionals are the prototypical examples of hypothetical sentences that are not (to speak the jargon of Jackson 1979) *robust* with respect to their antecedents, which means that it is possible to opt out once the truth of the protasis has been ascertained.

If the opt-out property induces any demarcation at all, then, it cannot be the one which Bennett points to. Rather, such a divide seems to cut the field across, setting squiggles and some corner conditionals in the subjunctive mood apart from arrows and other corner conditionals, both in the indicative and in the subjunctive mood.

(B) The zero property. Bennett says that a conditional "has the zero property if it is a conditional for which nobody could have any serious use while giving (its antecedent) a probability of zero". He maintains that the zero property characterises indicative conditionals in opposition to subjunctive ones. However, if indicative conditionals of the (13) sort are something for which anybody could have any use at all, this would happen precisely when their antecedents have probability zero.

11.3 Simplification of Disjunctive Antecedents

11.3.1 Background

Let us now take stock and examine more closely the three logical levels previously introduced. As the word "level" itself suggests, I assume that the progressive decrease in the number of operational properties observed in passing from the extensional disjunction to the intensional and then to the superintensional one corresponds to an underlying ordering: the first level, where idempotency is retained, should be the most basic or fundamental one, while the remaining levels should constitute subsequent steps towards greater generality and should therefore be characterised by the rejection of some basic properties of disjunction. I will also award the intermediate level a distinguished status, appointing the arrow conditional as the linguistic analogue of a metalinguistic derivability relation. In more precise terms, this means that when we set up an axiom system for our logic, it will be the intensional level that will provide a set of *equivalence formulae* (namely, the symmetrisation of the arrow conditional) for the resulting deductive system. Remark that such a role is played by the material conditional in most conditional logics, including the systems by Stalnaker and Lewis, and by the relevant conditional in most relevant logics, including **R**.

A question now arises quite naturally: how should these levels be linked to one another? An appealing *prima facie* thought would lead us to rank our disjunctions and conditionals in order of *inferential strength*, with connectives of upper levels ranking higher than their lower level counterparts. However, I think we should resist this easy temptation, which would commit us to accepting both (23) and (24) below:

(23) $(A \rightarrow B) \rightarrow (A \rightsquigarrow B)$.
(24) $(A > B) \rightarrow (A \rightarrow B)$.

(Remark that the principal connectives of both formulae, which intuitively express derivability claims, are arrow connectives! This is a consequence of the distinguished status I awarded to the intensional level.) I already discussed some reasons for disliking (23): given some modest assumptions in the underlying logic, it leads to an undesirable confusion between the squiggle and the horseshoe. (24), on the other side, closely resembles the principle known in the literature on conditionals as MP, or *conditional modus ponens*—indeed, it would be just MP if we had a horseshoe in place of the arrow. Although MP is a thesis of most conditional logics, it fails in some well-known basic conditional systems such as **CK** or **V** (Chellas 1975; Nute 1980), and it can be plausibly argued that it is an unwelcome principle for *ceteris paribus* conditionals.[13] However, the main reasons for my distrust in both (23) and (24) are the fact that, as argued above, the identity principle should hold for $\rightarrow$ but not for $\rightsquigarrow$, thus contradicting (23); and the fact that there are principles, of which more below, which hold for $>$ and not for $\rightarrow$, thus contradicting (24).

Taking up a suggestion advanced in Casari (1997), I prefer to choose a different way to connect together the levels of our construction. I already underlined the fact that in substructural logics there are two families of conjunction and disjunction connectives—the intensional and the extensional. Generally speaking, distribution of conjunction over disjunction, and of disjunction over conjunction, fails within the same family, but holds if the distributing connective is intensional and the connective which is distributed over is extensional. In other words: neither extensional disjunction (conjunction) distributes over extensional conjunction (disjunction), nor does intensional disjunction (conjunction) distribute over intensional conjunction (disjunction), but intensional disjunction (conjunction) *does* distribute over extensional conjunction (disjunction).[14]

My conditional logic simply adds one more superintensional level on top of the building: now we also have a noncommutative disjunction which is assumed to distribute from both sides over the conjunction of the level immediately below—i.e. the intensional level. Summing up, I assume that for $n \in \{1, 2\}$, the disjunction (conjunction) of level $n + 1$ distributes over the conjunction (disjunction) of level n. This ensures the required connection among the different levels.

Distribution of intensional connectives over extensional connectives of different name is commonplace in substructural logics, and is well-motivated in the light of the inferential content of such constants (in fact, you do not need either weakening or contraction to derive such principles in the context of the sequent calculus for classical logic).[15] But why should we assume distributivity of *superintensional* connectives over *intensional* ones? Is such a move triggered merely by an aesthetic

[13]It suffices to consider any subjunctive conditional with the opt-out property: "If Oswald hadn't killed Kennedy no one else would have", indeed, does not seem to imply "If Oswald didn't kill Kennedy no one else did".

[14]In most relevant logics, to be sure, distribution for extensional connectives is available, whereas it fails in linear logic and in Meyer (1966) **LR**, whose sequent version is obtained from the classical sequent calculus by dropping the rules of weakening. I argued in Paoli (2007) that nondistributive logics are the best motivated relevant logics.

[15]I tried to keep my preference for a proof-conditional (over a truth-conditional) semantics for logical constants in the background but, as you see, I failed. Observe, however, that what I said so far is independent from such a personal liking. For a defence of this view, see Paoli (2007).

desire of symmetry, or can it be justified by deeper philosophical and logical reasons? Well, it turns out that upholding such distribution patterns amounts to taking up an especially plausible form of the well-known and controversial principle of *simplification of disjunctive antecedents* (SDA: Nute 1980, 1984). If we denote by $\wedge$ (respectively, by $\vee$) the classical, truth-functional conjunction (disjunction), the standard form of this principle reads as follows:

(25) $(A \vee B > C) \supset (A > C) \wedge (B > C)$.

In many cases, this law appears to encode an intuitively plausible inference. For example, the following conditional with disjunctive antecedent from Nute (1984) seems to imply the conditionals that retain the same consequent and have as respective antecedents the members of the disjunction:

(26) If the world's population were smaller or agricultural productivity were greater, fewer people would starve.

Nonetheless, the validity of SDA has been the centre of a heated debate in conditional logic. On the "pro" side, it has been remarked that inferences like the one we just considered appear to be valid not in virtue of the meanings of the terms occurring therein, but in virtue of their *logical form*: and, if SDA happens to be invalid, what other valid argument schema could they possibly instantiate?

On the "con" side, however, it has been observed that SDA yields all the Undesirables (transitivity, monotonicity, contraposition) if paired with the seemingly innocent principle of *substitution of provable equivalents* (SPE) in the framework of even very weak conditional logics. By way of example, let us show that SDA together with SPE entails monotonicity: Let $\chi = A \wedge C, \psi = A \wedge \neg C$:

1. $A \equiv \chi \vee \psi$	**CPC** thesis
2. $(A > B) \supset (\chi \vee \psi > B)$	1, SPE
3. $(\chi \vee \psi > B) \supset (\chi > B) \wedge (\psi > B)$	SDA
4. $(A > B) \supset (\chi > B) \wedge (\psi > B)$	2, 3, transitivity$\supset$
5. $(\chi > B) \wedge (\psi > B) \supset (\chi > B)$	conj. simplification
6. $(A > B) \supset (\chi > B)$	4, 5, transitivity$\supset$

Furthermore, it has been contended that, even though most of its instances look unexceptionable, there are cases which are not that self-evident. For example (27) below does not seem to imply (28) (Nute 1984):

(27) If the US devoted more than half of its national budget to defence or to education, it would devote more than half of its national budget to defence.

(28) If the US devoted more than half of its national budget to defence, it would devote more than half of its national budget to defence; and if the US devoted more than half of its national budget to education, it would devote more than half of its national budget to defence.

A way out of the puzzle has been suggested by Loewer (1976), who claims that SDA, *per se*, never expresses a reliable mode of inference. Even its seemingly correct instances do not confirm its soundness, because the real logical form of their antecedents is $(A > C) \wedge (B > C)$, rather than $A \vee B > C$: hence such instances are actually instances of the identity principle. This solution, however, besides having a slight *ad hoc* flavour, borders on circularity: if asked when it is the case that a conditional with disjunctive antecedents should be formalised as a conjunction of conditionals, the supporter of this "translation lore" account cannot help but replying that this happens exactly when SDA fails.

A more convincing variant of this approach has been put forward by Humberstone (1978), who introduces a unary connective in the form of an antecedent forming operator ("If A"): thus, the binary conditional connective connects an antecedent whose logical form is "If A" and a standardly formed consequent. The difference between conditionals of the (26) and of the (27) type is that in (26) the antecedent *distributes* over the disjunction, while in (27) it does not; put differently, it has wide scope in the latter and narrow scope in the former. Although one cannot repeat here the same charges that had been levelled against the Loewer account, I observe that also in this proposal sentences having the same surface grammatical form are treated as having different logical forms. This, at the very least, shifts upon its propounder the burden of providing independent reasons—namely, independent from their accounting for the failure of SDA—for such a move.

Other writers prefer to retain SDA and to drop SPE. Nute (1980), for example, sets up systems of conditional logic where substitution of provable equivalents is not unconditionally valid; but "these systems are extremely cumbersome and there still is the extra-formal problem of justifying the particular choice of substitutions which are to be allowed in the logic" (Nute 1984, p. 416). As the last quotation shows, Nute later changed his mind and came to distrust SDA, attributing its prima facie appeal to the action of pragmatic conversational rules.

Finally, some linguists have recently suggested theories based on nonstandard natural language semantics accounts of disjunction or of the conditional (Alonso-Ovalle 2008; Klinedinst 2007). The merits of such proposals remain to be carefully assessed.

In sum, there is still no universally accepted account of the problem of simplification of disjunctive antecedents, as well as of the discrepancy between SDA and SPE.

11.3.2 A New Proposal

My approach proceeds along the following lines. Both SPE and SDA, if understood classically, are partly faulty. SPE is a principle of substitution of provably material equivalents; but material equivalence is no less ambiguous than the material conditional is. That allowing substitution of provably material equivalents is indeed wrong is borne out by line 1 of the above proof of monotonicity, which is a notorious paradox of material implication, used by C.I. Lewis in his proof that A entails $B \vee \neg B$ for A, B whatsoever. Therefore, since the conditional connective which mirrors at the language level the metalinguistic derivability relation in my logic is the arrow, I only endorse a substitution principle for provably *arrow* equivalents. The previous proof, as a result, breaks down at its very beginning. In the formal theory below, moreover, I will show that it is not only this specific proof of the Undesirables which fails—these principles, in fact, are demonstrably independent of the axiom system I shall set up.

So much for SPE. But even SDA needs to be rendered more precise, since it contains ambiguous classical connectives by the score. Once this has been done, the observed tension between cases which seem to disconfirm the principle and cases where no trouble is caused immediately disappears. To see why it is so, return for a while to (27) and (28) above. What kind of "if...then" and "either...or" are at issue here? Well, the conditional is a subjunctive one, hence necessarily a corner conditional; as to the disjunction, it is readily acknowledged that if the antecedent of (27) were to be asserted, it would not possible to do so because of a connection between the disjuncts, but only on the ground that a single disjunct is accepted. We have therefore an extensional disjunction. The logical forms of (27) and (28) are thus, respectively,

(29) $A \sqcup B > C$.
(30) $(A > C) \sqcap (B > C)$.

which are in turn respectively equivalent, via the interdefinability of $>$ and $\curlyvee$ and the De Morgan laws, to

(31) $(\neg A \sqcap \neg B) \curlyvee C$.
(32) $(\neg A \curlyvee C) \sqcap (\neg B \curlyvee C)$.

(32) would then follow from (31) if distribution of superintensional disjunction over extensional conjunction were permissible. But our discussion above suggests that it is not: there is no reason why a disjunction of level $n + 2$ should distribute over a conjunction of level n.

Another case where SDA seems to fail can be accounted for along similar lines. (33) below does not seem to imply (34):

(33) If the butler or the gardener did it, then if it wasn't the butler it was the gardener.
(34) If the butler did it, then if it wasn't the butler it was the gardener, and if the gardener did it, then if it wasn't the butler it was the gardener.

Here, both the disjunction and the conditional are obviously intensional. Arguing as above, it is soon realized that (34) would follow from (33) if distribution of intensional disjunction over intensional conjunction were permissible. But our discussion above suggests that it is not: there is no reason why a disjunction of level n should distribute over a conjunction of the same level.

A careful examination of the intuitively plausible instances of SDA—like (26)—reveals that the conditionals occurring therein are corner conditionals, while the disjunction is an intensional one. Hence, such instances are sound precisely because they call into play the distribution principle whose assumption we advocated at the beginning of this section. Here are other relevant examples drawn from the literature:

(35) If Thorpe or Wilson were to win the next general election, Britain would prosper (Fine 1975).
(36) If New Zealand had either not sent a rugby team to South Africa or had withdrawn from the Montreal games, then Tanzania would have competed (Ellis et al. 1977).

11.4 The Formal Theory

11.4.1 Syntax

In this subsection, I provide the informal intuitions of the preceding sections with a more formal clothing. I will extend the system **HL** of Paoli (2002), corresponding to a Hilbert-style axiomatisation of subexponential linear logic without lattice bounds, by adding superintensional connectives to it. The deductive system thus obtained will be dubbed **CHL**. It corresponds to a very weak conditional logic, a sort of substructural (and paraconsistent) version of the system **CE** of Nute (1980).

Definition 11.1 (The language of CHL). CHL is formulated in a propositional language £ containing a denumerable set Var(£) of variables and the connectives $0, 1$ (nullary), $\sqcap, \sqcup, \otimes, \rightarrow, \curlywedge$ (binary). Defined connectives are:

$$\begin{aligned} \neg p &= p \rightarrow 0 \\ p \rightsquigarrow q &= \neg p \sqcup q \\ p \oplus q &= \neg(\neg p \otimes \neg q) \\ p \curlyvee q &= \neg(\neg p \curlywedge \neg q) \\ p > q &= \neg p \curlyvee q \end{aligned}$$

$A \leftrightarrow B$ will be sometimes used as a metalinguistic abbreviation for the set $\{A \rightarrow B, B \rightarrow A\}$. We follow the convention according to which unary connectives bind stronger than binary ones, and $\sqcap, \sqcup, \otimes, \oplus, \curlywedge, \curlyvee$ bind stronger than $\rightsquigarrow, \rightarrow, >$. The class of formulae of £ (Fm(£)) is defined as usual. The $\langle \curlywedge \rangle$-free fragment of such language will be denoted by £$'$.

Definition 11.2 (Postulates of CHL). Here are the postulates of **CHL**:
Axioms for nullary and unary connectives:
0.1 $\neg\neg A \rightarrow A$
0.2 $(\neg A \rightarrow \neg B) \rightarrow (B \rightarrow A)$
0.3 1
0.4 $1 \rightarrow (A \rightarrow A)$
0.5 $0 \leftrightarrow \neg 1$

First level axioms:
1.1 $A \to (\neg A \rightsquigarrow B)$
1.2 $B \to (A \rightsquigarrow B)$
1.3 $\neg((A \to C) \rightsquigarrow \neg(B \to C)) \to ((\neg A \rightsquigarrow B) \to C)$
Second level axioms:
2.1 $A \to A$
2.2 $(A \to B) \to ((B \to C) \to (A \to C))$
2.3 $(A \to (B \to C)) \to (B \to (A \to C))$
2.4 $A \to (B \to A \otimes B)$
2.5 $(A \otimes B \to C) \leftrightarrow (A \to (B \to C))$
Third level axioms
3.1 $(A > B \otimes C) \leftrightarrow (A > B) \otimes (A > C)$
3.2 $(A \oplus B > C) \leftrightarrow (A > C) \otimes (B > C)$
Rules
$R1$ $A, A \to B \vdash B$
$R2$ $A, B \vdash A \sqcap B$
$R3$ $A \to B, B \to A \vdash (A > C) \to (B > C)$
$R4$ $A \to B, B \to A \vdash (C > A) \to (C > B)$

Observe that 3.2 encodes the nonproblematic version of SDA, while R3 and R4 formalise SPE—or, rather, substitution of provably *arrow* equivalents. The notions of proof (from assumptions) and derivability in **CHL** are defined as usual; by $\Gamma \vdash_{\mathbf{CHL}} A$ I will mean that the formula A is derivable in the calculus **CHL** from the assumptions in Γ. The relation $\vdash_{\mathbf{CHL}}$ is a finitary and substitution-invariant consequence relation. Therefore, we can abstractly identify **CHL** with the deductive system $\langle \mathrm{Fm}(\pounds), \vdash_{\mathbf{CHL}} \rangle$, something I will feel free to do hereafter.

Here is a list of additional postulates, named after the traditional labels they receive in the literature, from which one could draw to extend the third level of **CHL**:

RCK $\sqcap$: $A_1 \sqcap \ldots \sqcap A_n \to B \vdash (C > A_1) \sqcap \ldots \sqcap (C > A_n) \to (C > B)$ $(n \geq 0)$
RCK $\otimes$: $A_1 \otimes \ldots \otimes A_n \to B \vdash (C > A_1) \otimes \ldots \otimes (C > A_n) \to (C > B)$ $(n \geq 0)$
ID: $A > A$
CA: $(A > B \sqcap C) \leftrightarrow (A > B) \sqcap (A > C)$
MOD: $A \curlyvee A \to (B > A)$
CSO: $(A > B) \otimes (B > A) \to ((A > C) \to (B > C))$
CV: $(A > B) \sqcap (A \curlywedge B) \to (A \sqcap C > B)$
CS: $A \sqcap B \to (A > B)$
CEM: $(A > \neg B) \oplus (A > B)$
MP: $(A > B) \to (A \to B)$
TR: $(A > B) \otimes (B > C) \to (A > C)$
CONTR: $(A > \neg B) \to (B > \neg A)$
MON: $(A > B) \to (A \sqcap C > B)$

11.4.2 Algebra

The aim of this subsection is identifying a class of algebras which functions as an equivalent algebraic semantics for **CHL**. Let me remark that this semantics is meant to be, in the terminology of Copeland (1983), a typically *applied* semantics—as opposed to a pure, explicative semantics which yields deep insights on the logic which is being interpreted. Its only aim is showing that the above logic is consistent and that SDA can happily live therein together with SPE, while keeping the Undesirables from the door. To begin with, let us recall a fundamental concept from the algebraic semantics of substructural logics (see e.g. Galatos et al. 2007):

Definition 11.3. An $\mathbf{FL}_e$*-algebra* (also called *pointed commutative residuated lattice*) is an algebra

$$\mathbf{L} = \langle L, \otimes, \rightarrow, \sqcap, \sqcup, 0, 1 \rangle,$$

of type £′,[16] such that:

- $\langle L, \sqcap, \sqcup \rangle$ is a lattice;
- $\langle L, \otimes, 1 \rangle$ is an Abelian monoid;
- For every $a, b, c \in L$, we have that $a \otimes b \leq c$ iff $a \leq b \rightarrow c$, where $\leq$ denotes the induced order of the lattice reduct $\langle L, \sqcap, \sqcup \rangle$.

An $\mathbf{FL}_e$-algebra is called *involutive* iff it satisfies the identity

$$(p \rightarrow 0) \rightarrow 0 \approx p$$

The class of (involutive) $\mathbf{FL}_e$-algebras is a variety in its type: the residuation quasi-equations can be dispensed with in favour of a finite set of equations (Galatos et al. 2007). We now want to define an expansion of involutive $\mathbf{FL}_e$-algebras in the language £, in order to provide a suitable interpretation for the superintensional connectives.

Definition 11.4. A *ringoidal involutive* $\mathbf{FL}_e$*-algebra* is an algebra

$$\mathbf{L} = \langle L, \otimes, \rightarrow, \sqcap, \sqcup, \curlywedge, 0, 1 \rangle,$$

of type £, such that:

- $\langle L, \otimes, \rightarrow, \sqcap, \sqcup, 0, 1 \rangle$ is an involutive $\mathbf{FL}_e$-algebra;
- The term reduct $\langle L, \curlywedge, \oplus \rangle$ is a *ringoid*, i.e., for every $a, b, c \in L$,

[16] As it is customary to do in the tradition of abstract algebraic logic, I will not distinguish between logical languages and algebraic similarity types; therefore, I will use the same symbols for logical connectives and the corresponding operation symbols. Fm(£) will denote both the set of all formulas of £ (seen as a logical language) and the set of all terms of £ (seen as an algebraic similarity type), and likewise for Fm(£′).

$$a \curlywedge (b \oplus c) = (a \curlywedge b) \oplus (a \curlywedge c);$$
$$(b \oplus c) \curlywedge a = (b \curlywedge a) \oplus (c \curlywedge a).$$

While the variety of involutive $\mathbf{FL}_e$-algebras has been investigated in great detail (see e.g. Galatos et al. 2007 or Paoli 2002, where it is actually a term equivalent variant which is under scrutiny), the variety of ringoidal involutive $\mathbf{FL}_e$-algebras—hereafter denoted by $\mathbb{R}$—is new. Our first duty, therefore, is showing that it is not empty by providing appropriate examples. Here are some.

Example 11.1. Any lattice-ordered ring

$$\mathbf{R} = \langle R, +, \cdot, \sqcap, \sqcup, -, \mathbf{0} \rangle$$

gives rise to a ringoidal involutive $\mathbf{FL}_e$-algebra by taking, for any $a, b \in A$,

$$a \otimes b = a \oplus b = a + b$$
$$a \rightarrow b = b - a$$
$$a \curlywedge b = a \curlyvee b = a \cdot b$$
$$0 = 1 = \mathbf{0}.$$

The previous example does not yield, of course, any nontrivial finite algebra. The next one, due to Pierluigi Minari (and cited in Casari 1997), does. Remark that in this case the $\mathbf{FL}_e$-algebra reduct is actually a $\mathbf{FL}_{ew}$-algebra.

Example 11.2. It is possible to extend the three-element MV chain by ring-theoretical operations in such a way as to get the algebra $\mathbf{MV}_3^r$ with the following tables:

$\otimes$	0	$\frac{1}{2}$	1
0	0	0	0
$\frac{1}{2}$	0	0	$\frac{1}{2}$
1	0	$\frac{1}{2}$	1

$\sqcap$	0	$\frac{1}{2}$	1
0	0	0	0
$\frac{1}{2}$	0	$\frac{1}{2}$	$\frac{1}{2}$
1	0	$\frac{1}{2}$	1

$\sqcup$	0	$\frac{1}{2}$	1
0	0	$\frac{1}{2}$	1
$\frac{1}{2}$	$\frac{1}{2}$	$\frac{1}{2}$	1
1	1	1	1

$\curlywedge$	0	$\frac{1}{2}$	1
0	0	0	0
$\frac{1}{2}$	0	$\frac{1}{2}$	1
1	0	1	1

$\rightarrow$	0	$\frac{1}{2}$	1
0	1	1	1
$\frac{1}{2}$	$\frac{1}{2}$	1	1
1	0	$\frac{1}{2}$	1

11.4.3 Semantics

We now establish the required bridge between **CHL** and ringoidal involutive $\mathbf{FL}_e$-algebras:

Theorem 11.1. *The deductive system* **CHL** *is strongly and finitely algebraisable with equivalence formulas* $\{p \to q, q \to p\}$ *and defining equation* $\{1 \sqcap p \approx 1\}$, *and its equivalent algebraic semantics is the variety* $\mathbb{R}$ *of ringoidal involutive* $\mathbf{FL}_e$-*algebras.*

Proof. Me must show that:

1. $\vdash_{\mathbf{CHL}}$ can be faithfully interpreted into the equational consequence relation $\vDash_{\mathbb{R}}$ of $\mathbb{R}$, i.e. for any $\Gamma \subseteq \mathrm{Fm}(\pounds)$ and any $A \in \mathrm{Fm}(\pounds)$,

$$\Gamma \vdash_{\mathbf{CHL}} A \text{ iff } \{1 \sqcap B \approx 1 : B \in \Gamma\} \vDash_{\mathbb{R}} 1 \sqcap A \approx 1;$$

2. $\vDash_{\mathbb{R}}$ can be faithfully interpreted into $\vdash_{\mathbf{CHL}}$, i.e. for any set of £-equations $\Gamma \approx \Delta$ and any £-equation $A \approx B$,

$$\Gamma \approx \Delta \vDash_{\mathbb{R}} A \approx B \text{ iff } \{C \to D, D \to C : C \in \Gamma, D \in \Delta\} \vdash_{\mathbf{CHL}} \{A \to B, B \to A\};$$

3. The two interpretations are mutually inverse, i.e.

$$p \vdash_{\mathbf{CHL}} \{1 \sqcap p \to 1, 1 \to 1 \sqcap p\}$$
$$\{1 \sqcap p \to 1, 1 \to 1 \sqcap p\} \vdash_{\mathbf{CHL}} p$$
$$p \approx q \vDash_{\mathbb{R}} \{1 \sqcap (p \to q) \approx 1, 1 \sqcap (q \to p) \approx 1\}$$
$$\{1 \sqcap (p \to q) \approx 1, 1 \sqcap (q \to p) \approx 1\} \vDash_{\mathbb{R}} p \approx q$$

By Proposition 7.2 in Jansana (201+), it is enough to establish item 1 and the last two lines of 3. As to the first item, this is the content of a standard strong soundness and completeness theorem, and so it can be established as usual—via an inductive argument on the length of the derivations for the soundness part, and a Lindenbaum algebra argument for the completeness part. The second half of item 3. can be proved as follows. Suppose that $\mathbf{A} \in \mathbb{R}$, that $a, b \in A$ and that $a = b$. Then $a \leq b$, whence $1 \leq a \to b$, and $b \leq a$, whence $1 \leq b \to a$. Conversely, if $1 \leq a \to b, b \to a$, then $a \leq b$ and $b \leq a$, whereby $a = b$. □

The main point of the previous semantics was to show the independence of the principles of transitivity, monotonicity and contraposition, in order to prove that SPE and SDE, if appropriately disambiguated, can live together without forcing us to take the Undesirables aboard. The next proposition does the trick.

Proposition 11.1. *The principles ID, MOD, CS, CEM, MP, TR, CONTR, MON are all independent of* **CHL**.

Proof. We provide falsifying models for some instances of the mentioned principles, whence the result follows by Theorem 11.1. Consider the ringoidal involutive $\mathbf{FL}_e$-algebra $\mathbf{MV}_3^r$ of Example 11.2. A counterexample to ID is given by

$$p > p^{\mathbf{MV}_3^r}\left(\frac{1}{2}\right) = \frac{1}{2}$$

Counterexamples to MOD, CS, CEM are respectively given by

$$p \curlyvee p \to (q > p)^{\mathbf{MV}_3^r}\left(\frac{1}{2}, 1\right) = \frac{1}{2}$$

$$p \sqcap q \to (p > q)^{\mathbf{MV}_3^r}\left(1, \frac{1}{2}\right) = \frac{1}{2}$$

$$(p > \neg q) \oplus (p > q)^{\mathbf{MV}_3^r}\left(1, \frac{1}{2}\right) = 0$$

CONTR is falsified in any noncommutative lattice-ordered ring, as $(p > \neg q) \to (q > \neg p)$ is equivalent to $(\neg p \curlyvee \neg q) \to (\neg q \curlyvee \neg p)$. Finally, consider the ℓ-ring $\mathbf{Z}$ of the integers. Counterexamples to MP, TR, MON are respectively given by

$$(p > q) \to (p \to q)^{\mathbf{Z}}(+3, -1) = -1$$

$$(p > q) \otimes (q > r) \to (p > r)^{\mathbf{Z}}(+3, -1, +2) = -11$$

$$(p > q) \to (p \sqcap r > q)^{\mathbf{Z}}(+3, -1, +2) = -1$$

Acknowledgements A version of the present paper was presented at the 4th World Congress on Paraconsistency (Melbourne, Australia, July 2008). I thank the audience of my talk for their stimulating questions and comments. I also thank Charlie Donauhe and an anonymous referee for their precious suggestions.

References

Adams, V.M. 1970. Subjunctive and indicative conditionals. *Foundations of Language* 6: 89–94.

Allo, P. 2011. On when a disjunction is informative: Ambiguous connectives and pluralism. In *The realism antirealism debate in the age of alternative logics*, ed. S. Rahman et al., 1–23. Springer: Berlin.

Alonso-Ovalle, L. 2008. Alternatives in the disjunctive antecedents problem. In *Proceedings of the 26 west coast conference in formal linguistics*, ed. C.B. Chang and H.J. Haynie, 42–50. Somerville: Cascadilla Proceedings Project.

Anderson, A.R., and N.D. Belnap. 1975. *Entailment*, vol I. Princeton: Princeton University Press.

Bennett, J. 1988. Farewell to the phlogiston theory of conditionals. *Mind* 97: 509–527.

Bennett, J. 1995. Classifying conditionals: The traditional way is right. *Mind* 104: 331–354.

Bennett, J. 2003. *A philosophical guide to conditionals*. Oxford: Oxford University Press.

Bryant, C. 1981. Conditional murderers. *Analysis* 41: 209–215.

Casari, E. 1997. Conjoining and disjoining on different levels. In *Logic and scientific methods*, ed. M.L. Dalla Chiara et al., 261–288. Dordrecht: Kluwer.

Chellas, B. 1975. Basic conditional logic. *Journal of Philosophical Logic* 4: 133–153.

Copeland, B.J. 1983. Pure semantics and applied semantics. *Topoi* 2: 197–204.

Dowing, P. 1975. Conditionals, impossibilities and material implication. *Analysis* 35: 85–88.

Dudman, V.H. 1983. Tense and time in English verb clusters of the primary pattern. *Australian Journal of Linguistics* 3: 25–44.
Dudman, V.H. 1984. Parsing 'If' sentences. *Analysis* 44: 145–153.
Dudman, V.H. 1989. Vive la revolution. *Mind* 98: 591–603.
Edgington, D. 1995. On conditionals. *Mind* 104: 235–329.
Ellis, B., F. Jackson, and R. Pargetter. 1977. An objection to possible worlds semantics for counterfactual logics. *Journal of Philosophical Logic* 6: 355–357.
Fine, K. 1975. Review of Lewis (1973). *Mind* 84: 451–458.
Gibbard, A. 1981. Two recent theories of conditionals. In *Ifs*, ed. W.L. Harper et al., 211–247. Dordrecht: Reidel.
Galatos, N., P. Jipsen, T. Kowalski, and H. Ono. 2007. *Residuated lattices: An algebraic glimpse on substructural logics*. Amsterdam: Elsevier.
Humberstone, L. 1978. Two merits of the circumstantial operator language for conditional logic. *Australasian Journal of Philosophy* 56: 21–24.
Humberstone, L. 2011. *The connectives*. Cambridge: MIT Press.
Hunter, G. 1993. The meaning of 'if' in conditional propositions. *Philosophical Quarterly* 43: 279–297.
Jackson, F. 1979. On assertion and indicative conditionals. *Philosophical Review* 88: 565–589.
Jackson, F. 1987. *Conditionals*. Oxford: Blackwell.
Jansana, R. 201+. *A course in abstract algebraic logic*. e-notes available from the University of Barcelona website (Department of Philosophy).
Klinedinst, N. 2007. *Plurality and possibility*. Ph.D. thesis, UCLA.
Lewis, D. 1973. *Counterfactuals*. Oxford: Blackwell.
Lewis, D. 1976. Probabilities of conditionals and conditional probabilities. *Philosophical Review* 85: 297–315.
Loewer, B. 1976. Counterfactuals with disjunctive antecedents. *Journal of Philosophy* 73: 531–536.
Lowe, E.J. 1995. The truth about counterfactuals. *Philosophical Quarterly* 43: 41–59.
Mares, E. 2004. *Relevant logic*. Cambridge: Cambridge University Press.
Meyer, R.K. 1966. *Topics in modal and many-valued logics*. Ph.D. thesis, University of Pittsburgh.
Meyer, R.K. 1986. *A farewell to entailment*. RSSS research paper, ANU, Canberra.
Mares, E., and A. Fuhrmann. 1995. A relevant theory of conditionals. *Journal of Philosophical Logic* 24: 645–665.
Nute, D. 1980. *Topics in conditional logic*. Dordrecht: Reidel.
Nute, D. 1984. Conditional logic. In *Handbook of Philosophical Logic*, vol. II, ed. D. Gabbay and F. Guenthner, 387–439. Dordrecht: Reidel.
Paoli, F. 2002. *Substructural logics: A primer*. Dordrecht: Kluwer.
Paoli, F. 2005. The ambiguity of quantifiers. *Philosophical Studies* 124: 313–330.
Paoli, F. 2007. Implicational paradoxes and the meaning of logical constants. *Australasian Journal of Philosophy* 85: 553–579.
Priest, G. 2001. *An introduction to non-classical logics*. Cambridge: Cambridge University Press.
Read, S. 1988. *Relevant logic*. Oxford: Blackwell.
Restall, G. 2000. *An introduction to substructural logics*. London: Routledge.
Routley, R., V. Plumwood, R.K. Meyer, and R.T. Brady. 1982. *Relevant logics and their rivals*. Atascadero: Ridgeview.
Sainsbury, M. 1991. *Logical forms*. Oxford: Blackwell.
Sanford, D.H. 1989. *If P then Q*. London: Routledge.
Smiley, T. 1984. Hunter on conditionals. *Proceedings of the Aristotelian Society* 84: 113–122.
Stalnaker, R. 1968. A theory of conditionals. In *Studies in logical theory*, ed. N. Rescher, 98–112. Oxford: Blackwell.
Stalnaker, R. 1975. Indicative conditionals. *Philosophia* 5: 269–286.
Weatherson, B. 2001. Indicatives and subjunctives. *Philosophical Quarterly* 51: 200–216.
Weatherson, B. 2009. Conditionals and indexical relativism. *Synthese* 166(2): 333–357.

Part II
Applications

Chapter 12
An Approach to Human-Level Commonsense Reasoning

Michael L. Anderson, Walid Gomaa, John Grant, and Don Perlis

12.1 Introduction

Humans reason—of that there is no doubt. But what sort(s) of reasoning do we do? Clearly there are some among us who do mathematical reasoning, and do it well. And, it has been argued that all reasoning is an attempt to reach the ideal model of mathematics, i.e., to arrive at true conclusions (from given assumptions).

Perhaps for this reason, efforts to examine human reasoning have tended to be in formal logical dress, mimicking the rigor of mathematics, e.g., Aristotle, Leibniz,

M.L. Anderson (✉)
Institute for Advanced Computer Studies, University of Maryland, College Park, USA

Department of Psychology, Franklin and Marshall College, Lancaster, CA, USA
e-mail: michael.anderson@fandm.edu

W. Gomaa
Department of Computer and Systems Engineering, Alexandria University, Alexandria, Egypt

Department of Computer Science and Engineering, Egypt-Japan University of Science and Technology, New Borg El-Arab City, Alexandria-Egypt
e-mail: walid.gomaa@gmail.com

J. Grant
Institute for Advanced Computer Studies, University of Maryland, College Park, USA

Department of Computer Science, University of Maryland, College Park, USA

Department of Mathematics, Towson University, Towson, USA
e-mail: grant@cs.umd.edu

D. Perlis
Institute for Advanced Computer Studies, University of Maryland, College Park, USA

Department of Computer Science, University of Maryland, College Park, USA
e-mail: donperlis@gmail.com

K. Tanaka et al. (eds.), *Paraconsistency: Logic and Applications*, Logic, Epistemology, and the Unity of Science 26, DOI 10.1007/978-94-007-4438-7_12,

Boole, and Frege.[1] But there is evidence—recounted below—that human reasoning is not always aimed specifically at true conclusions.

The field of artificial intelligence also followed in this math-based mode, at least initially, despite many doubters. One such doubter, Marvin Minsky, pointed out an embarrassingly obvious difficulty: much of human reasoning tends to be non-monotonic. What we conclude depends not only on what we believe but also on what we do not believe. That is, we draw conclusions on the basis of not having certain beliefs.

Minsky's famous example is essentially this: Told Tweety is a bird, we may reasonably come to believe that Tweety can fly. But if we had been told Tweety is a bird and furthermore is a penguin, we would not have drawn that conclusion. That is, the original conclusion (Tweety can fly) was performed by an inferential process that can be halted by the presence of additional information. This completely flies (pardon the pun) in the face of mathematical reasoning, where only ironclad guaranteed true conclusions are of interest. Minsky's example highlights the fact that in everyday life very often we are content (indeed may have no other choice except) to seek a very highly plausible conclusion. The real world has far too many parameters for us to be able to have strict data on them all, so we end up reasoning as if we had beliefs such as "most birds fly", and so on. And this relaxing of the truth-demand into a plausibility demand opens the door to retractions in the face of further evidence.

Well, this did not stop the logic-based AI-ers from using logic. It simply encouraged them to find better logics, so-called non-monotonic logics, where an enlarged set of assumptions might lead to a different set of conclusions that happens to be missing one of the original ones. A number of proposals for such logics soon surfaced, and had high degrees of success, most notably those of McCarthy and of Reiter.[2]

Nevertheless, smart artificial systems did not spring up. It seems that being non-monotonic is not enough. In fact, it was pointed out a number of times that these new logics tended to be designed for the purpose of specifying the kinds of conclusions a smart reasoner ought to come to, but were not in general useful to system designers. The logics did not lend themselves to specifying ways for a system to actually arrive at these conclusions. The biggest roadblock was that the logics in question—like those before them—provided a characterisation of the set of all theorems that would follow from given axioms (or beliefs). This set typically is infinite, and the logics give little or no indication as to how and in what order these theorems are to be proven. Thus a system designer is stuck with the task of building an inference *engine* that produces, little by little, the theorems specified by one of the formal logics.

Yet even if one solves these problems, another surfaces. Very many of the theorems are of no use whatsoever to a given system's activity. Logics in general

---

[1]Note that we do *not* make the converse claim—that formal logic tends to aim at modelling reasoning processes (i.e., psychologism).

[2]While details are complicated, most such approaches aim at the inference of special additional "normal" or "typical" formulas—such as Flies(x) from Bird(x)—when *not* ruled out by axioms.

tend to have promiscuous rules of inference, concluding sentence after sentence without regard for their usefulness. This problem of *relevance* had long been recognised. But possibly an even worse aspect is that time (a *lot* of time) is being used up, both on these irrelevant results and on the enormous (even infinite) set of all theorems. Somehow a real-world reasoner must exercise some control over its reasoning so that time is not wasted without regard to real-world exigencies.

This is a major difficulty because time is highly significant in almost all endeavours, and because formal inference crunches on and on forever, oblivious to time. Clearly, on-board logic[3] must be able to take into account the fact that time passes as reasoning is going on, and what is important at one moment might not be so at another. In short, a reasoner ought to realise that "now" changes out from under it; it never stands still. So reasoning about time is slippery. As an example, the following makes perfect sense and is essential for effective on-board logic, yet is absurd from the point of view of spec logic:

$$\text{From } Now(t) \text{ infer } \neg Now(t).$$

That is, as soon as the time is known to be t, it no longer is t. Given a small unit or grain of time (e.g., a second, or a millisecond), the gist of the above can be approximated by this rule:

$$\frac{t : Now(t)}{t+1 : Now(t+1)}$$

The above "clock" rule is the essential feature of so-called active logics, a species of on-board logic. While it might not seem particularly revolutionary, it has major consequences. Three of the most important are as follows:

1. Reasoning can keep up with deadlines. Given a noon lunch appointment, one reasons at 11:30 a.m. that at 11:45 one should start walking to the restaurant. Then at 11:44 one reasons that it is time to stop reading the newspaper and put on one's coat. And by the time one's coat is on, one reasons that it is time to walk (because by the time that reasoning has been done it will be close enough to 11:45). Trivial enough, but impossible to do with spec logics. But the clock rule makes deadline sensitive reasoning possible.[4]

   Here is a much-simplified example of the above form of reasoning in active logic, involving a deadline. We have annotated each time-step in the reasoning

---

[3]Let us call a logic that is used by a real-world reasoning agent (human or otherwise) as it goes about its business an "on-board" logic (as opposed to a specification—"spec"—logic that characterises limiting behaviours such as the set of all sentences that (eventually) can be proven). Thus we are using the term "logic" quite broadly, to include any systematic method for drawing conclusions from premises.

[4]So-called tense logics and temporal logics express propositions about past, present, and future, but the present is not represented as evolving: *Now* does not change as theorems are proved, in contrast with the above Clock Rule. In other words, tense logics are also spec logics, rather than on-board logics.

with the actual time on the left; via the clock rule, the logic has effective access to this information as well, assuming it is started off with the correct time. In each step below we have placed the agent's relevant beliefs at that time, with any new ones listed first; among these is always the current time, $Now(t)$. And the last step shown has the newly inferred belief "Walk" as well. Note that beliefs of the form Now(t) are not inherited to the next step (see above clock rule) but that in general other beliefs—such as that one should start walking at 11:45—tend to be retained (precisely which beliefs are to be retained is a subtle issue; in particular cases there are useful heuristics but no single general principle). Note that, in general, a belief at one time is carried forward (remains a belief) at later times—for instance

$$Now(11:45) \rightarrow Walk$$

remains a belief indefinitely, whereas some special beliefs, such as knowledge of the current time, are dropped at the next step and replaced by a new belief (in this case due to the above clock rule, because time is always changing; but something similar can occur whenever there is reason to no longer hold a belief). Here is the example:

$$[11:30] : \mathrm{Now}(11:30), \mathrm{Now}(11:45) \rightarrow \mathrm{Walk}$$

$$[11:30:01] : \mathrm{Now}(11:30:01), \mathrm{Now}(11:45) \rightarrow \mathrm{Walk}$$

$$\ldots$$

$$[11:44:59] : \mathrm{Now}(11:44:59), \mathrm{Now}(11:45) \rightarrow \mathrm{Walk}$$

$$[11:45] : \mathrm{Now}(11:45), \mathrm{Now}(11:45) \rightarrow \mathrm{Walk}$$

$$[11:45:01] : \mathrm{Now}(11:45:01), \mathrm{Walk}, \mathrm{Now}(11:45) \rightarrow \mathrm{Walk}$$

At time 11:45 above, *modus ponens* goes to work on the then-current beliefs, and by 11:45:01 has inferred Walk. One simplifying assumption here is that it takes one "step" of time to apply an inference rule. Note that the belief that at 11:45 the Walk action should begin is still there among the beliefs, even though it is not likely to be useful; this can be "pruned" by a cleanup rule that drops conditionals of the form

$$Now(t) \rightarrow X$$

after time t has passed; after all, $Now(t)$ will never be true again after that time so the conditional will always remain true but never useful in concluding anything except at time t.

2. Inconsistency is a disaster for spec logics. They simply accept all sentences as theorems, making them useless. Paraconsistent logics adopt various means to avoid this "explosion" of consequences. But what is really needed is a paraconsistent logic with the ability not only to side-step a contradiction, but to notice it and consider what to do about it, possibly altering its status as a belief. After all, it might be an important clue to something amiss. Again, time comes to the rescue, providing a temporal "stratification" of theorems according to when they are proven, so that the time at which one sentence is proven (believed) allows

inferences at the next time-step to comment on the previous result, such as that it is in contradiction with other beliefs and should be abandoned:

$$\frac{t : P, \neg P}{t + 1 : Contra(P, t)}$$

However, active logic does not discover all inconsistencies; that is in general not computable in finite time. It simply scans the current knowledge base for an occurrence of a wff and its negation. If deeper inconsistencies remain, so be it: just as a human may unknowingly entertain contradictory beliefs, so with active logic. Only when a contradiction is noticed—such as in the form of a direct contradiction between a formula and its negation, is an agent (human or otherwise) in a position to do anything about it.

Also, once P and $\neg P$ are noticed and removed from the KB, there is no general method for adjudicating between them. In general, the reasoning agent may have to be content with uncertainty. In particular cases, there are heuristics that may be useful, such as deciding in favour of P if the evidence that produced it is more compelling than that for $\neg P$. That of course requires additional machinery.

3. Inconsistency is just one example of a situation needing some sort of change (e.g., distrust various sentences). But more generally, any manner of change might be called for in a given situation. Even a change in language may be needed, if for example that is a plausible way to resolve an inconsistency. For instance, given the beliefs "John is reading," and "John is wagging his tail," one might consider that the word John is being used to name two different entities. This might then prompt the introduction of two new names, John^1 and John^2. But to do this, the reasoning must be able to have breathing room, time to make such changes before the inference engine rushes ahead to all the infinitely many theorems that would arise from the two earlier beliefs that together are implausible: that a dog is reading.

So where the does this leave us? Are we closer to commonsense reasoning? We think so. One feature that we have identified, as a key to commonsense reasoning, is the ability to notice—and respond usefully to—anomalies. And it turns out that anomalies can easily be cast in the form of mismatches between expectations and observations, i.e., a contradiction—or close enough so that the Contra and Distrust rules can go into action. An evolving-time logic such as active logic provides just this capability.

12.2 Human Paraconsistency

Humans are very good at dealing with—reasoning and acting in the face of—uncertainty, change and even contradictions. In contrast, AI systems, especially those implemented with logic-based reasoning mechanisms, are notoriously bad at coping with these pervasive features of real environments. One widely-accepted conclusion from these observations has been that humans do not use logic-based mechanisms to implement their core reasoning abilities. And, indeed, there is a great deal of empirical evidence that seems to point in this direction. Humans often fail to achieve the ideal of valid logical deduction, and in many contexts we seem to utilise representational formats more suited to non- or extra-logical manipulations.

For instance, a large body of research has established that people are less likely to judge instances of *modus tollens* to be valid than instances of *modus ponens*.[5] Moreover, people are subject to some characteristic logical fallacies, such as the converse error (Example 12.1) and the inverse error (Example 12.2):

Example 12.1.
If the horses went to the watering hole, we would see their tracks.
We see their tracks.
∴ The horses went to the watering hole.

Example 12.2.
If the horses went to the watering hole, we would see their tracks.
The horses did not go to the watering hole.
∴ We will not see their tracks.

Interestingly, despite trouble with *modus tollens* in general, people have little trouble with that logical form in the following sort of case:

Example 12.3.
If the horses went to the watering hole, we would see their tracks.
We do not see their tracks.
∴ The horses did not go to the watering hole.

This pattern of results has suggested to many that what looks like logical reasoning is actually *causal* reasoning. Rather than building formal logical models from these sentences and judging the validity of the argument, we are in fact building causal models of the situations depicted, and judging the likelihood of the outcome. And, indeed, by those standards, arguments Examples 12.1 and 12.2 represent fairly plausible inferences.

Similarly, results from the Wason card selection task in Johnson-Laird and Wason (1970) apparently point to the use of inference mechanisms that are not logic-based. In this task, participants are shown four cards, e.g. $(A, K, 2, 7)$, given a rule of the form "If a card has a vowel on one side, it has an even number on the other" $(p \rightarrow q)$, and asked to choose the cards they need to turn over to check the validity of the rule. The majority of participants choose A and 2 (i.e. p and q), even though the logically correct choice is A and 7 (p and $\neg q$). To cite just two examples of how this evidence has been interpreted, Oaksford and Chater (1994) take it to indicate that decision making is instead driven by considerations of information yield (according to their analysis, turning over A and 2 yields more information about

[5]For an overview of the various findings reported in this paragraph, see Evans (1982).

the rule than does turning over any other two cards), while Cosmides (1989)—after noticing that participants make the logically correct choices when the abstract rule is replaced with one governing social conduct, e.g., "If you drink beer you must be over 21"—argues for the existence of mechanisms specialised for reasoning about social exchanges.

Such results—and there are many more like them—are of course deeply interesting, and assimilating them will be crucial to articulating a complete model of the mechanisms supporting human reasoning. And while we do not wish to question the existence and importance of the many different non-logical mechanisms that have been proposed to account for the vast amounts of available data on human reasoning and decision-making—including causal and other mental models (Gentner and Stevens 1983; Johnson-Laird 1983), Bayesian inference (Oaksford and Chater 2007), social exchange modules (Cosmides 1989), expected utility curves (Kahneman and Tversky 1979), frequency sensitivity (Gigerenzer 1994), and expected information gain (Oaksford and Chater 1994)—we would like to suggest that there is nevertheless room for continued empirical attention to human *logical* reasoning, for at least the following reasons.

First, and most obvious, from the fact that humans possess some inferential mechanisms that are not logic-based, it does not follow that we do not have and use some native logic-based reasoning abilities. It might be noted in support of this thought that people's vulnerability to fallacies like those presented in arguments Examples 12.1 and 12.2 largely disappears when the propositions involved are *not* causally related as they are in the examples. This suggests that causal-model-based mechanisms may be *interfering* with logic-based ones in circumstances in which both potentially apply.

Second, from the fact that logic-based AI is brittle while humans are not, it does not follow that human flexibility is necessarily or entirely the result of non-logical capacities. It may be that human logic takes a special form, or has certain features, or interacts with non-logical capacities in particular ways, and these attributes of human logic have simply not been captured in prevailing logic-based AI systems.

Third, even if it is proven that humans have no natural, native, logic-based inference mechanisms, the fact that humans can nevertheless reason logically would mean that our non- and extra- logical capacities can be harnessed to this end. Thus, investigating human logical reasoning, particularly in the face of contradiction and change, may help us understand what is special about our implementation of logic such that it supports the observed flexibility of human reasoning.

Fourth and finally, given the significant advantages of logic-based implementations in AI—including the fact that rule-based systems are relatively easy for humans to understand, and therefore to trust, and that changing their behaviour is as simple and quick as changing the rules that govern it (something that is not the case in systems that require extensive (re-)training)—it behooves us to consider how logic-based systems can be made more robust in the face of various perturbations. Human perturbation-tolerance can be a source of ideas and inspiration in this task.

Unfortunately, perhaps because human flexibility has been largely taken as an indication of non- or extra-logical mechanisms at work, there has been relatively little empirical work on human performance in the face of contradictory or changing information in specifically logical contexts. There has nevertheless been *some* work along these lines, enough to draw some preliminary conclusions that can be used to guide the development of more robust logic-based systems. We will first review the results, and then discuss what we take the implications to be.

In one interesting set of experiments Dean Sharpe and Lacroix (1999) asked adults and children how they resolve assertions of the form $p\&\neg p$, such as the response "yes and no" to the question "Was the movie good?" In this work, 24 adults and 48 children (ranging in ages from 3 to 8) were told a story about two characters having dinner. At the end of the meal, one asks the other, "Did you like your supper?", to which the other character replies "Yes and no. I liked my supper and I didn't like it." Participants were asked to explain what the second character meant by the response.

The vast majority of participants (around 70%, including some children as young as 4), dealt with the contradiction by reinterpreting the statement p (I liked my supper) to take advantage of the internal structure of the object "supper". That is, they took the character to be asserting that he liked one part of the supper, but didn't like a different part. In addition, two other strategies were employed. Two adults and nine children reinterpreted the meaning of p by drawing attention to the applicability of the predicate "like". These participants said things like: the supper was average, so he neither liked it nor didn't like it. In addition, four of the adults, but only one of the children simply denied p, explaining that he didn't like the supper, but was trying to be polite. There were no other resolution strategies employed. The authors summarise their main findings by noting that "adults and even preschoolers possess interpretive structures—particularly object structure—that are non-classical in the sense that they can be used to resolve apparent contradictions" (Sharpe and Lacroix 1999, p. 489).

A different set of experiments revealed some similar tendencies. Renee Elio (1997, 1998) asked what strategies people use to resolve logical contradictions of the form $\{p, p \rightarrow q, \neg q\}$. Participants were given premises like:

Example 12.4. A If the ignition key is turned the car will start.
B The ignition key was turned.
They were then told:

C The car did not start

and asked: assuming that C is true, which statement A or B do you think it is more plausible to disbelieve? What revision would you make to that statement to make it consistent with the other premises?

Overall, participants were more inclined (around 60% of the time) to doubt $p \rightarrow q$ than they were to doubt p,[6] and when they did so they usually (around 63% of the time) made the statement consistent by re-interpreting the meaning of p, typically by adding conditions. Thus, participants might revise the statement to read

[6]Although this preference was reversed when the initial statement was a definition such as: if a mineral is a diamond then it is made of compressed carbon.

"If the ignition is turned and the battery is not dead, then ..." Most of the remaining revisions (around 30%) involved reinterpretations of q, with the effect of turning the rule into a default, e.g. "If the ignition key is turned the car will *usually* start."

This last finding is related to an interesting discovery by Byrne (1989), that reasoners seem to tacitly treat *many* rules as defaults, and thus can be made to suppress valid inferences under certain conditions. In her studies she found that while participants were happy to accept as valid inferences like:

Example 12.5.
If she has an essay to write then she will study late in the library.
She has an essay to write.
∴ She will study late in the library.

they will suppress the logically valid inference if certain additional premises are added, as in the following.

Example 12.6.
If she has an essay to write then she will study late in the library.
If the library stays open then she will study late in the library.
She has an essay to write.
∴ She will study late in the library.

In the case of argument Example 12.6, participants' chance of accepting the conclusion that she will study late in the library drops from 96% to 38%.

So, what do these interesting findings mean?

1. Humans maintain control over their inferences, and don't necessarily come to all logically valid conclusions.
2. This control is *content based*, in that they do not manage inference by ceasing to apply valid rules to all applicable forms, but instead selectively block application of valid rules to *certain* formulas. As Byrne concludes: "The moral of these experiments is that in order to explain how people reason, we need to explain how the premises of the same apparent logical form can be interpreted in quite different ways."
3. Reinterpretation of the meanings of premises is the most commonly used strategy for dealing with contradictory formulas. People maintain consistency of beliefs by changing their meanings in appropriate ways.
4. People use only a few strategies to address inconsistencies; these strategies nevertheless suffice for the purposes of everyday reasoning.

Can these features be captured in a formal system? We think so, and active logic is intended as one proposal for how that might be done. For instance, feature 1 is captured by active logic's stepwise character—an active logic reasons in time and, through the use of rules like *contra*(), permits "inspection" of its beliefs at each step. This allows an active logic to decide whether to continue to trust certain beliefs, or cease using them in further inference. In conjunction with this, active logic allows sentences to be "superscripted", as in the earlier example of the two Johns. This is a formal device implementing features 2 and 3, above. Its effect is to give an active logic the freedom to resolve contradictions by giving sentences different interpretations. Exactly how all of this is effected by active logic is described in detail in the section on active logic below, and in Anderson et al. (2008). Before getting to that, however, we turn to a brief survey of some of the many other approaches to implementing AI reasoning systems. This will allow us to better highlight the unique, and we think valuable, features of active logic.

12.3 Formal Models of Human Reasoning

From the theoretical perspective any AI *reasoning system* typically consists of two main components: (1) a logical formalism for knowledge representation and (2) an inference engine to conclude new knowledge from existing knowledge. Based on the logical formalism and the theoretical and philosophical motivations behind the reasoning system, the inference mechanism can either be deductive, inductive, non-monotonic, default, defeasible, etc. An important issue in the implementation of the inference engine is the use of heuristics for typically the complexity of an algorithmic approach is prohibitively high. In the following subsections we survey some reasoning systems that take different approaches towards knowledge representation and inferencing.

General intelligence in human beings can be analyzed in terms of levels of description (see Newell 1990). Each level corresponds to a particular degree of abstraction or, more concretely, to a particular timescale of intelligent tasks. Every increase in the order of magnitude on the timescale would instantiate a new higher level of abstraction. Levels can be grouped into three bands (see Rosenbloom et al. 1991): (1) the neural band which corresponds to levels that do not exceed the order of few milliseconds; this band is the focus of the connectionist community, (2) the cognitive band which corresponds to levels starting with few milliseconds and up to levels with few seconds; this band is the focus of the cognitive science community, and (3) the rational band which corresponds to complex goal-oriented planning and actions which take at least the order of seconds; this band is the focus of the logicist and expert systems communities.

12.3.1 Soar

Soar (see Laird et al. 1987; Rosenbloom et al. 1991) is an implementation of a theoretical-based approach to general intelligence that focuses on the cognitive band. The relationship of Soar to other bands are investigated in Newell (1990), Rosenbloom (1989) and Rosenbloom et al. (1990). Soar assumes no distinction between human intelligence and machine intelligence, hence it has been extensively used both for developing artificial intelligence applications and cognitive models.

The architecture of Soar can be described by four levels of abstraction. First it uses an *associative parallel* memory to store long-term knowledge, and to identify and retrieve knowledge relevant to the current problem solving context.

This knowledge is stored as a set of productions of the form P: *condition* $\rightarrow$ *action*, where the correct action is performed when its preconditions hold. Memory access consists of the parallel execution of these productions. The result of this access is the retrieval of information into a short-term *working memory* that stores contextual information in the form of interrelated objects with attribute-value pairs. For example, an object representing a blue Ford car owned by Heather might look like

$$[Id = te12, type = car, model = Ford, color = blue, owner = Heather]$$

The second level of abstraction in Soar's architecture is the decision making mechanism which proceeds in two elaborate-decide cycles. During elaboration memory is accessed repeatedly and the corresponding relevant productions are executed in parallel. Then one or more of the retrieved actions is performed based upon preference knowledge about what actions are acceptable and/or desirable.

Above the decision making comes the determination of *goals*. Goals are set out whenever the decision procedure has reached a situation (called impasse) where alternatives do not exist any more or there are alternatives, but not enough discriminating information to choose among them (Rosenbloom et al. 1991). Along with the determination of a new goal, a new problem context is generated which allows the continuation of decision making. If in the new context another impasse is encountered, then a new sub-goal and context are generated and the whole process recurs.

The final layer of abstraction is learning. When Soar resolves an impasse it summarises and generalises all the reasoning that led to its resolution. This adds new knowledge to its long-term memory that will prevent the occurrence of such an impasse in similar future situations. Soar's learning mechanism can be used to learn new conceptual knowledge, learn new procedures, and correct its knowledge from the feedback obtained from its interactions with the surrounding environment.

12.3.2 Cyc

Cyc is a reasoning system that focuses on the construction of a vast knowledge base (KB) of trivial and commonsense knowledge (see Lenat et al. 1990; Lenat and Guha 1990). The rationale behind Cyc is as follows. The research and design of AI reasoning systems have largely been concentrating on the development of a logical formalism for knowledge representation and an efficient inference engine based on that formalism. However, little attention has been given to the construction of a real, or at least an approximation to a real, KB that grounds the whole enterprise in reality (the raw material over which the reasoning engine operates). This KB would encode commonsense knowledge about the world that we take for granted concerning things such as time, space, agenthood, life, death, etc.

The early systems lacked the kind and amount of knowledge that would make them effective. With modest-sized KBs (10^2–10^3 domain-specific assertions or rules), such systems sometimes showed very impressive performance in *narrow* task domains but notable problems remained. For example, consider an expert system that contains the following rules from Lenat et al. (1990):

if frog(x), then amphibian(x)

if amphibian(x), then lays_eggs_in_water(x)

if lays_eggs_in_water(x), then lives_near_lots_of(x,water)

if lives_near_lots_of(x,water), then $\neg$ *lives_in_desert(x)*

Given the assertion that Freda is a frog, the expert system can conclude various facts about Frida such as Frida is amphibian, lays eggs in water, lives near lots of water, etc. However, it can not answer simple commonsense questions, that would otherwise seem trivial to humans, such as: Does Freda lay eggs i.e., instead of asking about laying eggs in water? Is Freda sometimes in water? Is Freda a living being?, etc. Hence, such expert systems with complex detailed knowledge were very rigid, non-robust, and could easily fail when encountering a situation or question that is slightly different from the intended narrow domain.

Cyc is an attempt to overcome this brittleness. Its philosophy is to build a vast KB (size at least the order of millions of facts) containing general commonsense facts, domain-specific facts, general heuristics, specific heuristics, and heuristics for analogizing.

The construction of Cyc is, by its very nature, incremental. This includes the representation language, the inference engine, and of course the KB itself.

12.3.3 OSCAR

As opposed to Soar which is intended to simulate the cognitive band, OSCAR is constructed to simulate the *rational band* (Pollock 1992). It is an architecture for rational agents based upon an evolving philosophical theory of rational cognition (Pollock 1999). The general architecture is described in Pollock (1995). OSCAR's overall behaviour can be briefly described by the following cycle: (1) OSCAR has beliefs representing the surrounding environment, (2) it evaluates the current situation according to these beliefs, then (3) it engages in an activity to change the world to its liking and to update its belief system. The most distinguishing feature of OSCAR is that most of its rational cognition is performed by *epistemic cognition*, cognition about what to believe, as opposed to *practical cognition* which is cognition about what to do.

OSCAR is essentially a *defeasible* reasoner. Additionally, by providing it with the axiom schemas of first-order logic it becomes a *complete theorem prover* for that logic (that is OSCAR is able to deduce every valid first-order formula). Defeasible reasoning leads to conclusions that are not necessarily deductively valid. The truth of the premises along with a rationally compelling argument provide good support of the conclusion, even though it is still possible for the premises to be true and the conclusion false. Such premises are called *prima facie* reasons. Conclusions supported defeasibly might have to be withdrawn later in the face of new additional information (Pollock 1999). For instance, if something looks red to me, that gives me a prima facie reason for thinking that it is red. But if someone I trust insists that it is not red then that gives a rebutting defeater. This kind of defeater attacks the conclusion. Another kind of defeater would attack the relationship between the premises and the conclusion. For example, learning that there was red light illumination should weaken my belief that the object is red. The interested reader may consult Pollock (1987, 1989, 1991a,b) for further details.

12.3.4 SNePS

SNePS, the Semantic Network Processing System (Shapiro 1979; Shapiro and Rapaport 1987, 1992; Shapiro 1993), is a *logic-based* approach to natural language understanding and commonsense reasoning. Its ultimate goal is to acquire new knowledge through natural language interaction either with human agents or through media such as books, journals, radios, TVs, etc. SNePS should generally be able to represent everything expressible in natural language and should be able to reason in the presence of incomplete, circular, or inconsistent information.

Reasoning in SNePS is done through a formalism called SNePS logic SNePSLOG which is an enhanced version of first-order logic that is adapted to the natural language context (Shapiro 2000). For example, one of the features of SNePSLOG is the implementation of a new logical connective $andor(i, j)$, which can be used to express the fact that an object satisfies some properties among several alternatives. This is not easily expressible in first-order logic because it is neither *inclusive or* nor *exclusive or*. The general formal syntax of $andor(i, j)$ is:

$$andor(i, j)\{P_1, \ldots, P_n\}$$

is true if and only if at least i and at most j of the first-order properties $P_1, \ldots, P_n$ are true. Another improvement to first-order logic is the addition of the connective *thresh* which has the following syntactical form:

$$thresh(i, j)\{P_1, \ldots, P_n\}$$

and is true if and only if fewer than i or more then j of $P_1, \ldots, P_n$ hold. This connective could be used to capture equivalences among first-order properties. More connectives, quantifiers and other logical features are included in SNePSLOG (see Shapiro 2000).

SNePS memory is a *semantic network* modeled as a directed graph. Nodes in this graph represent concepts, individuals, general and specific rules, and propositions. The neighbours of any node in the semantic network can determine more complex

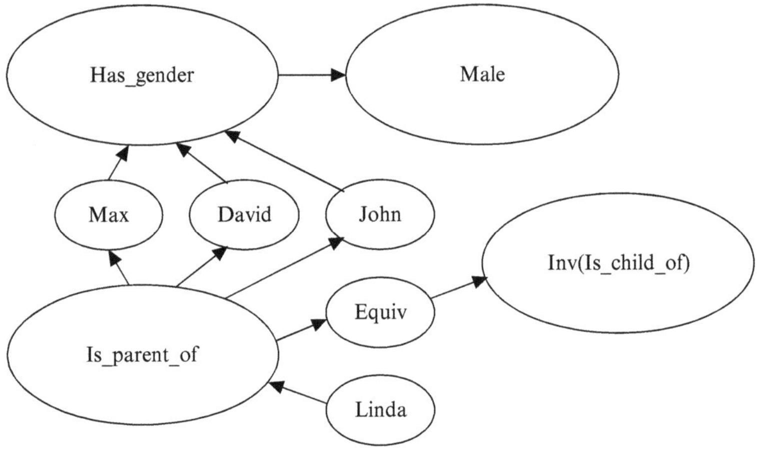

Fig. 12.1 An example of a SNePS network (Adapted from Shapiro et al. (1968))

structural properties of that node. For example, composite rules, propositions, and concepts can be formed by following a path of several nodes along the edges. Figure 12.1 shows an example of a SNePS semantic network. The nodes 'Max', 'David', and 'John' represent individuals, the node 'Male' represents a property, and the nodes 'Has_gender', 'Is_parent_of', 'Is_child_of', and 'Equiv' represent binary relations.

12.3.5 ACT-R

The ACT-R architecture is a simulation environment that supports the creation of cognitive models capable of predicting and explaining human behaviour (Anderson et al. 2004; Lebiere and Anderson 1993). The architecture is constrained by the theory of *rational analysis* which is an empirical program that aims at explaining the functions and purposes of cognitive processes (Anderson 1990, 1991; Oaksford and Chater 1999). According to rational analysis it is important to step back from the investigation of human methods and mechanisms to ask about the environment within which these mechanisms are applied (Gray et al. 2006). In the context of ACT-R, each component of the cognitive system is optimised with respect to environmental demands, given computational limitations (Taatgen et al. 2006). According to this pragmatic approach *truth* is not a fundamental notion in ACT-R, though it is a derivative one: useful demand-based knowledge (either sensed directly from the surrounding environment or extracted from the current beliefs given the contextual environment) is usually true (weaker than defeasible reasoning described above in OSCAR); however, true knowledge is not necessarily useful (deducing Fermat's Last Theorem or solving the Continuum Hypothesis are not useful in everyday activities). This is in contrast to purely logical-based systems built upon (presumed) true premises that are acted upon by sound reasoning rules irrespective of *usefulness*, which is not a logical notion. As will be seen below, this notion of usefulness/utility upon which ACT-R is based is manifested in the design of its memory.

ACT-R has two kinds of memory: declarative memory for facts and procedural memory for rules. Declarative memory is defined by items called *chunks*. Chunks have different levels of *activation* which reflect both their general access pattern and their relevance to the current context. Chunks that are frequently accessed receive a high activation. This activation decays stochastically over time if the chunk is not used. Procedural memory is defined by a set of *production rules*. Similar to the use of activation in declarative memory, each production rule has an associated *utility* value that determines its usefulness in reaching the desired goal. Selection of productions is based on the values of this attribute which are updated stochastically through the use of learning mechanisms.

12.4 Active Logic

In contrast with most of the systems outlined above, active logic was explicitly designed to capture some of the non-classical aspects of human commonsense reasoning, including time-awareness, control of inference, paraconsistency and non-monotonicity, including the ability to re-interpret the meanings of formulas. We have provided a detailed semantics (for a propositional version of active logic) in Anderson et al. (2008), but we offer some of the highlights here.

Formulas in active logic are expressed in a sorted first-order language $\mathcal{L}$ with two parts $\mathcal{L}_w$, a propositional language in which are expressed facts about the world, and $\mathcal{L}_a$, a first-order language used to express facts about the agent, including the agent's beliefs, for instance that the agent's time is now t, that the agent believes P, or that the agent discovered a contradiction in its beliefs at a given time.

$\mathcal{L}_w$ is a propositional language consisting of the following symbols:

- A set S of sentence symbols (propositional or sentential variables) $S = \{S_i^j : i, j \in N\}$ (N is the set of natural numbers).
- The propositional connectives $\neg$ and $\rightarrow$
- Left and right parentheses (and)

$Sn_{\mathcal{L}_w}$ is the set of sentences of $\mathcal{L}_w$ formed in the usual way. These represent the propositional beliefs of the agent about the world. For instance S_1^0 might mean "John is happy". For later use we assume there is a fixed lexicographic ordering for the sentences in $Sn_{\mathcal{L}_w}$.

$\mathcal{L}_a$, contains the unary predicate symbol *Now*, used to express the agent's time, the binary predicate symbol *Contra*, used to indicate the existence of a direct contradiction in its beliefs at a given time, and the binary predicate symbol *Bel*, which expresses the fact that the agent had a particular belief at a given time. $\mathcal{L}_a$ contains only the connective $\neg$; hence statements such as $Bel(\theta, t) \rightarrow Bel(\theta, t+1)$ are not in the language.

All inferences in active logic depend on the knowledge base (KB) of the agent. The agent's knowledge base at time t, KB_t, is a finite set of sentences from $\mathcal{L}$, that is, $KB_t \subseteq Sn_{\mathcal{L}}$. In the case of KB_0 we allow only formulas of $Sn_{\mathcal{L}_w}$ whose superscripts are all 0.

For $\mathcal{L}_w$, we use a fairly standard notion of interpretation $h : Sn_{\mathcal{L}_w} \rightarrow \{T, F\}$ over the sentences in $\mathcal{L}_w$ that extends an $\mathcal{L}_w$-truth assignment h as follows:

$$h(\neg\varphi) = T \iff h(\varphi) = F$$

$$h(\varphi \rightarrow \psi) = F \iff (h(\varphi) = T \text{ and } h(\psi) = F)$$

We also stipulate a standard definition of consistency for $\mathcal{L}_w$: a set of $\mathcal{L}_w$ sentences is *consistent* iff there is some interpretation h in which all the sentences are true. Notationally we write the usual $h \models \Sigma$, to mean that all the sentences of Σ are assigned T by h.

The interpretation for $\mathcal{L}_a$ is somewhat more unusual. The symbol for the interpretation is H_{t+1}^{Σ}; it is an interpretation at time $t+1$ based on Σ, where Σ is to be understood formally as any set of sentences from $\mathcal{L}$. For current purposes, the most important aspects of the interpretation are as follows:

- The predicate symbol *Now* has the following semantics: $H_{t+1}^{\Sigma} \models Now(s) \iff s = t+1$ and $Now(t) \in \Sigma$; otherwise $H_{t+1}^{\Sigma} \models \neg Now(s)$.
- The predicate symbol *Contra* has the following semantics: $H_{t+1}^{\Sigma} \models Contra(\sigma, s) \iff$ either $s < t$ and $Contra(\sigma, s) \in \Sigma$ or $s = t$ and $\exists\sigma, \neg\sigma \in \Sigma$; otherwise $H_{t+1}^{\Sigma} \models \neg Contra(\sigma, s)$.
- The predicate symbol *Bel* has the following semantics: $H_{t+1}^{\Sigma} \models Bel(\theta, s) \iff$ either $s < t$ and $Bel(\theta, s) \in \Sigma$ or $s = t$ and $\theta \in \Sigma$; otherwise $H_{t+1}^{\Sigma} \models \neg Bel(\theta, s)$.

For this version of active logic, we assume that the sentences in $\mathcal{L}_a$ are consistent, but allow for the possibility of inconsistency in the set of $\mathcal{L}_w$ sentences. We use the term Γ to refer to the potentially *inconsistent* set of $\mathcal{L}_w$ sentences in Σ: $\Gamma = \Sigma \cap Sn_{\mathcal{L}_w}$.

In order to model the sentences in Γ, active logic uses an "apperception function". The notion of an apperception function is intended to help capture, at least roughly, how the world might seem to an agent with a given inconsistent belief set Γ. For a real agent, only some logical consequences are believed at any given time, since it cannot manage to infer all the potentially infinitely many consequences in a finite time, let alone in the present moment. Moreover, even if the agent has contradictory beliefs, the agent still has a view of the world, and there will be limits on what the agent will and won't infer. This is in sharp distinction to the classical notion of a model, where (1) inconsistent beliefs are ruled out of bounds, since then there are no models, and (2) all logical consequences of the KB are true in all models.

The idea is simple: suppose S_i^0, $S_i^0 \rightarrow S_j^0$ and $\neg S_j^0$ are all in Γ, we imagine that the agent might not realise, at first, that the two instances of S_i are in fact instances of the same sentence symbol. That is, it might seem to the agent that the world is one in which, say, S_i^1 is true, and so is $S_i^2 \rightarrow S_j^0$.

The apperception functions we define can make changes only to Γ. An apperception function does not change $\Sigma - \Gamma$. We use the same notation ap when the apperception function is applied to an occurrence of a sentence symbol, a sentence, or a set of sentences. We start by defining a function that changes the superscripts of sentence symbols to 0. This is used to recover the original direct contradictions that were modified by the assignment of superscripts.

Definition 12.1. For any sentence $\phi \in Sn_{\mathscr{L}_w}$, let $z(\phi)$ be the sentence ϕ with all superscripts reset to 0. If $\Sigma \subseteq Sn_{\mathscr{L}_w}$, then $z(\Sigma) = \{z(\phi) | \phi \in \Sigma\}$.

Definition 12.2. An apperception (awareness) ap is a function $ap\colon \Sigma \rightarrow \Sigma'$ where Σ and Σ' are sets of $\mathscr{L}$-sentences. An ap is represented as a finite sequence of nonnegative integers: $\langle n_1, \ldots, n_p \rangle$. The effect of ap on Σ is as follows:

1. Let Σ be a set of $\mathscr{L}$-sentences and let $\Gamma = \Sigma \cap \mathscr{L}_w$. Using the lexicographic order given earlier, let the kth sentence symbol in Γ be S_i^j. The effect of the $ap = \langle n_1, \ldots, n_p \rangle$ is to change S_i^j to $S_i^{n_k}$ if $1 \leq k \leq p$, otherwise S_i^j is unchanged.
2. $ap(\Sigma) = (\Sigma - \Gamma) \cup ap(\Gamma)$. ($ap$ does not change $\Sigma - \Gamma$).

Example 12.7. Let $\Sigma = \{Now(5), Bel(S_2^0, 4), \neg S_2^1, S_2^1, S_1^0 \rightarrow S_5^4\}$. In this case $\Gamma = \{\neg S_2^1, S_2^1, S_1^0 \rightarrow S_5^4\}$. Writing the elements lexicographically yields $ord(\Gamma) = \{S_2^1, \neg S_2^1, S_1^0 \rightarrow S_5^4\}$. Consider $ap = \langle 1, 3, 2, 16, 7 \rangle$. Then $ap(\Sigma) = \{Now(5), Bel(S_2^0, 4), S_2^1, \neg S_2^3, S_1^2 \rightarrow S_5^{16}\}$.

The purpose of the apperception functions is to get rid of inconsistencies in Σ. Hence we are interested only in apperception functions that output consistent sets. The set of apperception functions that do this depends on Σ.

Definition 12.3. Let AP denote the class of all apperception functions. $AP^{\Sigma} = \{ap \in AP | ap(\Sigma) \text{ is consistent}\}$.

It turns out that AP^{Σ} is never empty (Anderson et al. 2008).

At this point we are ready to define the notion of *active consequence* at time t—the active logic equivalent of logical consequence. Here again, the full technical details are given in Anderson et al. (2008), but we outline some of the more important elements here. We start by defining the concept of *1-step active consequence* as a relationship between sets of sentences Σ and Θ of $\mathscr{L}$, where $\Sigma \subseteq KB_t$ and Θ is a potential subset of KB_{t+1}. When we define this notion we want to make sure that Θ contains only sentences required by Σ and the definition of H_{t+1}^{Σ}. This is the reason for the next definition.

Definition 12.4. Given Σ and $ap \in AP^{\Sigma}$, define
$dcs(\Gamma) = \{\phi \in \Gamma | \exists \psi \in \Gamma \text{ such that } z(\phi) = \neg z(\psi) \text{ or } \neg z(\phi) = z(\psi)\}$.
$ap^z(\Gamma) = ap(\Gamma) - dcs(\Gamma)$.

The meaning of Definition 12.4 is that we are removing direct contradictions from $ap(\Gamma)$ while ignoring the superscripts.

Definition 12.5. Let $\Sigma, \Theta \subseteq Sn_{\mathcal{L}}$. Then Θ is said to be a 1-*step active consequence* of Σ at time t, written $\Sigma \models_1 \Theta$ if and only if $\exists ap \in AP^{\Sigma}$ such that

i. If $\sigma \in \Theta \cap Sn_{\mathcal{L}_w}$ then $ap^z(\Gamma) \models \sigma$ (σ is a classical logical consequence of $ap^z(\Gamma)$), and
ii. If $\sigma \in \Theta \cap Sn_{\mathcal{L}_a}$ then $H_{t+1}^{(\Sigma-\Gamma)\cup z(\Gamma)} \models \sigma$.

Definition 12.6.

i. Let $\Sigma, \Theta \subseteq Sn_{\mathcal{L}}$. Then Θ is said to be an n-*step active consequence* of Σ at time t, written $\Sigma \models_n \Theta$, if and only if

$$\exists \Delta \subseteq Sn_{\mathcal{L}}\colon \Sigma \models_{n-1} \Delta \; and \; \Delta \models_1 \Theta. \tag{12.1}$$

ii. We say that Θ is an *active consequence* of Σ, written $\Sigma \models_a \Theta$, if and only if $\Sigma \models_n \Theta$ for some positive integer n.

Next we give some examples to illustrate the concept of active consequence.

Example 12.8.

i. Let $\Sigma = \{Now(t), S_1^0, S_1^0 \rightarrow S_4^0, S_{12}^0\}$ and $\Theta = \{Now(t+1), S_4^0, S_{12}^0\}$. Let $ap \in AP^{\Sigma}$ be the identity function. It is easy to see that $\{S_4^0, S_{12}^0\}$ are logical consequences of $\{S_1^0, S_1^0 \rightarrow S_4^0, S_{12}^0\}$. Also by definition $H_{t+1}^{\Sigma} \models Now(t+1)$. Hence $\Sigma \models_1 \Theta$.
ii. Let $\Sigma = \{S_1^0, S_2^0, S_2^0 \rightarrow \neg S_1^0\}$ and $\Theta = \{Contra(S_1^0, t+1)\}$. We will see that $\Sigma \models_2 \Theta$. Let $\Delta = \{S_1^1, \neg S_1^2\}$. Then $\Sigma \models_1 \Delta$, through the apperception function $ap(\Sigma) = \{S_1^1, S_2^2, S_2^2 \rightarrow \neg S_1^2\}$. Then $\Delta \models_1 \Theta$ by the second part of the definition, regardless of the apperception function applied in this step.

Note that in Example 12.8(ii), it is not the case that $\Sigma \models_1 \{Contra(S_1^0, t)\}$ even though the conditions for the later appearance of the relevant direct contradiction were already in place at time t. This underlines the fact that in active logic it can take time for consequences to appear in the *KB*. Apperception functions give active logic agents control over which inferences to make, and which to suppress. They allow the agent to have inconsistent beliefs while still having a consistent world model. Moreover, this allows us to see how an agent with inconsistent beliefs could avoid vacuously concluding *any* proposition, and also reason in a directed way, by applying inference rules only to an appropriately apperceived subset of its beliefs.

For instance, consider the following active logic inference:

Definition 12.7. If $\varphi, \neg\varphi \in KB_t$, where $\varphi \in Sn_{\mathcal{L}_w}$, then the *direct contradiction inference rule* is defined as follows:

$$\frac{t : \varphi, \neg\varphi}{t+1 : Contra(\varphi, t)}$$

This inference is sound based on the definition and interpretation of *Contra*. And because of this, along with apperception functions, the following inference is *unsound*:

Definition 12.8. Let $\Sigma \subseteq Sn_{\mathscr{L}_w}$ be inconsistent. Let $\psi \in Sn_{\mathscr{L}_w}$. We define the *explosive rule* with respect to the language $\mathscr{L}_w$ as follows.

$$\frac{t : \Sigma; Inconsistent(\Sigma)}{t+1 : \psi}$$

The explosive inference rule is unsound. For consider the case where ψ is $\neg(S_1^0 \rightarrow S_1^0)$. No apperception function *ap* that turns Σ into a consistent set can logically derive ψ. Hence $ap(\Sigma) \not\models_1 \psi$.

This shows that active logic is paraconsistent. We hope that this approach to paraconsistency can shed some light on focused, step-wise, resource-bounded reasoning more generally. More details on the semantics for active logic, and many more examples of its use, can be found in Anderson et al. (2008).

12.5 Comparison with Reasoning Systems and Formalisms

Active logic possesses several interesting properties. It has a temporal component so that inference occurs in time: for a set of formulas Γ at time t deduce formula ϕ at time $t + 1$. Active logic is paraconsistent as both ϕ and $\neg\phi$ may hold at some time t. Active logic is also non-monotonic because a formula ϕ that holds at time t does not necessarily hold at time $t + 1$; this happens in particular when ϕ and $\neg\phi$ are replaced by the *Contra* formula.

We are not aware of any other logic system that possesses such a temporal component as well as paraconsistency and non-monotonicity. SOAR, Cyc and ACT-R do not appear to incorporate any of these features, and while OSCAR is non-monotonic, it is neither time-tracking nor paraconsistent. The closest of the above systems to having the distinctive features of active logic is SNePS, but there are some important differences between the two approaches. For instance although SNePS incorporates a time-tracking feature, in a SNePS-based agent *NOW* is a meta-logical variable, rather than a logical term fully integrated into the SNePS semantics. The variable *NOW* is implemented so that it does, indeed, change over time, but this change is the result of actions triggering an external time-variable update. In active logic, in contrast, reasoning itself *implies* the passage of time. Perhaps in part because of this difference, SNePS is a monotonic logic, whereas active logic is non-monotonic, leveraging the facts that beliefs are held at times, and beliefs can be held about beliefs, to easily represent such things as "I used to believe P, but now I believe $\neg P$" using the *Bel* operator. SNePS is also able to represent beliefs about beliefs, but there is no indication that this ability is leveraged by SNePS to guide belief updates. Rather, all beliefs are about states holding over time, so that belief change is effected by allowing beliefs to expire, rather than by formally retracting them. This is a strategy similar to that employed by the situation calculus (which does not itself incorporate a changing *Now* term) (McCarthy and Hayes 1969). Finally, although SNePS is a paraconsistent logic, in SNePS contradictions imply nothing at all, whereas in active logic contradictions imply *Contra*, a meta-level operator that can trigger further reasoning.

Nevertheless, although there are few examples of implemented systems with the features of active logic, we know that a substantial amount of work has been done on non-monotonic paraconsistent logics. While these logics are not really comparable to active logics, we provide here information on some such systems.

An early influential paraconsistent non-monotonic logical system was presented in Priest (1989). The logic **LP** has three truth values: True, False, and Both. The connectives and entailment in **LP** are defined as in classical logic, but on account of the third truth value, **LP** is paraconsistent. **LP** is then extended to $\mathbf{LP_m}$ with consistency as a default assumption and a notion of default consequence relation $\models_m$ is defined using minimal models. $\mathbf{LP_m}$ is a non-monotonic paraconsistent system.

Another such system is a combination of LEI (Logic of Epistemic Inconsistency) and IDL (Inconsistent Default Logic), called IDL&LEI. We refer to Martins et al. (2002) for details about it including a multiple world semantics. Formulas in LEI are divided into two groups: the irrevocable formulas and the plausible formulas; the latter are distinguished by a question mark, as in α?. No contradictions are allowed involving any irrevocable formula; contradictions are allowed only for plausible formulas. LEI is paraconsistent. Non-monotonicity is obtained by adding default rules using IDL. The IDL&LEI system has both an elegant syntax and a multiple world semantics.

Finally we mention the work in Arieli and Avron (1998) where a non-monotonic paraconsistent logic uses Belnap's four-valued logic with a notion of logical consequence based on minimal preferential models. The approach here is primarily semantical. (Actually, it turns out that a four-valued semantics is available also for IDL.) The recent paper by Arieli (2007) uses quantified Boolean formulas in the context of multiple-valued logics to represent several non-monotonic paraconsistent logics. This paper also contains many references to recent related work.

12.6 Conclusions

As shown by many psychological experiments, the logic used by humans is substantially different from classical logic, and for just this reason may be more useful to commonsense reasoning. Hence logic-based AI systems should be attuned to, and where possible implement, these non-classical features. We have described several AI reasoning systems, as well as active logic, a logic designed to capture features such as time-awareness, control of inference, paraconsistency, and non-monotonicity, that we think are important to human commonsense reasoning.

Acknowledgements We would like to thank the referees for many helpful comments and suggestions.

References

Anderson, J. 1990. *The adaptive character of thought*. Hillsdale: Lawrence Erlbaum.

Anderson, J. 1991. Is human cognition adaptive? *Behavioral and Brain Sciences* 14(3): 471–517.

Anderson, J.R., D. Bothell, M.D. Byrne, S. Douglass, C. Lebiere, and Y. Qin. 2004. An integrated theory of mind. *Psychological Review* 111(4): 1036–1060.

Anderson, M.L., W. Gomaa, J. Grant, and D. Perlis. 2008. Active logic semantics for a single agent in a static world. *Artificial Intelligence* 172: 1045–1063.

Arieli, O. 2007. Paraconsistent reasoning and preferential entailments by signed quantified boolean formulae. *ACM Transactions on Computational Logic* 8(3): Article 18.

Arieli, O., and A. Avron. 1998. The value of four values. *Artificial Intelligence* 102(1): 97–141.

Byrne, R. 1989. Suppressing valid inferences with conditionals. *Cognition* 31: 61–83.

Cosmides, L. 1989. The logic of social exchange: Has natural selection shaped how humans reason? Studies with the Wason selection task. *Cognition* 31: 187–276.

Elio, R. 1997. What to believe when inferences are contradicted: The impact of knowledge type and inference rule. In *Proceedings of the nineteenth annual conference of the cognitive science society*, 211–216. Hillsdale: Lawrence Erlbaum.

Elio, R. 1998. How to disbelieve p → q: Resolving contradictions. In *Proceedings of the twentieth annual conference of the cognitive science society*, 315–320. Mahwah: Lawrence Erlbaum.

Evans, J.S.B. 1982. *The psychology of deductive reasoning*. London: Routledge Keegan Paul.

Gentner, D., and A. Stevens. 1983. *Mental models*. Hillsdale: Lawrence Erlbaum.

Gigerenzer, G. 1994. Why the distinction between single-event probabilities and frequencies is important for psychology (and vice versa). In *Subjective probability*, ed. G. Write and P. Ayton, 129–161. New York: Wiley.

Gray, W., C. Sims, W.T. Fu, and M. Schoelles. 2006. The soft constraints hypothesis: A rational analysis approach to resource allocation for interactive behavior. *Psychological Review* 113(3): 461–482.

Johnson-Laird, P. 1983. *Mental models*. Cambridge: Cambridge University Press.

Johnson-Laird, P., and P. Wason. 1970. A theoretical insight into a reasoning task. *Cognitive Psychology* 1: 134–148.

Kahneman, D., and A. Tversky. 1979. Prospect theory: An analysis of decision under risk. *Econometrica* 47(2): 263–291.

Laird, J., A. Newell, and P. Rosenbloom. 1987. Soar: An architecture for general intelligence. *Artificial Intelligence* 33(1): 1–64.

Lebiere, C., and J.R. Anderson. 1993. A connectionist implementation of the ACT-R production system. In *Proceedings of the fifteenth annual conference of the cognitive science society*, 635–640. Hillsdale: Lawrence Erlbaum.

Lenat, D., and R.V. Guha. 1990. Cyc: A midterm report. *AI Magazine* 11(3): 32–59.

Lenat, D., R.V. Guha, K. Pittman, D. Pratt, and M. Shepherd. 1990. Cyc: Toward programs with commonsense. *Communications of the ACM* 33(8): 30–49.

Martins, A.T., M. Pequeno, and T. Pequeno. 2002. A multiple worlds semantics to a paraconsistent nonmonotonic logic. In *Paraconsistency, the logical way to the inconsistent*, ed. W.A. Carnielli, M.E. Coniglio, and I.M.L. D'Ottavio, 187–212. New York: Marcel Dekker.

McCarthy, J., and P. Hayes. 1969. Some philosophical problems from the standpoint of artificial intelligence. In *Machine intelligence*, ed. B. Meltzer and D. Michie, 463–502. Edinburgh: Edinburgh University Press.

Newell, A. (1990). *Unified theories of cognition*. Cambridge: Harvard University Press.

Oaksford, M., and N. Chater. 1994. A rational analysis of the selection task as optimal data selection. *Psychological Review* 101: 608–631.

Oaksford, M., and N. Chater (ed.). 1999. *Rational models of cognition*. Oxford: Oxford University Press.

Oaksford, M., and N. Chater. 2007. *Bayesian rationality: The probabilistic approach to human reasoning*. Oxford: Oxford University Press.

Pollock, J. 1987. Defeasible reasoning. *Cognitive Science* 11: 481–518.
Pollock, J. 1989. *How to build a person*. Cambridge: Bradford/MIT.
Pollock, J. 1991a. A theory of defeasible reasoning. *International Journal of Intelligent Systems* 6: 33–54.
Pollock, J. 1991b. Self-defeating arguments. *Minds and Machines* 1: 367–392.
Pollock, J. 1992. How to reason defeasibly. *Artificial Intelligence* 57: 1–42.
Pollock, J. 1995. *Cognitive carpentry*. Cambridge: Bradford/MIT.
Pollock, J. 1999. Rational cognition in OSCAR. In *Proceedings of ATAL-99*, ed. N. Jennings and Y. Lesperance. Berlin: Springer.
Priest, G. 1989. Reasoning about truth. *Artificial Intelligence* 39(2): 231–244.
Rosenbloom, P.S. (1989). A symbolic goal-oriented perspective on connectionism and soar. In *Connectionism in perspective*, ed. R. Pfeifer, Z. Schreter, F. Fogelman-Soulie, and L. Steels. Amsterdam: Elsevier.
Rosenbloom, P.S., A. Newell, and J.E. Laird. 1990. Towards the knowledge level in soar: The role of the architecture in the use of knowledge. In *Architectures for intelligence*, ed. K. VanLehn. Hillsdale: Lawrence Erlbaum.
Rosenbloom, P., J. Laird, A. Newell, and R. McCarl. 1991. A preliminary analysis of the soar architecture as a basis for general intelligence. *Artificial Intelligence* 47: 289–325.
Shapiro, S.C. 1979. The SNePS semantic network processing system. In *Associative networks: The representation and use of knowledge by computers*, ed. N. Findler, 179–203. New York: Academic.
Shapiro, S.C. 1993. Belief spaces as sets of propositions. *Journal of Experimental and Theoretical Artificial Intelligence* 5: 225–235.
Shapiro, S.C. 2000. SNePS: A logic for natural language understanding and commonsense reasoning. In *Natural language processing and knowledge representation: Language for knowledge and knowledge for language*, ed. Ł. Iwańska and S.C. Shapiro, 175–195. Menlo Park: AAAI/MIT.
Shapiro, S.C., and W. Rapaport. 1987. SNePS considered as a fully intensional propositional semantic network. In *The knowledge frontier*, ed. N. Cercone and G. McCalla, 263–315. New York: Springer.
Shapiro, S.C., and W. Rapaport. 1992. The SNePS family. *Computers and Mathematics with Applications* 23: 243–275.
Shapiro, S.C., G.H. Woodmansee, and M.W Kreuger. 1968. A semantic associational memory net that learns and answers questions (SAMENLAQ). Technical report, Computer Science Department, University of Wisconsin, Madison, WI.
Sharpe, D., and G. Lacroix. 1999. Reasoning about apparent contradictions: Resolution strategies and positive–negative asymmetries. *Journal of Child Language* 26(2): 477–490.
Taatgen, N.A., C. Lebiere, and J.R Anderson. 2006. Modeling paradigms in ACT-R. In *Cognition and multi-agent interaction: From cognitive modeling to social simulation*, ed. R. Sun, 29–52. Cambridge: Cambridge University Press.

Chapter 13
Distribution in the Logic of Meaning Containment and in Quantum Mechanics

Ross T. Brady and Andrea Meinander

13.1 Introduction

Distribution of conjunction over disjunction, in the form $A\&(B \vee C) \rightarrow (A\&B) \vee (A\&C)$, holds in the vast majority of logics. The converse form, $(A\&B) \vee (A\&C) \rightarrow A\&(B \vee C)$, is easily derived using standard lattice properties of conjunction and disjunction[1] and does not require any additional distribution axiom. The alternative form $(A \vee B)\&(A \vee C) \rightarrow A \vee (B\&C)$ is derivable from $A\&(B \vee C) \rightarrow (A\&B) \vee (A\&C)$ by substitution and lattice properties and its converse $A \vee (B\&C) \rightarrow (A \vee B)\&(A \vee C)$ is derivable from lattice properties alone. So, when we talk of distribution we usually refer to the form $A\&(B \vee C) \rightarrow (A\&B) \vee (A\&C)$.

Distribution in the form $A\&(B \vee C) \supset (A\&B) \vee (A\&C)$ holds in classical logic, the form $A\&(B \vee C) \rightarrow (A\&B) \vee (A\&C)$ holds in intuitionist logic and also in relevant logics, with appropriate interpretational adjustments for the connectives. This can all be seen from the appropriate truth-functional semantics for these logics: the standard semantics for classical logic, the Kripke worlds semantics for intuitionist logic and the Routley-Meyer worlds semantics for a vast array of relevant logics. So, it would seem that distribution holds quite generally.

[1]The lattice properties are the meet and join properties of conjunction and disjunction which follow from the following axioms: $A\&B \rightarrow A$, $A\&B \rightarrow B$, $(A \rightarrow B)\&(A \rightarrow C) \rightarrow (A \rightarrow B\&C)$, $A \rightarrow A \vee B$, $B \rightarrow A \vee B$, $(A \rightarrow C)\&(B \rightarrow C) \rightarrow .(A \vee B) \rightarrow C$. With respect to an ordering based on the '$\rightarrow$', conjunction represents the greatest lower bound and disjunction represents the least upper bound.

R.T. Brady (✉)
Department of Philosophy, La Trobe University, Melbourne, Australia
e-mail: Ross.Brady@LaTrobe.edu.au

A. Meinander
Departments of Philosophy and Physics, University of Helsinki, Helsinki, Finland

K. Tanaka et al. (eds.), *Paraconsistency: Logic and Applications*, Logic, Epistemology, and the Unity of Science 26, DOI 10.1007/978-94-007-4438-7_13,

However, there are two main areas of dissent. One is quantum logic, which consists, roughly speaking, of the rules of classical logic without the rule form of distribution, $A\&(B \vee C) \Rightarrow (A\&B) \vee (A\&C)$. It is obtained from a modelling of quantum mechanics using Hilbert-spaces, which also models an ortho-modular law. It is argued that we need to drop distribution to make sense of various quantum phenomena (see Putnam 1975). We will investigate three alternative logics for quantum mechanics, determine which one is preferable, and then see whether distribution should hold or not.

The other is of more recent origin. It concerns the logic MC of meaning containment, the axiomatisation of which is given in the next section. (See Brady 1996, 2006 for much detail concerning the logic.) Its content semantics captures the meaning containment aspect of the logic, but since it is a relevant logic, it has a truth-functional worlds semantics as well. However, as argued by Rush and Brady (2007), the content semantics is better as a semantics. The issue then is whether distribution can be said to hold in its content semantics, as opposed to its truth-functional semantics for which distribution can be seen to hold, as mentioned above. However, Restall (2007), in his review of Brady (2006), states 'I found most unsatisfactory the short argument ... intended to show that the distribution principle ... is motivated on grounds of meaning containment'. We will determine whether distribution should hold here and, if so, in what form and under what conditions.

Another important and related concern is Mares and Goldblatt's truth-functional semantics for the quantified relevant logic QR (Mares and Goldblatt 2006). The quantified logic QR does not have the quantified distribution principle $\forall x(A \vee B) \rightarrow (A \vee \forall xB)$, where x is not free in A, called 'extensional confinement', and its semantics is more straightforward than that for the corresponding quantified logic RQ, which does include extensional confinement. And, we think the Mares and Goldblatt semantics is the best truth-functional semantics for quantified relevant logic achieved thus far. The dropping of extensional confinement also occurs in the logic QMC, used by Brady and Rush (2008) to prevent classicality spreading from atomic statements of Peano arithmetic to all quantified arithmetical statements. In addition, extensional confinement is also invalid in quantified intuitionist logic (see Dummett 1977, pp. 202–204). However, for these three logics, the sentential forms of distribution all hold.

The issue of distribution has been recently discussed by Paoli, where he gives five reasons for rejecting sentential distribution (Paoli 2007, pp. 572–577). Some of these reasons are in response to Belnap (1993), who argues in favour of distribution, and some involve the replacement of conjunction and/or disjunction by their corresponding intensional connectives, fusion and fission. Moreover, Paoli and Belnap's arguments focus on rather strong logics such as R, BCK and linear logic. We do not have the space here to deal with the arguments in this discussion, but some idea of our positions regarding these arguments can be obtained from the conclusion of this paper.

We endeavour to maintain a fairly broad perspective but, in doing so, we wish to focus on the logic MC of meaning containment and its quantificational extension MCQ, which includes extensional confinement. This paper will examine the rationale behind the use of distribution in logics in some generality, and in a broad range of both proof-theoretic and semantic perspectives. We will examine the logical treatment of quantum mechanics separately as its underpinnings are rather different. Unfortunately, space does not allow us to do justice to linear logic, for which extensional distribution fails but some intensional forms of distribution hold. This will need to done at a future time.

13.2 The Logics MC and MCQ

Before axiomatizing MC, we briefly give three key properties that help to pin it down as a logic. Firstly, as mentioned above, it is a *relevant logic*, which means that if $A \to B$ is a theorem of the logic then A and B share a sentential variable. This can be tightened up to *depth relevance*, which requires a shared variable of A and B to be of the same depth. Roughly, the depth of a variable occurrence in a formula is the number of nested '$\to$'s one passes through in order to reach the variable occurrence (see Brady 2006, pp. 162–165 for details of this). Secondly, it is *paraconsistent* in that $A, \sim A \Rightarrow B$ is not a derived rule of the logic. This enables applications of the logic to include contradictions, without the logic becoming trivial, i.e. with all its formulae being provable. The same applies to models of the logic with regard to validity. Thirdly, it is *metacomplete*. As such, MC has a primary focus on its entailments $A \to B$ in the sense that all its non-entailment theorems can be systematically built up from its entailment theorems (see Brady 2006, pp. 155–159 for details of metacompleteness and its key properties).

However, we still need to add some further characterising feature to these properties in order to fully pin down the logic. Such a feature is that its axioms and rules mimic set-theoretic containment, which ties in well with meaning containment, when meaning is represented set-theoretically, and this does happen in the canonical model for the content semantics (see Brady 2006, pp. 63–80). Moreover, we still need to examine distribution to see if its corresponding set-theoretic property holds, and this will show up in our examination of Venn Diagrams.

We now axiomatise MC, which includes distribution (A8). We also add the quantifiers, giving us MCQ, as quantified distribution will be discussed as well. So, it will also include the quantified distribution laws (QA3) and (QA6).

13.2.1 The System MC

Primitives:

$\&, \vee, \sim, \to$ (connectives)
$p, q, r, \ldots$ (sentential variables)

Axioms:

1. $A \to A$.
2. $A\&B \to A$.
3. $A\&B \to B$.
4. $(A \to B)\&(A \to C) \to (A \to B\&C)$.
5. $A \to A \vee B$.
6. $B \to A \vee B$.
7. $(A \to C)\&(B \to C) \to (A \vee B \to C)$.
8. $A\&(B \vee C) \to (A\&B) \vee (A\&C)$.
9. $\sim\sim A \to A$.
10. $(A \to \sim B) \to (B \to \sim A)$.
11. $(A \to B)\&(B \to C) \to (A \to C)$.

Rules:

1. $A, A \to B \Rightarrow B$.
2. $A, B \Rightarrow A\&B$.
3. $A \to B, C \to D \Rightarrow (B \to C) \to (A \to D)$.

Meta-rule:

1. If $A \Rightarrow B$ then $C \vee A \Rightarrow C \vee B$.

13.2.2 The System MCQ

Quantificational Primitives:

$\forall, \exists$ (quantifiers)
$a, b, c, \ldots$ (free individual variables)
$x, y, z, \ldots$ (bound individual variables)
$f, g, h, \ldots$ (predicate variables)

Quantificational Axioms:

1. $\forall xA \to A^a/_x$.
2. $\forall x(A \to B) \to (A \to \forall xB)$.
3. $\forall x(A \vee B)(A \vee \forall xB)$.
4. $A^a/_x \to \exists xA$.
5. $\forall x(A \to B) \to (\exists xA \to B)$.
6. $A\&\exists xB \to \exists x(A\&B)$.

Note that, in distinguishing free and bound individual variables, x can only occur bound in the A of QA2, QA3 and QA6 and in the B of QA5.

Quantificational Rules:

1. $A^a/_x \Rightarrow \forall xA$, where a does not occur in A.

Quantificational Meta-rule:

1. If $A^a/_x \Rightarrow B^a/_x$ then $\exists x A \Rightarrow \exists x B$.

The Meta-rule and Quantificational Meta-rule are both subject to the proviso that, in the respective derivations $A \Rightarrow B$ and $A^a/_x \Rightarrow B^a/_x$, the Quantificational Rule must not generalise on any free individual variable in A or in $A^a/_x$, respectively.

13.3 Distribution in Proof Theory

We now examine distribution in the two major proof-theoretic contexts, viz. natural deduction and Gentzen sequent calculus, giving some general comments at the end of this section.

13.3.1 Natural Deduction

We use Fitch-style natural deduction, as it is standardly used for relevant logics (see Anderson and Belnap 1975, pp. 6–32, 271–274). We start by setting out a proof of distribution $A\&(B \vee C) \supset (A\&B) \vee (A\&C)$ in a Fitch-style system for classical logic, without subscripts.

$$
\begin{array}{l l}
\left| \; A\&(B \vee C) \right. & \\
\hline
\left| \; A \right. & \&\text{E} \\
\left| \; B \vee C \right. & \&\text{E} \\
\left| \; \left| \; B \right. \right. & \\
\left| \; \left| \; A\&B \right. \right. & \&\text{I} \\
\left| \; \left| \; (A\&B) \vee (A\&C) \right. \right. & \vee\text{I} \\
\left| \; \left| \; C \right. \right. & \\
\left| \; \left| \; A\&C \right. \right. & \&\text{I} \\
\left| \; \left| \; (A\&B) \vee (A\&C) \right. \right. & \vee\text{I} \\
\left| \; (A\&B) \vee (A\&C) \right. & \vee\text{E} \\
A\&(B \vee C) \supset (A\&B) \vee (A\&C) & \supset\text{I}
\end{array}
$$

We can see how the standard $\&E$, $\&I$, $\vee I$ and $\vee E$ rules, together with the standard $\supset I$ rule, work to yield distribution. Note that the $\&I$ rule takes its premises A and B from two distinct subproofs. The above argument also holds for intuitionist logic (see Heyting 1966; Dummett 1977).

Let us examine the derivation of distribution in a Fitch-style natural deduction system for relevant logics. As explained in Anderson and Belnap (1975), to ensure relevance, the conclusion of each subproof should be derived using its hypothesis. To ensure that this is the case, subscripts $\{k\}$ are introduced to keep track of hypotheses of subproof depth k. Index sets are used on each step in a subproof to keep a record of all hypotheses used in its derivation. To this end, the $\rightarrow I$ rule has the following proviso:

$\rightarrow I$: From a proof of B_a on hypothesis $A_{\{k\}}$ to infer $A \rightarrow B_{a-\{k\}}$, provided $k \in a$.

Further, to maintain relevance, the two premises of $\&I$ must have the same index set a, as follows:

$$\&I : \text{From } A_a \text{ and } B_a \text{ to infer } A\&B_a.$$

Further still, the two inferential premises of $\vee E$ must also have the same index set.

$$\vee E : \text{From } A \rightarrow C_a, B \rightarrow C_a \text{ and } A \vee B_b \text{ to infer } C_{a \cup b}.$$

There are various provisos applying to this rule for relevant logics weaker than the logic R, but these will not impact on the derivation of distribution.

However, with these restrictions to the three rules, the classical proof of distribution fails to go through, primarily due to the restriction on $\&I$. To overcome this problem, Anderson and Belnap introduce a special distribution rule:

> $\&\vee$: From $A\&(B \vee C)_a$ to infer $(A\&B) \vee C_a$. (Anderson and Belnap 1975, pp. 273–274)

Since we are working with the more standard form of distribution, we replace it by the deductively equivalent form:

$\&\vee$: From $A\&(B \vee C)_a$ to infer $(A\&B) \vee (A\&C)_a$.[2]

Using this rule, the derivation of distribution is trivial and this applies to all the relevant logics in Brady (1984), where a wide range of such natural deduction systems are set out.

In the normalised natural deduction system for the depth relevant logic DW in Brady (2006a),[3] distribution is essentially absorbed into the structure of the natural deduction system to enable the normalisation to be proved. Signs T and F are placed in front of formulae to help capture negation and disjunctively interpreted commas are placed between signed formulae, so that the T$\vee$E rule can be applied within the same subproof, as follows:

---

[2]To derive $(A\&B) \vee (A\&C)$ from $A\&(B \vee C)$ using the Anderson and Belnap version of the $\&\vee$ rule, first derive $A, (A\&B) \vee C, C \vee (A\&B), A\&(C \vee (A\&B))$ and then apply $\&\vee$ again. The derivation of $(A\&B) \vee C_a$ from $A\&(B \vee C)_a$ using our version of $\&\vee$ follows from $(A\&B) \vee (A\&C)_a$, $A\&B \rightarrow (A\&B) \vee C_\emptyset$, and $A\&C \rightarrow (A\&B) \vee C_\emptyset$, by applying $\vee E$.

[3]The logic DW is MC without A11. $(A \rightarrow B)\&(B \rightarrow C) \rightarrow (A \rightarrow C)$.

$$
\begin{array}{|ll}
TA \vee B_a & \\
TA, TB_a & \text{T}\vee\text{E} \\
\ldots & \\
\ldots & \\
TC, TC_a & \\
TC_b & \text{,E}
\end{array}
$$

Also, the T & I rule is applied within the same subproof, as follows:

$$
\begin{array}{|ll}
TA_a & \\
\ldots, TB, \ldots_a & \\
\ldots, TA\&B, \ldots_a & \text{T\&I}
\end{array}
$$

Note that TB (or TA) can be inside a sequence of commas. This enables distribution to be proved thus:

$$
\begin{array}{ll}
\left| \; TA\&(B \vee C)_{\{1\}} \right. & \\
\hline
\left| \; TA_{\{1\}} \right. & \text{T\&E} \\
\left| \; TB \vee C_{\{1\}} \right. & \text{T\&E} \\
\left| \; TB, C_{\{1\}} \right. & \text{T}\vee\text{E} \\
\left| \; TA\&B, TA\&C_{\{1\}} \right. & \text{T\&I} \\
\left| \; T(A\&B) \vee (A\&C), T(A\&B) \vee (A\&C)_{\{1\}} \right. & \text{T}\vee\text{I} \\
\left| \; T(A\&B) \vee (A\&C)_{\{1\}} \right. & \text{,E} \\
TA\&(B \vee C) \rightarrow (A\&B) \vee (A\&C)_{\emptyset} & \text{T}\rightarrow\text{I}
\end{array}
$$

Here, deductions take place along suitably defined threads of proof within a subproof rather than in separate subproofs as in the classical case. The same applies to the logic MC, as set out in Brady (201+a).

A similar story applies to the quantified distribution axioms (QA3) and (QA6), which we will call universal and existential distribution, respectively. For classical logic, the '⊃' versions are derivable in a Fitch-style natural deduction system based on introduction and elimination rules, as follows:

$$
\begin{array}{ll}
\left| \; A\&\exists x B \right. & \\
\hline
\left| \; A \right. & \text{\&E} \\
\left| \; \exists x B \right. & \text{\&E} \\
\left| \quad \left| \; B^a/_x \right. \right. & \\
\left| \quad \left| \; A\&B^a/_x \right. \right. & \text{\&I} \\
\left| \quad \left| \; \exists x(A\&B) \right. \right. & \exists\text{I} \\
\left| \; \exists x(A\&B) \right. & \exists\text{I} \\
A\&\exists x B \supset \exists x(A\&B) & \supset\text{I}
\end{array}
$$

Note again that $\&I$ is applied to premises from two distinct subproofs, whilst the $\exists I$ and $\exists E$ rules are standardly applied. However, the derivation of $\forall x(A \vee B) \supset (A \vee \forall x B)$ is not straightforward as the $\forall I$ rule cannot be applied to generalise on a free variable in a hypothesis. Instead, we can prove $(\sim A \& \exists x {\sim} B) \supset \exists x(\sim A \& {\sim} B)$, and then apply $\sim$-rules to contrapose it back to $\forall x(A \vee B) \supset (A \vee \forall x B)$. For intuitionist logic, however, its negation properties are not adequate enough to do this and $\forall x(A \vee B) \rightarrow (A \vee \forall x B)$ fails to go through, given that $A \& \exists x B \rightarrow \exists x(A \& B)$ is provable as above.

For relevant logics, in Brady (1984), the corresponding rules $\forall\vee$ and $\exists\&$ (below) are added to introduction and elimination rules for the quantifiers $\forall$ and $\exists$, because the $\&I$ rule, as used in the above classical proof, is not allowable in these natural deduction systems.

$$\forall\vee: \text{ From } \forall x(A \vee B)_a \text{ to infer } A \vee \forall x B_a.$$

$$\exists\&: \text{ From } A \& \exists x B_a \text{ to infer } \exists x(A \& B)_a.$$

These can be used to trivially derive (QA3) and (QA6), respectively.

Again, when normalizing the natural deduction system for DWQ, as can be seen in Brady (201+b) and below, both forms of distribution are absorbed into the structure of the natural deduction system to enable the normalisation to be proved. The same points will apply for MCQ.

$T\forall x(A \vee B)_{\{1\}}$	
$TA \vee B^a/x_{\{1\}}$	T∀E
$TA, TB^a/x_{\{1\}}$	T∨E
$TA, T\forall x B_{\{1\}}$	T∀I
$TA \vee \forall x B, TA \vee \forall x B_{\{1\}}$	T∨I
$TA \vee \forall x B_{\{1\}}$	,E
$T\forall x(A \vee B) \rightarrow (A \vee \forall x B)_{\emptyset}$	T→I

Here, an exception is made in the application of $T\forall$I to allow it to generalise on a variable that only occurs in the one signed formula to which the rule is applied.

$TA \& \exists x B_{\{1\}}$	
$TA_{\{1\}}$	T&E
$T\exists x B_{\{1\}}$	T&E
$,_a TB^a/x_{\{1\}}$	T∃E
$,_a TA \& TB^a/x_{\{1\}}$	T&I
$,_a T\exists x(A \& B)_{\{1\}}$	T∃I
$T\exists x(A \& B)_{\{1\}}$	$,_a$E
$T\forall x(A \vee B) \rightarrow (A \vee \forall x B)_{\emptyset}$	T→I

Here, the '$,_a$' is used to signify that an existential instantiation has taken place with introduced variable a. It is eliminated in the same subproof after the a is eliminated from the signed formula. Again, T&I is applied with one of its premises inside a comma, this time an existential comma.

13.3.2 Gentzen Sequent Calculus

We start with the derivation of $A\&(B \vee C) \supset (A\&B) \vee (A\&C)$ in the classical Gentzen sequent calculus:

$$\dfrac{\dfrac{\dfrac{\dfrac{\dfrac{\dfrac{A \vdash A}{A, B \vdash A} \qquad \dfrac{\dfrac{B \vdash B}{B, A \vdash B}}{A, B \vdash B}}{A, B \vdash A\&B}}{A, B \vdash (A\&B) \vee (A\&B)} \qquad \dfrac{\dfrac{\dfrac{A \vdash A}{A, C \vdash A} \qquad \dfrac{\dfrac{C \vdash C}{C, A \vdash C}}{A, C \vdash C}}{A, C \vdash A\&C}}{A, C \vdash (A\&B) \vee (A\&C)}}{A, (B \vee C) \vdash (A\&B) \vee (A\&C)}}{A\&(B \vee C) \vdash (A\&B) \vee (A\&C)}}{\vdash A\&(B \vee C) \supset (A\&B) \vee (A\&C)}$$

In deriving $A\&(B \vee C) \vdash (A\&B) \vee (A\&C)$ from $A, B \vee C \vdash (A\&B) \vee (A\&C)$, we used the Ketonen form of the (& $\vdash$) rule on the left, which replaces a comma by a '&' (see Ketonen 1944). This can be derived from the original Gentzen rules by two applications of Gentzen's left-side rule $\& - IA$ (available in Gentzen 1969, pp. 83–84), which we express as (& $\vdash$) in current terminology, and an application of the contraction rule on the left, currently expressed as (W$\vdash$). As shown by Gentzen, this proof also holds for intuitionist logic, replacing '$\supset$' by '$\rightarrow$' in the conclusion.

We can instead apply the Ketonen form of the ($\vdash$ $\vee$) rule on the right, which replaces a comma by a '$\vee$', to derive distribution (also see Ketonen 1944). It too can be derived from the original Gentzen rules, by applying the right-side rules ($\vdash$ $\vee$) twice and contraction ($\vdash$W). However, this proof does not satisfy the intuitionist requirement of at most one formula to the right of the '$\vdash$'.

The feature of both these derivations of distribution is the use of contraction either directly or indirectly through the use of Ketonen forms. Given that the Gentzen proof, with a ceiling on the use of contraction, is a decision procedure, it can be seen that there is no proof of distribution without the use of contraction. To see this, we examine again the passage from $A, B \vee C \vdash (A\&B) \vee (A\&C)$ to $A\&(B \vee C) \vdash (A\&B) \vee (A\&C)$ in the above proof. The only rules that can be applied to yield $A\&(B \vee C) \vdash (A\&B) \vee (A\&C)$ are the original Gentzen rules (& $\vdash$) and ($\vdash$ $\vee$), giving the following possible previous lines: $A \vdash (A\&B) \vee (A\&C)$; $B \vee C \vdash (A\&B) \vee (A\&C)$; $A\&(B \vee C) \vdash A\&B$; $A\&(B \vee C) \vdash A\&C$. A simple truth-table test will show that none of these inferences hold and so none of these can be proved in the Gentzen system either.

A similar point can be made for the weaker relevant logics also. In Brady (1996a), left-handed Gentzenisations of a wide range of relevant logics are given, including a proof of distribution for the basic relevant logic B, using signed formulae, an extensionally interpreted comma and an intensionally interpreted colon (Brady 1996a, p. 406). There are no turnstiles '$\vdash$', the sign F represents what would normally be called the right-hand side and the sign T represents what would normally be called the left-hand side of a Gentzen proof. The colon $TA : TB$ is interpreted as co-tenability, $\sim(A \rightarrow \sim B)$, and, being a left-handed system, each step is interpreted negatively. The axioms are of shape $TA : FA$, thus interpreted as $\sim(A \rightarrow \sim\sim A)$, a negated theorem. The theorems are of form $Tt : FA$, establishing the formula A as a theorem of the logic.

Let us examine the proof of distribution of (Brady 1996a, p. 406), below:

$$\dfrac{\dfrac{\dfrac{\dfrac{\dfrac{\dfrac{\dfrac{\dfrac{TA{:}FA}{TA{:}(FA,F(A\&C))}\text{(Ke)}}{T(A\&(B\vee C)){:}(FA,F(A\&C))}\text{(T\&)} \qquad \dfrac{\dfrac{\dfrac{\dfrac{TA{:}FA}{TA{:}(FA,FB)}\text{(Ke)}}{TA{:}(FB,FA)}\text{(Ce)}}{T(A\&(B\vee C)){:}(FB,FA)}\text{(T\&)} \qquad \dfrac{\dfrac{\dfrac{TB{:}FB}{TB{:}(FB,FC)}\text{(Ke)} \quad \dfrac{\dfrac{TC{:}FC}{TC{:}(FC,FB)}\text{(Ke)}}{TC{:}(FB,FC)}\text{(Ce)}}{T(B\vee C){:}(FB,FC)}\text{(T}\vee\text{)}}{T(A\&(B\vee C)){:}(FB,FC)}\text{(T\&)}}{T(A\&(B\vee C)){:}(FB,F(A\&C))}\text{(F\&)}}{T(A\&(B\vee C)){:}(F(A\&B),F(A\&C))}\text{(F\&)}}{T(A\&(B\vee C)){:}(F(A\&B),F((A\&B)\vee(A\&C)))}\text{(F}\vee\text{)}}{T(A\&(B\vee C)){:}(F((A\&B)\vee(A\&C)),F((A\&B)\vee(A\&C)))}\text{(F}\vee\text{)}}{T(A\&(B\vee C)){:}F((A\&B)\vee(A\&C))}\text{(We)}}{Tt{:}(T(A\&(B\vee C)){:}F((A\&B)\vee(A\&C)))}\text{(TtiI)}}{Tt{:}F(A\&(B\vee C)\rightarrow(A\&B)\vee(A\&C))}\text{(F}\rightarrow\text{)}$$

Note the key use of the extensional contraction rule (We) to derive the third-last step $TA\&(B \vee C) : F(A\&B) \vee (A\&C)$ from $TA\&(B \vee C) : (F(A\&B) \vee (A\&C), F(A\&B) \vee (A\&C))$. All of the rules applied to the axioms down to this step correspond to original classical Gentzen rules for weakening, permutation, and conjunction and disjunction rules on the left and right. So, there is an easy translation between this proof and its corresponding classical proof, ignoring the subsequent Tt introduction rule used to state the theorem.

One could have also applied the (We) rule to $TA\&(B \vee C)$ with the above steps following the classical Gentzen proof of $A\&(B \vee C) \supset (A\&B) \vee (A\&C)$, given above, but with the Ketonen step replaced by the original Gentzen steps. As in the classical case, one can see that with the removal of the (We) rule (and the Tt rules), a proof of distribution is not possible in the weaker relevant logics, including the logic DJ, which has the same theorems as MC. Note that, for the stronger logics T and R, there is a general intensional contraction rule (see Brady 1996a, p. 419) and also that these logics are undecidable (see Urquhart 1984).

We now consider the addition of quantifiers and the proofs of the quantified forms of distribution, $\forall x(A \vee B) \supset (A \vee \forall xB)$ and $A\&\exists xB \supset \exists x(A\&B)$. We start with the classical Predicate Calculus:

$$\dfrac{\dfrac{\dfrac{\dfrac{\dfrac{\dfrac{A \vdash A}{A \vdash A, B^a/_x} \qquad \dfrac{\dfrac{B^a/_x \vdash B^a/_x}{B^a/_x \vdash B^a/_x, A}}{B^a/_x \vdash A, B^a/_x}}{A \vee B^a/_x \vdash A, B^a/_x}}{\forall x(A \vee B) \vdash A, B^a/_x}\ [\text{a is new to } A \text{ and } B]}{\forall x(A \vee B) \vdash A, \forall xB}}{\forall x(A \vee B) \vdash A \vee \forall xB}}{\vdash \forall x(A \vee B) \supset (A \vee \forall xB)}$$

$$\dfrac{\dfrac{\dfrac{\dfrac{\dfrac{\dfrac{A \vdash A}{A, B^a/_x \vdash A} \qquad \dfrac{\dfrac{B^a/_x \vdash B^a/_x}{B^a/_x, A \vdash B^a/_x}}{A, B^a/_x \vdash B^a/_x}}{A, B^a/_x \vdash A \& B^a/_x}}{A, B^a/_x \vdash \exists x(A \& B)}\ [\text{a is new to } A \text{ and } B]}{A, \exists xB \vdash \exists x(A \& B)}}{A \& \exists xB \vdash \exists x(A \& B)}}{\vdash A \& \exists xB \supset \exists x(A \& B)}$$

Note the usage of the Ketonen disjunction rule in passing from $\forall x(A \vee B) \vdash A, \forall xB$ to $\forall x(A \vee B) \vdash A \vee \forall xB$ and the Ketonen conjunction rule in passing from $A, \exists xB \vdash \exists x(A\&B)$ to $A\&\exists xB \vdash \exists x(A\&B)$. Also, without the contraction rule, embodied in these Ketonen forms, neither of these quantified distributions would be provable. This can be seen from the non-provability of the various possible preceding steps in lieu of $\forall x(A \vee B) \vdash A, \forall xB$ and $A, \exists xB \vdash \exists x(A\&B)$ in the above proofs. These would be due to the application of ($\forall \vdash$) or ($\vdash \vee$) to obtain $\forall x(A \vee B) \vdash A \vee \forall xB$, or ($\& \vdash$) or ($\vdash \exists$) to obtain $A\&\exists xB \vdash \exists x(A\&B)$. None of these preceding steps are valid in Predicate Calculus. As a result of this, for intuitionist logic, $\forall x(A \vee B) \rightarrow (A \vee \forall xB)$ is not derivable, as at most one formula can appear on the right, but $A\&\exists xB \rightarrow \exists x(A\&B)$ is derivable by the above proof.

The Gentzenisation for quantified relevant logic appears in (Brady 2003 pp. 350–351), where the same four signed quantifier rules are added to the sentential Gentzenisations of Brady (1996a). Similarly to the sentential case, the derivations down to $T\forall x(A \vee B) : FA \vee \forall xB$ and down to $TA\&\exists xB : F\exists x(A\&B)$ correspond to steps in the classical derivations of $\forall x(A \vee B) \vdash A \vee \forall xB$ and $A\&\exists xB \vdash \exists x(A\&B)$, respectively. Similarly to the case of classical predicate logic, with the removal of the (We) rule (and the Tt rules), proofs of these two forms of distribution are not possible in the weaker quantified relevant logics, including the logic MCQ. Furthermore, in Brady (1996b,c), Gentzenisations of distributionless relevant logics are shown to be much simpler due to the fact that the extensional comma is not needed and not included.

13.3.3 General Comments

As can be seen from the Natural Deduction section, the key to the proof of sentential distribution in Fitch-style natural deduction depends on allowing the &I rule to apply to premises in different subproofs (for classical and intuitionist logic) or on enforcement (for relevant logics in general) or on defining threads of proof within a subproof such as to allow T&I to apply to premises in the same thread for normalised deduction (for weak relevant logics). A similar dependency applies for existential distribution. Apart from the roundabout classical proof of universal distribution and the enforced proof, the normalised proof requires a special allowance for $T\forall$I to apply adjacent to a comma to enable a straightforward proof to go through. Summing up, the straightforward derivation of the forms of distribution using connective and quantifier rules essentially depend on structural properties of the natural deduction proofs.

A similar conclusion can be drawn for the Gentzen sequent calculus. The proof of sentential distribution essentially depends on the extensional contraction rule (and, to a lesser extent, weakening), and also for the quantified forms of distribution. Again, extensional contraction is a structural rule, as opposed to a connective or quantifier rule. This dependency of sentential distribution on structural rules has also been pointed out by Belnap 1993, pp. 31–32.

Another differentiation between logics with and without distribution has been made by Restall and Paoli (2005). They use special directed graphs to encode proofs in logics without distribution and show that they have a natural geometry that distributive logics do not have.

13.3.4 Schroeder-Heister's Proof-Theoretic Semantics

We continue in this vein by examining some research by Schroeder-Heister on what he calls 'proof-theoretic semantics'. He introduces the concept in Schroeder-Heister (1991) for logical constants and in Schroeder-Heister (2006) provides validity concepts based on proof theory rather than on truth-functional semantics. His philosophical justification for this is that meaning is usage and the usage here consists of the application of proof-theoretic rules. Thus, the meaning of each logical concept will be understood in terms of the rules governing the concept. His main focus is on natural deduction, but the points we need to make carry over to Gentzen sequent calculi as well.

In a tutorial Schroeder-Heister (2007) introduced the uniqueness of the logical connectives for classical logic. For example, $A \supset B \vdash A \supset' B$ and $A \supset' B \vdash A \supset B$ are both derivable from their common introduction and elimination rules. Thus, '$\supset$' is said to be unique as $(A \supset B) \equiv (A \supset' B)$ is derivable, establishing the equivalence of $A \supset B$ with any $A \supset' B$ which has the same introduction and elimination rules. What will concern us here is the uniqueness of conjunction and disjunction, obtained as follows:

$$\begin{array}{|l} A\&B \\ \hline A \\ B \\ A\&'B \end{array} \qquad \begin{array}{|l} A\&'B \\ \hline A \\ B \\ A\&B \end{array} \qquad \begin{array}{|l} A \vee B \\ \hline \quad \begin{array}{|l} A \\ \hline A \vee' B \end{array} \\ \quad \begin{array}{|l} B \\ \hline A \vee' B \end{array} \\ A \vee' B \end{array} \qquad \begin{array}{|l} A \vee' B \\ \hline \quad \begin{array}{|l} A \\ \hline A \vee B \end{array} \\ \quad \begin{array}{|l} B \\ \hline A \vee B \end{array} \\ A \vee B \end{array}$$

Hence, $(A\&B) \equiv (A\&'B)$ and $(A \vee B) \equiv (A \vee' B)$ are both derivable. Thus, '&' and '$\vee$' are uniquely determined by their introduction and elimination rules, which determine their meaning in proof-theoretic semantics. Due to these equivalences, they are also inter-substitutible into any context. The same applies to intuitionist and relevant logics as $A\&B \leftrightarrow A\&'B$ and $A \vee B \leftrightarrow A \vee' B$ are similarly derivable.

However, this uniqueness is established independently of distribution, which then means that distribution is independent of the meaning of conjunction and disjunction in this "semantics". This takes further what we have been saying above, viz. that distribution essentially depends on structural considerations of the natural deduction system over and above the introduction and elimination rules. In proof-theoretic semantics, distribution cannot be absorbed into the introduction and elimination rules for conjunction and disjunction, as it combines both of these, and thus relies on structural adjustment for its provability. This enables one to have what we could call different "proof-theoretic modellings", both with and without distribution, according to whether the structural adjustment is present or absent.

13.4 Distribution in Semantics

We next examine distribution in the context of truth-functional semantics, Venn diagrams, algebraic semantics and content semantics, to see what light these various styles of semantics can bring to this discussion.

13.4.1 Truth-Functional Semantics

Let us first examine the standard semantics of classical logic. Distribution follows from the truth-conditions for conjunction and disjunction, as follows:

$I(A\&B) = T$ iff $I(A) = T$ and $I(B) = T$.

$I(A \vee B) = T$ iff $I(A) = T$ or $I(B) = T$.

$I(A \supset B) = T$ iff, if $I(A) = T$ then $I(B) = T$.

Thus, $I(A\&(B \vee C) \supset (A\&B) \vee (A\&C)) = T$

iff, if $I(A\&(B \vee C)) = T$ then $I((A\&B) \vee (A\&C)) = T$

iff, if $I(A) = T$ and $I(B \vee C) = T$ then $I(A\&B) = T$ or $I(A\&C) = T$

iff, if $I(A) = T$ and $(I(B) = T$ or $I(C) = T)$

then $(I(A) = T$ and $I(B) = T)$ or $(I(A) = T$ and $I(C) = T)$.

However, this last line is just distribution in the logic of the meta-theory. As is standard practice, the meta-logic of truth-functional semantics is classical and this applies to these three connectives in particular, making distribution true, for all interpretations, and hence valid. So, whether distribution holds or not in truth-functional semantics entirely depends on whether or not it holds in its meta-logic. This is just circular, as the issue of whether distribution holds in classical logic or not is just passed from the object logic to the meta-logic, which is assumed to be classical and in which distribution holds. The semantics does not tell us why or how distribution holds and thus does not do its proper semantical job.

A similar situation applies to intuitionist logic and relevant logic. The Kripke semantics for intuitionist logic has the following truth-conditions:

$I(A\&B, a) = T$ iff $I(A, a) = T$ and $I(B, a) = T$.

$I(A \vee B, a) = T$ iff $I(A, a) = T$ or $I(B, a) = T$.

$I(A \rightarrow B, a) = T$ iff, for all $b \geq a$, if $I(A, b) = T$ then$I(B, b) = T$.

Thus, $I(A\&(B \vee C, a) \rightarrow (A\&B) \vee (A\&C), a) = T$

iff, for all $b \geq a$, if $I(A\&(B \vee C), b) = T$ then $I((A\&B) \vee (A\&C), b) = T$

iff, for all $b \geq a$, if $I(A, b) = T$ and $(I(B, b) = T$ or$I(C, b) = T)$

then $(I(A, b) = T$ and $I(B, b) = T)$ or $(I(A, b) = T$ and $I(C, b) = T)$.

The truth of distribution at world a depends on whether distribution holds at such worlds b, which in turn depends on whether distribution holds in the meta-logic. Since the meta-logic is again assumed to be classical, distribution holds for intuitionist logic as well.

For relevant logics, we will first use the Routley-Meyer semantics, which evaluates distribution in a very similar way to that of the Kripke semantics for intuitionist logic. Being of the form $A \rightarrow B$, we only need the following lemma:

$A \rightarrow B$ is valid iff, for all model structures M, for all interpretations I on M, for all worlds b in M, if $I(A, b) = T$ then $I(B, b) = T$.

So, again, the validity of distribution depends on whether distribution holds at all worlds b, which in turn depends on whether distribution holds in the meta-logic, which is classical.

Also, in Read's homophonic semantics for the relevant logic R in Read (1988), the meanings of the connectives δ in the object language are maintained in the metalanguage to provide their truth-conditions, that is, '$\delta A_1 \ldots A_n$' is true iff δ (A_1 is true, ..., A_n is true) (see Read 1988, p. 156). So, again, distribution will be valid for the logic R iff it is in the meta-logic, which it is, as the meta-logic is also R.

We now consider the quantified forms of distribution. For the standard semantics of classical predicate logic, we need to add the truth-conditions:

$$I(\forall xA) = T \text{ iff, for all } a\text{-variants } I' \text{ of } I \text{ with the domain } D, I'(A^a/_x) = T.$$

$$I(\exists xA) = T \text{ iff, for some } a\text{-variant } I' \text{ of } I \text{ with the domain } D, I'(A^a/_x) = T.$$

For these to work, we need to choose a so that it does not occur in A.

$$\begin{aligned}
&\text{Then, } I(\forall x(A \vee B) \supset (A \vee \forall xB)) = T \\
&\quad \text{iff, if } I(\forall x(A \vee B)) = T \text{ then } I(A \vee \forall xB) = T \\
&\quad \text{iff, if, for all } a\text{-variants } I' \text{ of} I,\ I'(A \vee B^a/_x) = T \text{ then } I(A) = T \text{ or} \\
&\qquad I(\forall xB) = T \\
&\quad \text{iff, if, for all } a\text{-variants } I' \text{ of } I,\ I(A) = T \text{ or } I'(B^a/_x) = T \\
&\qquad \text{then } I(A) = T \text{ or, for all } a\text{-variants } I' \text{ of } I,\ I'(B^a/_x) = T.
\end{aligned}$$

Note that, for any a-variant I' of I, $I'(A) = I(A)$, since a does not occur in A.

As for the case of sentential distribution, this last line is just universal distribution in the logic of the meta-theory. Again, because of the classicality of the meta-logic, universal distribution is true for any interpretation, and hence valid. Also, this is circular as the issue of whether universal distribution holds in classical predicate logic or not is again passed from the object logic to the meta-logic. The same applies for existential distribution since:

$$\begin{aligned}
&I(A\&\exists xB \supset \exists x(A\&B)) = T \\
&\quad \text{iff, if } I(A) = T \text{ and, for some } a\text{-variant } I' \text{ of } I, I'(B^a/_x) = T \\
&\qquad \text{then, for some } a\text{-variant } I' \text{ of } I,\ I(A) = T \text{ and } I'(B^a/_x) = T.
\end{aligned}$$

For intuitionist logic, variable domain semantics should be used so that if $a \geq b$ then $D_a \supseteq D_b$, where D_a is the domain of a. For existential distribution:

$$\begin{aligned}
&I(A\&\exists xB \to \exists x(A\&B), b) = T \\
&\quad \text{iff, for all } c \geq b,\ \text{if } I(A, c) = T \text{ and, for some } a\text{-variant } I' \text{ of } I, \\
&\quad I'(B^a/_x, c) = T \\
&\qquad \text{then, for some } a\text{-variant } I' \text{ of } I,\ I(A, c) = T \text{ and } I'(B^a/_x, c) = T.
\end{aligned}$$

This holds because a can be taken from the domain of c and the same a-variant can be made to apply in both instances, and because existential distribution holds in the meta-logic for the world c and domain D_c.

However, for universal distribution:

$$I(\forall x(A \vee B) \to (A \vee \forall xB, b)) = T$$

$$\text{iff, for all } c \geq b, \text{ if, for all } d \leq c, \text{ for all } a\text{-variants } I' \text{ of } I, I(A, d) = T \text{ or } I'(B^a/_x, d) = T$$

$$\text{then } I(A, c) = T \text{ or, for all } d \leq c, \text{ for all } a\text{-variants } I' \text{ of } I, I'(B^a/_x, d) = T.$$

As explained by Dummett, the universal quantifier is evaluated over all worlds $d \leq c$ with domains $D_d \supseteq D_c$ and this ensures that universal distribution is invalid (Dummett 1977, pp. 202–204). Universal distribution would be valid if universal quantification was evaluated using a constant domain, and this would follow using classical meta-logic (Dummett 1977, p. 202).

For relevant logic, Mares & Goldblatt's alternative truth-functional semantics for QR (Mares and Goldblatt 2006) is much more intuitive and also much neater than the original semantics in Fine (1988). The evaluation of universal quantification $\forall xA$ is based on an LA proposition entailing A for each individual. This semantics does not validate universal distribution, called 'extensional confinement' in their paper. Because of the De Morgan negation properties and their quantificational counterparts, unlike intuitionist logic, existential distribution is also invalid. However, Mares and Goldblatt do consider adding universal distribution to their semantics but it is at the expense of adding the Boolean difference between propositions for that purpose. As was also mentioned earlier, Brady and Rush (2008) drop universal and existential distribution to prevent classicality spreading from atomic statements of Peano arithmetic to all quantified arithmetical statements. In particular, this allows m = n to be classical, without spreading classicality all the way to the Gödel sentence, where it is inappropriate for the concerns of Brady and Rush (2008) and our subsequent work.

13.4.2 Venn Diagrams

The quickest and easiest representation of distribution is that given by Venn diagrams. We consider three circles representing the membership of sets X, Y and Z. They are drawn so that each of the eight possible intersections, called segments, are shown. Each of these segments represents a union of X or $\overline{X}$ with Y or $\overline{Y}$ and with Z or $\overline{Z}$, where $\overline{X}$, $\overline{Y}$, and $\overline{Z}$ are the respective complements.

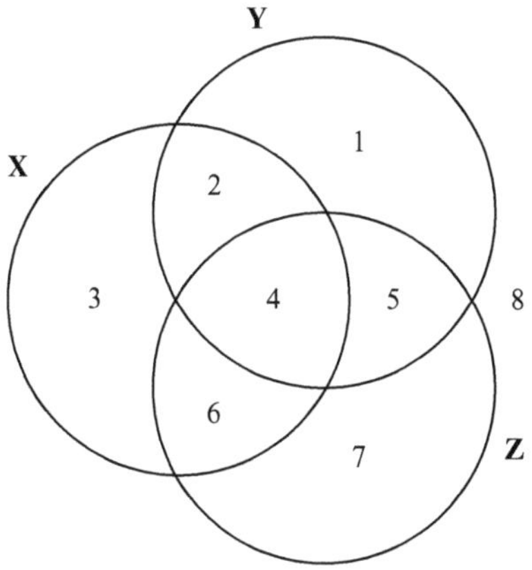

Instead of working with conjunction and disjunction, we work with intersection and union, respectively. So, what we need to examine is:

$$X \cap (Y \cup Z) = (X \cap Y) \cup (X \cap Z).$$

Using the segment numbers, both sides of the identity are represented by the totality of the segments 2, 4 and 6. Each of these three segments represents an intersection, as follows:

2: $X \cap Y \cap \overline{Z}$,
4: $X \cap Y \cap Z$,
6: $X \cap \overline{Y} \cap Z$.

So, what the Venn diagram is telling us is that $X \cap (Y \cup Z)$ and $(X \cap Y) \cup (X \cap Z)$ are both identical with $(X \cap Y \cap \overline{Z}) \cup (X \cap Y \cap Z) \cup (X \cap \overline{Y} \cap Z)$, a disjoint union.

How do we justify this latter identity in logical terms? We need to create the segments to do this and we start by dividing up the elements of a single class into two in accordance with their presence or absence in a second class. We thus identify Y with $Y \cap (Z \cup \overline{Z})$ and distribute to obtain $(Y \cap Z) \cup (Y \cap \overline{Z})$, thus splitting Y up into two segments based on Z. Similarly $Z = (Y \cap Z) \cup (\overline{Y} \cap Z)$, from which $Y \cup Z = (Y \cap Z) \cup (Y \cap \overline{Z}) \cup (\overline{Y} \cap Z)$. In order to obtain the left-hand side, $X \cap (Y \cup Z)$, we distribute the X over this union to obtain $(X \cap Y \cap \overline{Z}) \cup (X \cap Y \cap Z) \cup (X \cap \overline{Y} \cap \overline{Z})$. We can see the strong usage of the Law of Excluded Middle in the form $X \cup \overline{X} = U$, where U is the universe, and the distribution of intersection over union in this justification. The right-hand side, $(X \cap Y) \cup (X \cap Z)$, also uses the LEM and distribution, as follows:

$$X \cap Y = X \cap Y \cap (Z \cup \overline{Z}) = (X \cap Y \cap Z) \cup (X \cap Y \cap \overline{Z}),$$

$$X \cap Z = X \cap Z \cap (Y \cup \overline{Y}) = (X \cap Z \cap Y) \cup (X \cap Z \cap \overline{Y}) \text{ and}$$

$$(X \cap Y) \cup (X \cap Z) = (X \cap Y \cap \overline{Z}) \cup (X \cap Y \cap Z) \cup (X \cap \overline{Y} \cap Z)$$

Thus, despite being an obvious medium for displaying distribution, the use of Venn diagrams assumes distribution and the LEM in the setting up of the segments in the diagrams in order for distribution to be shown.

13.4.3 Algebraic Semantics

We just consider sentential algebraic semantics in a general way so as to make our point. The basic properties of conjunction and disjunction are generally represented by a lattice structure. This captures the following axioms: $A\&B \to A$, $A\&B \to B$, $(A \to B)\&(A \to C) \to (A \to B\&C)$, $A \to A \vee B$, $B \to A \vee B$, $(A \to C)\&(B \to C) \to (A \vee B \to C)$. If distribution holds in the logic, then the lattices would be distributive. So, lattices are distributive or not in accordance with the presence or absence of distribution in the logic.

13.4.4 Content Semantics

A particular kind of algebraic semantics is the content semantics for the relevant logic MC of meaning containment. Unlike the algebraic semantics for relevant logics which, like truth-functional semantics, caters for a broad range of logics, content semantics by and large does the job of pinning down a particular logic, that of MC. In the content semantics, unlike algebraic semantics generally, there is real set-theoretic intersection and containment, used to capture disjunction and entailment, respectively. Conjunction is captured by an analytically closed union, since plain union is inadequate to deal with entailments from conjunctions. The axioms and rules of the logic, given earlier, can be seen to reflect the set-theoretic properties of containment, which suits this kind of semantics. Nevertheless, in developing the content theory in Brady (1996, 2006) using the standard lattice properties of conjunction and disjunction, the justification of distribution is not entirely satisfactory, as pointed out by Restall (2007).

Setting this out, the elements of the content semantics consist of the logical contents $c(A)$ of formulae A. For the connectives, $c(A\&B) = c(A)\overline{\cup}c(B)$, $c(A \vee B) = c(A) \cap c(B)$, $c(\neg A) = c(A)*$ and $c(A \to B) = c(c(A) \supseteq c(B))$, where '$\cap$' and '$\supseteq$' are understood set-theoretically, '$\overline{\cup}$' is an analytically closed set-theoretic union and '$*$' is defined in Brady (2006) in terms of a corresponding theory of ranges. However, negation will be of less concern to us here. Note that we can have contents of containment statements themselves, which are seen as a specific type of statement shape in comparison to the general statements that the formulae can represent. Distribution of '$\overline{\cup}$' over '$\cap$' for contents is introduced in Brady (2006, pp. 18–19) on the grounds that it is a property of set-theoretic containment, which is what the axioms were seen to reflect. However, there is a certain circularity in the justification for this, given in footnote 12, in that the analyticity concept, used in defining contents does embody the distribution property itself. Thus, as we saw in the Proof-Theoretic Semantics, the introduction and elimination rules for conjunction and disjunction do not enable distribution to be established, and the same applies to the lattice properties.

13.4.5 General Comment

So, the general message from the various semantics is that of circularity, and that these forms of semantics do not do their proper semantic job of determining whether distribution should hold or not. Even the content semantics that does provide rationale for the rejection of a number of key non-theorems of MC (see Brady 2006, pp. 29–30) and provides support for its axioms and rules (see Brady 2006, pp. 13–38) does not provide non-circular support for distribution, despite it being a standard set-theoretic property of containment. That leads us to Venn diagrams as a simple visual means of representing set-theoretic containment properties, but even they do not escape the charge of circularity.

13.5 Distribution in Quantum Mechanics

The theory of quantum mechanics gives rise to a Hilbert space (i.e. a special kind of vector space), which is a technical term for a collection of objects governed by certain mathematical rules. In this case it is basically the collection of all possible states of a physical system, equipped with the mathematical rules that dictate how the states behave together. A subspace is any part of this space which in itself has the required properties of a vector space (which then will make it a Hilbert space). The subspaces of the quantum mechanical Hilbert space have a one-to-one correspondence with statements regarding the system. For this reason we are interested in the subspaces and the relations between them, and we call this quantum logic.

Opinions are divided as to whether or not there is a need for a separate logic for quantum mechanics. There are those who feel that no new logic is necessary (see Kochen and Specker 1965; Ludwig 1983), rather that all information gained from quantum mechanics is described by classical logic. Others believe that if we can talk reasonably about two different atomic statements in quantum mechanics, then we should naturally be able to talk also of their conjunction or disjunction. In this case the rules of classical logic lead to problems, as we shall see below, and one needs a new logic. A quantum logic can be realised in different ways, either by a two-valued logic, a many-valued logic, or a fuzzy logic. We will restrict ourselves to the traditional two-valued logic initiated by Birkhoff and von Neumann (1936), and the proposal for a 3-valued logic in Reichenbach (1994).

13.5.1 Two-Valued Quantum Logic

The study of standard two-valued quantum logic was initiated by Birkhoff and von Neumann (1936). They came to the conclusion that statements in quantum mechanics correspond to the closed subspaces of a Hilbert space, i.e. closed with respect to the strong topology. (The technical details of this topology are not necessary to the following discussion.) The relation between these is pictured in mathematical terms as an orthomodular lattice (see below) with the inclusion relation as the 'less than' relation. This is in contrast to classical theories, in which statements correspond to subsets of a set. This would give the structure of a Boolean lattice, in other words classical logic. Rather than being derived from the theory of quantum mechanics, orthomodularity is a property which has been shown to hold for the structure of subspaces of a Hilbert space. Since this is the structure of the propositions of quantum mechanics, these propositions form an orthomodular lattice. If one disregards orthomodularity one has a structure called an ortholattice (see below), which gives what is sometimes termed minimal quantum logic or orthologic. (See, e.g. Dalla et al. 2004.)

We give some definitions. A *lattice* is a set P with a partial order relation '$\leq$', where any two elements x and y have a greatest lower bound or "meet", $x \wedge y$, and a least upper bound, or "join", $x \vee y$.

An *ortholattice* is a lattice which has a least and a greatest element with respect to '$\leq$', i.e. it is bounded, and an *orthocomplementation* x', which is a unary operation on P satisfying (1)–(3), for all x and y in P:

1. if $x \leq y$ then $y' \leq x'$,
2. $x' = x$,
3. (a) $x \vee x'$ and (b) $x \wedge x'$ exist and are equal to the greatest and least element, respectively.

The greatest and least element express truth and falsity, respectively, while the orthocomplementation serves as negation, and so (3)(a) is in effect the Law of Excluded Middle.

The *orthomodular law* states that:

4. if $x \leq y$, then $(x \vee (x' \wedge y)) = y$.

An *orthomodular lattice* is one that satisfies (1)–(4).

The statements we are concerned with, the so called *experimental propositions*, are statements regarding the value of some observable, in other words some measurable quantity such as position or momentum. Every observable is related to a mathematical operator. When the operator acts in a calculation on the state of a quantum mechanical system, it gives the value of the associated observable. In this way, theory can be related to experiment. We restrict ourselves to yes-no questions, since any question can be expressed with a suitable combination of these, and it allows us to compare answers to different questions in a rigorous manner.

Thus, an experimental proposition S is a statement such as "the particle has energy 100 eV". When we perform a suitable experiment on the system, that is make a measurement of the observable in question, the energy, we then get a value which either is or is not 100 eV, giving us either a 'yes' or 'no' to that statement. However, this is not so cut and dried as it may initially appear, as we may get different readings at different times. So, we use the probability $P(S)$ to express the probability of a 'yes' as opposed to a 'no' answer to the proposition S. If we are certain to always get a 'yes', i.e. $P(S) = 1$, we say the proposition is true. Naturally if we always get a 'no', i.e. $P(S) = 0$, the statement is false. However, there are many situations in which we have some probability less than 1 and greater than 0 of getting a 'yes', i.e. $0 < P(S) < 1$. If we wish to have a two-valued logic, these statements, which can hardly be called true since they don't always hold, are given the other value. Thus, our two values are actually true and not true, rather than true and false. In what follows, we use the values 1 and 0 for true and not-true, respectively.

The state of a quantum mechanical system can be described as a vector in a Hilbert space. If the so-called state vector lies within the subspace associated with the experimental proposition in question, then the proposition is true. If it lies orthogonal to the subspace, then the proposition is false. (This occurs when the projection of the orthogonal vector on the subspace is zero, which translates into a zero probability of getting a 'yes' to the experimental proposition related to the subspace.) If the vector lies neither within nor orthogonal to the subspace, then there is a probability of this happening, but no certainty. This results in the 0-value in this two-valued logic.

Now, consider a one particle system (prepared in a certain way), and the experimental proposition 'the spin of the particle in the x-direction is up or down'. If we actually measure the spin in the x-direction, we will always get a result which is either up or down. There is no other possibility. So this proposition is always true. However, sometimes we will get up, and sometimes down, and there is no way of knowing beforehand which it will be. So neither "spin up" nor "spin down" are true independently. This is unavoidable in the two-valued quantum logic, and follows directly from the theory of quantum mechanics. Once we actually measure the spin of a particle, it will be completely that which we measure, so that a second measurement of the same particle will give the same result. Here, this is so, because the particle is not disturbed between the two measurements. However, if we disturb the system after the measurement, one or the other will be true, but not necessarily before. Therefore, quantum logic does not have the disjunction property. One can have a disjunction which is true without either one of the disjuncts being true on its own.

Failure of distribution hinges on this property of disjunction. This can be demonstrated in the following way. Let $[x+]$ and $[x-]$ denote that the x-component of spin is up and that it is down, respectively, and correspondingly for the z-component. Now $([x+]\&[z+]) \vee ([x+]\&[z-])$ is a statement which is always false, since both disjuncts are always false. This is because, in both cases, any time one of the conjuncts, say $[x+]$, is true the other, $[z+]$ or $[z-]$, has only a 0.5 probability of being true, so it is given value 0. This is due to the quantum mechanical principle that spin cannot be measured in two orthogonal directions at the one time, which can be represented by the non-commutativity of their operators in a Hilbert space. Thus the whole conjunction has value 0. The following equivalence is logically correct if we assume classical logic:

$$\begin{aligned}([x+]\&[z+]) \vee ([x+]\&[z-]) &\equiv [x+]\&([z+] \vee [z-]) \\ &\equiv [x+]\&\top \\ &\equiv [x+],\end{aligned}$$

where $\top$ is a constant with truth-value 1.

The first equivalence applies the law of distribution. Since the following statements are all equivalent, the disjunction $[z+] \vee [z-]$ being always true, for the reason explained above for the x-component, this means that $[x+]$ should always be false, which obviously isn't the case since there are circumstances in which it is true. Thus we have a contradiction. This shows that a consistent two-valued logic for quantum mechanics must be one where distribution is not valid.

13.5.2 Reichenbach's Three-Valued Logic

One way of dealing with statements that are neither certainly true nor certainly false is with probability logic, but this doesn't take into account the inherent indeterminism in quantum mechanics. In probability logic the assumption is that anything can in principle be determined if one has enough information about a system. In quantum mechanics, however, no amount of information would make it possible to determine, say, position and momentum of a particle simultaneously. In order to take into account the complete impossibility of answering certain questions, Reichenbach (1994) proposes a 3-valued logic. In addition to true and false, he assigns a third value, indeterminate, to statements about unobserved values which cannot be measured together with another already determined value. This approach tacitly assumes one is considering simultaneous values of the observables, since there is nothing restricting measurement of the second value at a later time, other than that it may affect the value of the previously measured observable.

In terms of the Hilbert space described above, the statements assigned the indeterminate value are the ones which lie neither within nor orthogonal to the subspace in question, and which in the two-valued approach were given the value 0. In addition, and essentially, they concern values which have an operator that does not commute with that of an already measured value. (If $AB - BA = 0$, A and B are said to commute. Else their order is not arbitrary, i.e. they do not commute). Specifically they are statements concerning simultaneous values of certain physical observables such as position and momentum, spin in two different directions, energy and time, and so on. These are examples of things which cannot be measured simultaneously, and so a conjunction or disjunction of these can never be verified. Only properties corresponding to operators that commute can be measured simultaneously. An example would be the three components of momentum in the x, y, and z directions.

The connectives in Reichenbach's logic include three different implications and three negations. Distributivity holds in this logic, but the Law of Excluded Middle does not, since in effect the indeterminate value is the middle value. The idea of the 3-valued logic is that unobservable values exist and so must be taken account of in a quantum logic.

13.5.3 Classical Logic

Among physicists advocation for classical logic is strong, in part because quantum logic has failed to be particularly fruitful in regards to physics. The arguments for classical logic made by Griffiths (2002) hold that certain statements are nonsensical, that certain questions simply cannot be asked or even expressed within the theory of quantum mechanics, and are thus meaningless. A similar notion of validity of a quantum mechanical statement was already developed by Kochen and Specker (1965). The meaningless statements do not correspond exactly to statements assigned the indeterminate value by Reichenbach. Many of those statements can be perfectly meaningful, i.e. they can be both expressed and measured. It's just that the outcome of the measurement is uncertain and a previously measured value may be disturbed. In particular, the meaningless statements in Griffiths' approach are all conjunctions or disjunctions. Atomic statements don't have this problem, since any single value can always be both expressed and measured.

For example, Reichenbach assigns the value indeterminate to certain atomic statements which at a given time cannot be measured without disturbing other already determined values. Griffiths is not actually concerned with assigning truth-values, so these atomic statements are perfectly legitimate in his view, only trying to combine them with certain others will not work. And this is because the mathematical language cannot express such a combination.

Let us now consider the well-known 2-slit experiment, which demonstrates that it is impossible to determine position and momentum simultaneously. This is related to the infamous particle-wave duality. If we let photons travel through two slits in a wall to a screen behind the wall, they will exhibit an interference pattern such as that from waves. However, if we decide to determine the path which the particles take by placing a sensor at one of the slits, the interference pattern on the screen disappears, and the photons act as point-particles. The mathematics describing wave motion is linked to the momentum operator, while the mathematics describing motion like that of a point-particle is linked to the position operator. These two operators do not commute, meaning that the result is different depending on which one is taken first in a calculation. Therefore, on a fundamental mathematical level they cannot be applied at the same point in time, since then there would be no well-defined order, and thus no unambiguous result. In other words it is not meaningful to speak of momentum and position simultaneously, and the 2-slit experiment shows that in fact we can only ever have access to one of the two at a given time.

This hinges on technicalities of the Hilbert space of the system. A Hilbert space can be divided into a so-called spectrum, which is different depending on the observable, or operator, in question. In order to consider two observables simultaneously, their spectra must be compatible. One cannot combine a subspace of one spectrum with a subspace of another incompatible one, and so such a statement would be meaningless.

The meaningful statements form a Boolean sublattice of the larger orthomodular lattice and so indeed if one restricts oneself to such statements they obey classical logic. The final result of all approaches is the same; they all agree with experimental results. Statements which are always true or always false are meaningful and their values coincide in all three approaches, while statements that are meaningless in Griffiths' interpretation end up with truth value 0 in the Birkhoff and von Neumann approach, and are indeterminate in Reichenbach's approach, so either way this is consistent with the fact that they will never be verified by experiment. The only difference really is at what point in the argument one rules out certain events, but the end result is the same.

Birkhoff and von Neumann single out the atomic propositions of quantum theory and their internal relations. These are expressed as sentences in a logical language, and the logical connectives are applied to them. Essentially Reichenbach does the same, although he takes into account the fact that some things cannot be measured. However, he believes these things can be asked, just not answered as true or false. Griffiths, on the other hand, adds "logical" connectives on a mathematical theory level, where certainly some combinations of statements cannot even be expressed because of the mathematics involved. By restricting discussion to only those compound sentences which this theory itself expresses, the result is that all those statements which would violate classical logic are forbidden by the fact that the language of the (mathematical) theory does not express them. Thus no new logic is needed.

13.5.4 General Comments

We will present arguments to enable us to pick one of these three logical alternatives, also taking into account Putnam's influential work on the rejection of distribution. We will then, in the conclusion show how this logic fits into the logic MC of meaning containment and determine what form distribution should then take.

Birkhoff and von Neumann's rejection of distribution can be seen to follow from their pooling together of all the probabilities less than 1 into one value, viz. 0. This will mean that statements with probabilities of $^1/_2$ and 0, say, would be regarded as logically equivalent, and it is not surprising that oddities would follow from this. In the example given in the section on Two-valued Quantum Logic, $[z+] \vee [z-]$ takes the value 1 because of its probability and $[x+]$ also takes 1 because it is taken to be contingently true, but $[x+]\&[z+]$ takes 0 as $[z+]$ and hence $[x+]\&[z+]$ has a probability of $^1/_2$. And, $[x+]\&[z-]$ takes 0 for the same reason. So, instead of the probability of their disjunction being summed to 1, the disjunction takes 0, refuting distribution. Indeed, it would be better to use probability theory totally than to regard non-truth as falsity, as this value covers too large a range of statements.

So, we proceed with what has become the standard philosophical account in which distribution fails, which will also throw some light back on Birkhoff and von Neumann. This is based on the series of papers in Putnam (1975), especially his paper "The Logic of Quantum Mechanics". Although he refers to Birkhoff and von Neumann, he does not refer to their pooling of non-truth into the one false value. However, the failure of distribution does derive from the mathematical modelling, i.e. the Hilbert spaces. On p. 178, he defines the disjunction space $S(p \vee q)$ as the span of the spaces $S(p)$ and $S(q)$, and the conjunction space $S(p\&q)$ as the intersection of the spaces $S(p)$ and $S(q)$. The disjunction is "always true" if its space spans the whole Hilbert space and the conjunction is false if $S(p)$ and $S(q)$ are orthogonal, yielding a single point space for the conjunction. This suffices to falsify the consequent of distribution, whilst maintaining the truth of its antecedent.

We can also construct a sublattice of an ortholattice exhibiting this failure of distribution, in the case of electron spin in the orthogonal x- and y-directions:

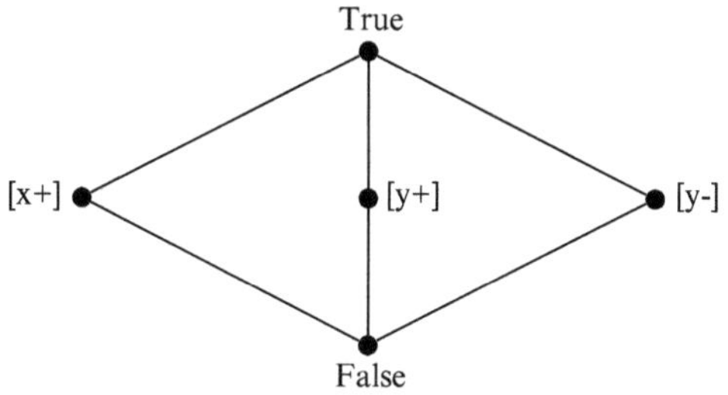

$[y+] \vee [y-]$ is necessarily true on this account, with $[x+]$ a contingent truth implying it, and $[y+]\&[y-]$ is necessarily false and thus implies $[x+]$. None of $[x+]$, $[y+]$ and $[y-]$ imply each other, as they are different spin directions or different spin readings. And, this lattice is of the standard shape for rejecting distribution. To see this, $[x+]\&([y+] \vee [y-]) = [x+]\&\mathit{True} = [x+]$, whilst $([x+]\&[y+]) \vee ([x+]\&[y-]) = \mathit{False} \vee \mathit{False} = \mathit{False}$.[4]

However, there are also problems with this account. Firstly, a logic is a general reasoning system, enabling one to deduce conclusions from premises, usually covering a wide domain of argument. Putnam is modelling quantum logic using Hilbert spaces which correspond one-to-one with quantum mechanical statements, and orthomodular quantum logic is what one gets from this process. This is at best a logic specifically for quantum mechanics, but how does this fit into ordinary reasoning, which we would expect to be able to use not only for quantum statements but also quite generally? Nevertheless, we should not use mathematical modelling

[4]Thanks to Bob Meyer and Greg Restall for pointing out this lattice to the first author.

as a justification for the failure of distribution. And, sure, there are many types of non-distributive modelling, exemplified by the above lattice. To pursue this question further we pass on to another problem.

Secondly, there is a strict adherence to the Law of Excluded Middle even when both disjuncts are unmeasured, or indeed immeasurable. In Putnam's general explanation of the workings of Hilbert spaces (Putnam 1975, pp. 178–179), and in his examples (Putnam 1975, pp. 183, 186) he always has a disjunction, which is held to be true on the basis of the certainty of probabilities or the spanning of a Hilbert space or that of some constraint, and which is conjoined with a true statement. This true statement is such that it cannot be measured together with any of the disjuncts, because of quantum mechanical constraints. What one is saying here is that the disjunction holds even when none of its disjuncts can be measured. One can consider a non-quantum example of a coin that is in my pocket now and that will be tossed next week. There is no current determination of its being heads or tails, and the disjunction that it is heads or tails will be unsupported by either disjunct. Similarly, in quantum mechanics, the Law of Excluded Middle is said to apply, even when neither disjunct can be determined in order to give support for it.

To assess this properly, we need to go back to general logical application and natural deduction. When we apply a logic, we start with a logical system such as MCQ (for discussion, see the beginning of this paper) and add non-logical axioms which capture concepts and state truths in relation to these concepts. In order to establish a disjunction we would naturally apply a disjunction introduction rule. Putting this into the context of quantum mechanics, we use the statements occurring in the above ortholattice diagram. $[x+]$ is true and is thus included in the logical application, but neither $[y+]$ nor $[y-]$ is included as they cannot be shown. Thus, $[y+] \vee [y-]$ cannot be derived using disjunction introduction, and would have to be independently assumed, just like the LEM in our earlier discussion in Brady and Rush (2008), where it is argued that it is a reasonable assumption.

There may be grounds for the truth of $[y+] \vee [y-]$, based on the fact that the measurement $[x+]$ distorts the system in such a way that the spin in the y-direction cannot be measured, even though one of $[y+]$ or $[y-]$ still holds true. However, if one of them is true then surely that one can be conjoined with $[x+]$ to yield a truth, but this is not the quantum mechanical conjunction that talks of their immeasurability, this is a simple matter of putting two truths together into a conjunction. Thus, there is an equivocation here between a disjunction being true because one of its disjuncts holds whilst another disjunction in the same argument is false because of the unmeasurability of the disjuncts. So, one can only have it one way or the other. If one of $[y+]$ or $[y-]$ holds then one of $[x+]\&[y+]$ or $[x+]\&[y-]$ holds, and distribution is saved. If $[y+]$ and $[y-]$ are both unmeasurable due to $[x+]$, then $([x+]\&[y+]) \vee ([x+]\&[y-])$ should fail as should $[y+] \vee [y-]$, again saving distribution.

The same sort of equivocation occurs in the case where the disjunction $[y+] \vee [y-]$ has a probability of 1, in the presence of $[x+]$ being true. Because distribution holds in probability theory, as demonstrated, $([x+]\&[y+]) \vee ([x+]\&[y-])$ still has a probability of 1. But these are not the unmeasurable conjunctions of quantum mechanics as they are based on the genuineness of $[y+] \vee [y-]$ having a probability of 1. In the unmeasurable case, $[y+] \vee [y-]$ would have a probability of 0, in the context of $[x+]$, as neither can be measured. In the event of measurement of the spin in the y-direction being undertaken, $[y+] \vee [y-]$ would again have a probability of 1, but this is a different scenario at a different point in time. (The above coin-tossing example makes a similar point.)

What is somewhat special is Putnam's account of the 2-slit experiment (Putnam 1975, pp. 180–181). Let A_1 be the statement that the photon passes through slit 1 and A_2 be the statement that the photon passes through slit 2. Let R be the statement that the photon hits a tiny region on the photographic plate at the back of the slits. In his classical calculation that the probability $P(R/A_1 \vee A_2)$ is the average of $P(R/A_1)$ and $P(R/A_2)$, he uses distribution in the form $P((A_1 \vee A_2)\&R) = P((A_1\&R) \vee (A_2\&R))$ and recognises that $P(A_1\&A_2) = 0$, i.e. that the photon cannot pass through both slits. He states however that this is not the observed probability and that quantum mechanics gets it right. Indeed, if the tiny region is at the centre of the plate with respect to the slits, then the observed probability is zero (or very close to it) as the interference pattern has the effect of cancelling out the waves coming through the two slits. However, there is still an equivocation here between the two slits considered separately, to set up the disjunction $A_1 \vee A_2$, this yielding the classical calculation, and the two slits considered together, which enables the interference pattern to be created. This wave-generated phenomenon behaves very differently from the pattern of particles just passing through each slit singly.

So, generally speaking, there is an equivocation between the circumstances in which a disjunction holds, whether it is true in accordance with the Law of Excluded Middle, its probability is 1, or it exhausts a set of possibilities under consideration (an extended form of the LEM), and the quantum circumstances which make each disjunct false, when taken in conjunction with a true statement such that both these statements cannot be checked by a measurement taken at the same time. The upshot of all this is that distribution can be saved but the LEM is thrown into doubt as a general logical principle, as has been argued elsewhere in Brady (2006, 2008) and Brady and Rush (2008).

What is noticeable about quantum logic, with or without the orthomodular law, is that it is only slightly changed from classical logic, since distribution is a relatively minor property, leaving the bulk of classical logic intact. Our alternative of dropping of the LEM, however, would rip the heart out of classical logic, as can be seen by its central status in conjunctive normal forms (though distribution does play a lesser role here as well). What we need is a logic like MC which dispenses with the LEM in a natural way, but still provides the basic entailments.

This still leaves us with the question of which of the three alternatives considered above are best for quantum mechanics. The third account using classical logic is too restrictive in its application to certain quantum mechanical conjunctions and disjunctions, in that it considers them meaningless or outside the scope of logic. From a logical point of view, this is the least attractive of the options. Thus, having argued against Birkhoff and von Neumann's and Putnam's advocacy of quantum logic, we are left with Reichenbach's 3-valued logic. The important point here is that the third value 'indeterminate' is evaluated such that the disjunction of two indeterminates is also indeterminate. However, there are a number of 3-valued logics in Reichenbach (1994), the main differences being in the evaluation of negation (see Reichenbach 1994, pp. 150–153). The one that suits our purposes is what he calls 'diametrical negation', which takes truth to falsity, falsity to truth, and indeterminate to indeterminate, as set out on p. 151. This last valuation would enable both $[y+]$ and $[y-]$ to be indeterminate, when $[x+]$ is measurably true. So, given diametrical negation, the LEM, in the form $A \vee \neg A$, with A set as indeterminate, is also indeterminate and the LEM is invalid in the 3-valued logic.

The 3-valued logic that is determined by Reichenbach's matrices (Reichenbach 1994, p. 151), together with diametrical negation, is the familiar 3-valued logic of Kleene, which can be seen in the textbook Priest (2001, pp. 119–120). We set it out below:

$\neg$	
1	0
i	i
0	1

$\wedge$	1	i	0
1	1	i	0
i	i	i	0
0	0	0	0

$\vee$	1	i	0
1	1	1	1
i	1	i	i
0	1	i	0

$\supset$	1	i	0
1	1	i	0
i	1	i	i
0	1	1	1

[$A \supset B$ is defined in the classical way as $\neg A \vee B$.]

Only the value 1 is designated, and thus a formula A is *valid* iff A takes the value 1 for all assignments to its sentential variables. However, it can easily be seen, by assigning i to each variable of a formula A, that all formulae are invalid. However, rules (or deductions) can still be semantic consequences, defined as follows: A rule $A_1, \ldots, A_n \Rightarrow B$ is a *semantic consequence* iff, for all assignments to the sentential variables of $\{A_1, \ldots, A_n, B\}$, whenever all of $A_1, \ldots, A_n$ take the value 1, B also takes the value 1. For example, *Modus Ponens* in the form: $A, A \supset B \Rightarrow B$ is a semantic consequence, as is Distribution in the rule-form: $A\&(B \vee C) \Rightarrow (A\&B) \vee (A\&C)$. And, quantum logic is standardly formalised as a set of rules, often using the lattice ordering relation in place of '$\Rightarrow$' (see Dunn 1981).

13.6 Conclusions

Firstly, we show how the transition is made from Kleene's 3-valued logic to the logic MCQ of meaning containment with quantification. We do need to consider a logic which applies generally and not just to quantum mechanics and Kleene's 3-valued logic is very restricted in not having any valid formulae. To achieve our aim, one needs at least to add an inference connective '$\rightarrow$' so that basic inferences such as $A \rightarrow A$, $A\&B \rightarrow A$, $A \rightarrow A \vee B$, etc. can be valid. Further, one should add an extra value to the logic to account for theories that over-specify their concepts or overstate their facts to the point of being inconsistent. This is quite possible, and would complement the under-specification of concepts (or understatement of their facts) in theories such as Peano arithmetic (due to Gödel's Theorem) and axiomatic set theories such as that of Zermelo–Fraenkel. (For some discussion of truth-value gluts and gaps, see Priest 2001, Chap. 7). To achieve both of these ends, we add the value 'b', for both true and false, to the values 1, i and 0 of the Kleene matrices, creating a 4-valued logic FDE, initially due to Smiley (see Anderson and Belnap 1975, § 15). All the formulae of this logic are first-degree entailments, i.e. they are of the form $A \to B$, where A and B are $\to$-free. However, the irrelevant classical rule, $A\&\neg A \Rightarrow B$, from Kleene's 3-valued logic, will fail to transform to a valid formula with '$\to$' instead of '$\Rightarrow$' in FDE, and this also extends to other irrelevant rules with undesignated premises. Beyond these rules, there must be at least a variable in common, this property being in conformity with that of FDE.

We set out the matrices for the logic FDE below, with 'n' replacing 'i' (see also Priest 2001, p. 144):

$\neg$	
1	0
b	b
n	n
0	1

$\wedge$	1	b	n	0
1	1	b	n	0
b	b	b	0	0
n	n	0	n	0
0	0	0	0	0

$\vee$	1	b	n	0
1	1	1	1	1
b	1	b	1	b
n	1	1	n	n
0	1	b	n	0

$\to$	1	b	n	0
1	1	0	0	0
b	1	1	0	0
n	1	0	1	0
0	1	1	1	1

Smiley designates just the value 1, but b (or even n) can be designated as well, without changing the set of valid formulae. Validity and semantic consequence is defined as for Kleene's logic above.

This system FDE not only provides a set of valid formulae of great intuitive appeal, as can be seen from the exposition of tautological entailments in § 15 of Anderson and Belnap (1975), but it is also the common first-degree fragment of all relevant logics from the basic sentential system B of the Routley-Meyer semantics right through to the logic R of relevant implication. This of course includes the logic MC, a depth relevant logic obtained from B by adding A10 and A11 in place of the rule form of A10 (see earlier). So, MC is a conservative extension of FDE through the addition of formulae beyond those of first-degree entailments. We then, by making what is now the third addition of quantification to the original Kleene 3-valued logic, arrive at the logic MCQ of meaning containment with quantification. So, the rules of the Kleene 3-valued logic are still embedded in MCQ, when applied to the statements taking either of the 3 values: 1, i and 0, but what we have achieved is a well-rounded entailment logic.

Secondly, we need to determine the status of distribution within the logic MCQ. As we saw earlier, distribution is obtained from structural rules within the proof-theory, rather than from the connective rules for conjunction and disjunction, and so it is an additional property over and above what is basically needed to understand these two connectives. We also saw that the semantic justification of distribution is circular. Nevertheless, it was then argued that there is no need to drop distribution in the context of quantum mechanics. The upshot of all this is that distribution does not follow from the meaning of conjunction and disjunction and thus should not be expressed as a meaning containment. So, $A\&(B \vee C) \to (A\&B) \vee (A\&C)$ should not be in the logic. Thus, the set-theoretic containment properties captured in the content semantics of MC would be intensional properties rather than extensional.

We next need to consider distribution in the rule-form: $A\&(B \vee C) \Rightarrow (A\&B) \vee (A\&C)$. If this is to be a deductively valid inference, its conclusion should follow as a matter of certainty from the premise, but not necessarily as an entailment. Its proof can involve a number of rule applications. Indeed, it can be proved from the two-premise version of the meta-rule MR1 of MC, viz.

$$\text{If } A, B \Rightarrow D \text{ then } C \vee A, C \vee B \Rightarrow C \vee D \ (*).$$

It is hard to see how one can accept MR1 and reject the above two-premise version of the disjunctive rule. The proof is as follows:

$(C \vee A)\&(C \vee B) \Rightarrow C \vee (A\&B)$, by (*) and R2, yielding one version of Distribution.

The Anderson and Belnap version of the distribution rule is then immediate:

$A\&(B \vee C) \Rightarrow (A\&B) \vee C$, using $A \to C \vee A$. (see Anderson and Belnap 1975, p. 158)

This is then applied twice in the following argument:

$A\&(B \vee C) \Rightarrow A\&(C \vee B) \Rightarrow (A\&C) \vee B \Rightarrow A\&((A\&C) \vee B) \Rightarrow A\&(B \vee (A\&C)) \Rightarrow (A\&B) \vee (A\&C)$, yielding the other (standard) version of Distribution. (See also footnote 2.)

Thus, we include distribution in rule-form in our logic.

Moreover, it is hard to think of any counter-examples, apart from the rather artificial one discussed in the previous section on quantum logic, where non-truth was pooled into one false value. The other case in that section involved equivocation and thus cannot be seriously considered. However, the sublattice of an ortholattice exhibited there provides a modelling of a non-distributive logic, but one still needs to provide suitable statements for the three nodes in the middle, which have the unusual property of not implying each other, yet when any two are conjoined, this conjunction implies the third, and when any two are disjoined, this disjunction is implied by the third.

The quantified forms of distribution, given as QA3 ($\forall x(A \vee B) \to (A \vee \forall xB)$) and QA6 ($A\&\exists xB \to \exists x(A\&B)$), would also not be in the logic. For, if they were in the logic, then their sentential forms would be derivable by restricting the domain to be finite. Also, the arguments we used against sentential distribution will equally apply here. So, we next need to consider whether the rule-forms of QA3 and QA6 hold. Let us consider the case where their domains are infinite. This issue came up in Brady and Rush (2008), where QA3 and QA6 were argued against on the grounds that the quantifiers are generally intensional, but in QA3 and QA6 they were being supported using the extensional connectives, conjunction and disjunction. The reason that the quantifiers are intensional in general is that the domains can be non-recursive, in which case quantified statements cannot be supported by mathematical induction or by an extensional element-by-element approach. This means that they can only be supported through the meanings of the concepts involved. In a natural deduction setting, this would be through the use of a universal introduction rule, which is based on an arbitrary element from the domain. So, by this account, the failure of QA3 and QA6 is due to the oddity of combining $\forall$ with $\vee$ or $\exists$ with &. This also applies to the rule-forms, both in Brady and Rush (2008) and here.

Examples of the failure of QA3 based on intuitionism can be found in Dummett (1977, p. 31) and in Beall and Restall (2006, p. 65), and these examples will similarly do for us as well. More generally, let $\forall x(A \vee Fx)$ be derived in the most general way, just by the universal introduction rule from $A \vee Fx$. The domain may be non-recursive and, in which case, these x-instances could not all be individuated, leaving $\forall$I as the appropriate rule. Then, neither of the disjuncts A nor Fx need follow, as such a theory need not be prime. In this case, neither A nor $\forall xFx$ need follow, and ditto for $A \vee \forall xFx$. The failure of $\forall x(A \vee B) \Rightarrow A \vee \forall xB$ under a particular interpretation, whether recursive or not, would also require it to fail in the logic MCQ, where just the fundamental meaning of universal quantification is captured, leaving recursion or non-recursion as a specific property which may induce additional axioms and rules.

Nevertheless, as for intuitionism, the rule $A\&\exists xB \Rightarrow \exists x(A\&B)$ seems to hold regardless of whether the quantifier is interpreted recursively or non-recursively. Whatever element makes $\exists xB$ hold, would also make $\exists x(A\&B)$ hold, given A. This seems to follow from the meaning of the existential quantifier as 'at least one of, without needing to know which one'. Further, this rule is derivable from the two-premise version of the meta-rule QMR1:

$$\text{If } A, B^a/_x \Rightarrow C^a/_x \text{ then } A, \exists xB \Rightarrow \exists xC.$$

Again, as for the sentential case, it is hard to see how one can accept QMR1 and reject the above two-premise version of the existential rule. For proof of $A\&\exists xB \Rightarrow \exists x(A\&B)$, simply replace C by $A\&B$.

The upshot of all this is that the rule-form $\forall x(A\vee B) \Rightarrow A\vee\forall xB$ is not included, whilst $A\&\exists xB \Rightarrow \exists x(A\&B)$ and the sentential $A\&(B \vee C) \Rightarrow (A\&B) \vee (A\&C)$ are included as rules in the logic. However, $\forall x(A \vee B) \Rightarrow A \vee \forall xB$ can be assumed much like the Law of Excluded Middle and the Disjunctive Syllogism. (See Brady and Rush (2008) for discussion of 'reasonable assumptions' and 'technical assumptions', especially concerning the LEM and the DS.) Clearly, if the domain is finite, $\forall x(A \vee B) \Rightarrow A \vee \forall xB$ would follow from the sentential form $A\&(B \vee C) \Rightarrow (A\&B) \vee (A\&C)$, with some help from MR1. So, there is at least a solid block of cases where they hold. In other cases, it would be up to the individual circumstances as to whether they hold or not. Moreover, it is interesting to note that if the LEM and the DS are assumed for A then $\forall x(A \vee B) \Rightarrow A \vee \forall xB$ follows.[5]

[5]This is shown as follows: $\neg A, A \vee B^a/_x \Rightarrow B^a/_x$, where a does not occur in B, by the DS, and hence $\neg A, \forall x(A\vee B) \Rightarrow \forall xB$. By the two-premise metarule (*) and the LEM, $A\vee\forall x(A\vee B) \Rightarrow A \vee \forall xB$, and hence $\forall x(A \vee B) \Rightarrow A \vee \forall xB$.

Acknowledgements We acknowledge the help of Prof. John Bigelow of Monash University for initiating the current enterprise, by pointing out his concerns about distribution from the point of view of quantum mechanics. We also acknowledge help from the audiences at the Australasian Association for Logic conference at Auckland and the World Congress on Paraconsistency at Melbourne, 2008, especially Greg Restall, Bob Meyer, Francesco Paoli and Graham Priest. We also thank the referee for diligent work on the paper, which found quite a number of omissions and inaccuracies. The first author also acknowledges the financial assistance of the Australian Research Council grant DP0556114 in carrying out this research.

References

Anderson, A.R., and N.D. Belnap. 1975. *Entailment, the logic of relevance and necessity*, vol 1. Princeton: Princeton University Press.

Beall, J.C., and G. Restall. 2006. *Logical pluralism*. Oxford: Oxford University Press.

Belnap, N.D. 1993. Life in the undistributed middle. In *Substructural logics*, ed. K. Dosen and P. Schroeder-Heister, 31–41. Oxford: Oxford Science Publications.

Birkhoff, G., and J. von Neumann. 1936. The logic of quantum mechanics. *Annals of Mathematics* 37: 823–843.

Brady, R.T. 1984. Natural deduction systems for some quantified relevant logics. *Logique et Analyse* 27: 355–377.

Brady, R.T. 1996. Relevant implication and the case for a weaker logic. *Journal of Philosophical Logic* 24: 151–183.

Brady, R.T. 1996a. Gentzenizations of relevant logics with distribution. *The Journal of Symbolic Logic* 61: 402–420.

Brady, R.T. 1996b. Gentzenizations of relevant logics without distribution I. *The Journal of Symbolic Logic* 61: 353–378.

Brady, R.T. 1996c. Gentzenizations of relevant logics without distribution II. *The Journal of Symbolic Logic* 61: 379–401.

Brady, R.T. (ed.). 2003. *Relevant logics and their rivals*, vol. 2. Aldershot: Ashgate.

Brady, R.T. (2006). *Universal logic*. Stanford: CSLI Publications.

Brady, R.T. 2006a. Normalized natural deduction systems for some relevant logics I: The logic DW. *The Journal of Symbolic Logic*, 71: 35–66.

Brady, R.T. 2008. Negation in metacomplete relevant logics. *Logique et Analyse*, 51: 331–354.

Brady, R.T. 201+a. *Normalized natural deduction systems for some relevant logics II: Other logics and deductions*. Presented to the Australasian association for logic conferences, University of Adelaide, 2003, and University of Otago, 2004.

Brady, R.T. 201+b. *Normalized natural deduction systems for some relevant logics III: Modal and quantificational logics*. Presented to the Australasian association for logic Conference, ANU, 2002, and to the 1st World Congress on Universal Logic, Montreux, 2005.

Brady, R.T., and P.A. Rush. 2008. What is wrong with Cantor's diagonal argument? *Logique et Analyse* 51: 185–219.

Dalla Chiara, M.L., R. Giuntini, and R. Greechie. 2004. *Reasoning in quantum theory*. Dordrecht: Kluwer.

Dummett, M. 1977. *Elements of intuitionism*. Oxford: Oxford University Press.

Dunn, J.M. 1981. *Quantum mathematics*. Philosophy of science association 1980, vol. 2. Princeton: Princeton University Press.

Fine, K. 1988. Semantics for quantified relevance logic. *Journal of Philosophical Logic* 17: 22–59.

Gentzen, G. 1969. Investigations into logical deduction. In *The collected papers of gerhard gentzen*, ed. M. Szabo. Amsterdam: North-Holland.

Griffiths, R.B. 2002. *Consistent quantum theory*. New York: Cambridge University Press.

Heyting, A. 1966. *Intuitionism, an introduction*. Amsterdam: North-Holland.

Ketonen, O. 1944. Untersuchungen zum Prädikatenkalkül. *Annales Academiae Scientiarum Fennicae*, A1, 23: 1–71.
Kochen, S., and E.P. Specker. 1965. Logical structures arising in quantum theory. In *The theory of models*, ed. J. Addison, L. Henkin, and A. Tarski, 177–189. Amsterdam: North-Holland.
Ludwig, G. 1983. *Foundations of quantum mechanics*, vol. 1. Berlin: Springer.
Mares, E.D., and R. Goldblatt. 2006. An alternative semantics for quantified relevant logic. *The Journal of Symbolic Logic* 71: 163–187.
Paoli, F. 2007. Implicational paradoxes and the meaning of logical constants. *Australasian Journal of Philosophy* 85: 553–579.
Priest, G. 2001. *An introduction to non-classical logic*. Cambridge: Cambridge University Press.
Putnam, H. 1975. *Mathematics, matter and method: Philosophical papers*, vol. 1. Cambridge: Cambridge University Press.
Read, S. 1988. *Relevant logic: A philosophical examination of inference*. Oxford: Blackwell.
Reichenbach, H. 1944. *Philosophic foundations of quantum mechanics*. Berkeley/Los Angeles: University of California Press.
Restall, G. 2007. Review of R.T. Brady, Universal logic. *Bulletin of Symbolic Logic* 13: 544–547.
Restall, G., and F. Paoli. 2005. The geometry of nondistributive logics. *Journal of Symbolic Logic* 70: 1108–1126.
Rush, P.A., and R.T. Brady. 2007. *What is wrong with truth-functional semantics?* Presented to the congress of logic, methodology and the philosophy of science, Beijing, 2007, and to the Australasian association for logic conference, University of Melbourne, 2007.
Schroeder-Heister, P. 1991. Uniform proof-theoretic semantics for logical constants (abstract). *The Journal of Symbolic Logic* 56: 1142.
Schroeder-Heister, P. 2006. Validity concepts in proof-theoretic semantics. *Synthese* 148: 525–571.
Schroeder-Heister, P. 2007. *Proof theoretic semantics*, Tutorial given to the World Universal Logic Conference. Xian, China.
Urquhart, A. 1984. The undecidability of entailment and relevant implication. *The Journal of Symbolic Logic* 49: 1059–1073.

Chapter 14
Wittgenstein on Incompleteness Makes Paraconsistent Sense

Francesco Berto

14.1 "A Completely Trivial and Uninteresting Misinterpretation"

Wittgenstein's comments on Gödel's First Incompleteness Theorem in the *Remarks on the Foundations of Mathematics* were dismissed by early commentators, such as Kreisel, Anderson, Dummett, and Bernays, as an unfortunate episode in the career of a great philosopher. It appears that Wittgenstein had in his sights only the informal account of the Theorem, presented by Gödel in the introduction of his celebrated 1931 paper, and was misguided by it (not that he was the only one: because of the misunderstandings it originated, Helmer said that exposition "without any claim to complete precision"—see Gödel (1931, p. 597)—is the only mistake in Gödel's paper). It is claimed that Wittgenstein erroneously considered essential the natural language interpretation of the Gödel sentence, whose undecidability within (the modified system considered by Gödel, taken from) Russell and Whitehead's *Principia mathematica* is at the core of the First Theorem, as claiming "I am not provable". On the contrary, Gödel's proof can be phrased in syntactic terms in which no such interpretation of the formulas is required.

Commentators were particularly struck by the fact that Wittgenstein seems to take the Gödel formula as a paradoxical sentence, not too different from the usual Liar—and Gödel's proof itself, therefore, as the deduction of an inconsistency:

> 11. Let us suppose I prove the unprovability (in Russell's system) of P; then by this proof I have proved P. Now if this proof were one in Russell's system – I should in this case have proved at once that it belonged and did not belong to Russell's system. – That is what comes of making up such sentences. But there is a contradiction here! – Well, then there is a contradiction here. Does it do any harm here? (Wittgenstein 1953, p. 51e)

F. Berto (✉)
Department of Philosophy and Northern Institute of Philosophy, The University of Aberdeen, Aberdeen, UK
e-mail: f.berto@abdn.ac.uk

K. Tanaka et al. (eds.), *Paraconsistency: Logic and Applications*, Logic, Epistemology, and the Unity of Science 26, DOI 10.1007/978-94-007-4438-7_14,

Zermelo, Perelman, and probably Russell himself made similar mistakes in the interpretation of the First Theorem, in the years following the publication of Gödel's results. It is usually maintained that the error rests on a confusion between a theory and its metatheory, or between syntax and semantics (see Perelman (1936), who claimed that Gödel had just discovered a new logical paradox; see also Dawson (1984) on Russell, and on Zermelo's letter to Gödel on this issue[1]), which makes it impossible to understand the difference between the truth predicate, inexpressible (by Tarski's theorem) within the theory to which the First Theorem applies, and the provability predicate, which, on the contrary, is (weakly) expressible (Anderson explicitly charges Wittgenstein with such confusion: Anderson 1958, p. 486). Until a few years ago, the discussion on Wittgenstein's remarks seemed to be concluded by the trustworthy verdict of Gödel himself, who, in a letter to Abraham Robinson, stated that Wittgenstein "advance[d] a completely trivial and uninteresting misinterpretation" (see Dawson 1984, p. 89) of the First Theorem.

However, in recent years some commentators have argued that it is possible to extract interesting philosophical theses from the comments of the *Bemerkungen*. Floyd and Putnam (2000) have claimed that Wittgenstein's intuitions anticipate some metamathematical acquisitions concerning the non-standard models of arithmetic. Wittgenstein's further remarks on Gödel, recently published in cd-rom format within the Bergen project, according to Rodych (2002), show that he did not consider the self-referential natural language interpretation of the Gödel sentence essential to the proof of the First Theorem—on the contrary, he "correctly understood the number-theoretic nature of Gödel's proposition" (see Rodych 2002, p. 380). And the debate is nowadays lively and rapidly evolving, with authoritative commentators taking a stance on Wittgenstein's "real thoughts" in the most important international reviews—from the *Journal of Philosophy* to *Dialectica* and *Erkenntnis*; see also Hintikka (1999), Rodych (1999), Rodych (2003), and Floyd (2001).

I also believe that no significant philosophical idea is past its use-by date. And in this paper I will show that it is possible to provide an interpretation of Wittgenstein's position on Gödel's results in the light of contemporary mathematical logic—that is to say, precisely from the point of view from which the comments of the *Bemerkungen* were most severely attacked. My interpretation, however, shall not follow the line of the latest commentators. In particular, I will not take the direction of non-standard models, suggested by Floyd and Putnam—although I will deal with other models of arithmetics, which definitely deserve to be called "non-standard". If we read Wittgenstein's stance on Gödel's First Theorem as conforming to the single, simple argument to be exposed below, then interesting facts will follow for the philosophical significance of the incompleteness results. The "single argument" will also allow me to vindicate and support two other ideas which harmed Wittgenstein's reputation among mathematicians and logicians: (1) his plain rejection of Hilbert's very idea of a *metamathematics*; and (2) the view that we should not dramatise

[1]In the correspondence between the two mathematicians, as Dawson points out, Zermelo "failed utterly to appreciate Gödel's distinctions between syntax and semantics" (Dawson 1984, p. 80).

the possibility that a calculus turns out to be inconsistent (a dramatisation which, according to some, puzzled Wittgenstein precisely since he began to pay attention to the role of consistency proofs within Hilbert's strategy, see Marconi (1984)). Wittgenstein's ideas on contradiction and consistency proofs were dismissed as absurdities by the same commentators who found his remarks on the First Theorem outrageous, see Anderson (1958) and Bernays (1959).

The "single argument", therefore, will capture several fundamental intuitions at the core of Wittgenstein's philosophy of mathematics—although, eventually, it will not capture them *all*. For instance, I will have to exploit some ideas on the formalisation of deductive theories, and some notions of model theory, which constitute established acquisitions of the current logico-mathematical practice, but which Wittgenstein would probably have rejected. Furthermore, I will not trust Wittgenstein's own declarations, according to which his remarks should not have any strictly mathematical import. On the contrary, my interpretation will entail a strong revisionism with respect to classical logic and classical mathematics.

14.2 Metamathematics Is Just Mathematics

At the core of the "single argument" is the idea that, in maintaining an interpretation of Gödel's proof that made of it a paradoxical derivation, Wittgenstein was consequent upon his bold move of rejecting the standard distinction between theory and metatheory (therefore, between formalised arithmetic and metamathematics).

Logicians have learned precisely from Gödel's results (and from Tarski's, on the undefinability of truth) to be much more careful than they had been before in distinguishing between theory and metatheory and between syntax and semantics; we may therefore forgive Gödel's contemporaries for being careless on this. Unlike Zermelo and Perelman, however, Wittgenstein knowingly refused several aspects of such distinctions. During his entire philosophical career he never had second thoughts on his rejection of Hilbert's metamathematics. This is expressed in the Philosophical Remarks and, most explicitly, in a paragraph of the Philosophical Grammar whose title is precisely "There is no metamathematics":

> I said earlier "calculus is not a mathematical concept"; in other words, the word "calculus" is not a chess piece that belongs to mathematics.
>
> There is no need for it to occur in mathematics. – If it is used in a calculus nonetheless, that doesn't make the calculus into a metacalculus; in such a case the word is just a chessman like all the others. Logic isn't metamathematics either; that is, work within the logical calculus can't bring to light essential truths about mathematics. Cf. here the "decision problem" and similar topics in modern mathematical logic. [...]
>
> (Hilbert sets up rules of a particular calculus as rules of metamathematics) (Wittgenstein 1953, pp. 296–297.)

That is to say: Hilbert's metamathematics is, in fact, nothing but mathematics. It is not a metacalculus, because there are no metacalculi: it is just one more calculus.

It would take too much space here to discuss Wittgenstein's motivations for discarding Hilbert's conception of metamathematics. Roughly, they are closely connected to a rejection of the Platonic idea that mathematical sentences describe an independently existing domain—the "realm of numbers". If we follow this line, the claim that Gödel's proof actually is the derivation of a paradox follows ineluctably. Contrary to what Bernays claimed, the discussion of Gödel's results in the *Bemerkungen* does *not* "suffer from the defect that Gödel's quite explicit premises of the consistency of the considered formal system is ignored" (Bernays 1959, p. 523). Bernays' charge just begs the question against Wittgenstein, for the consistency of the relevant system is precisely what is called into question by Wittgenstein's reasoning. For a particularly clear statement of this issue, see Rodych (2002, pp. 384–385). Let us see why.

14.3 Prose vs. Proof

Here is, to begin with, a standard exposition of the First Incompleteness Theorem, which will help us in the following. An exemplary case of a theory to which Gödel's Theorems apply is provided by Peano arithmetic, PA.[2] This can be obtained by simply adding to the ordinary axioms of first-order predicate logic with identity the following principles:

$$\text{(PA1) } \forall x(Succ(x) \neq 0)$$
$$\text{(PA2) } \forall xy(Succ(x) = Succ(y) \rightarrow x = y)$$
$$\text{(PA3) } \forall x(x + 0 = x)$$
$$\text{(PA4) } \forall xy(x + Succ(y) = Succ(x + y))$$
$$\text{(PA5) } \forall x(x \times 0 = 0)$$
$$\text{(PA6) } \forall xy(x \times Succ(y) = (x \times y) + x)$$
$$\text{(PA7) } \alpha[x/0] \rightarrow (\forall x(\alpha[x] \rightarrow \alpha[x/Succ(x)]) \rightarrow \forall x\alpha[x]).$$

The theory holds in the so-called standard model of arithmetic (be it $\mathbb{N}$), i.e., the model constituted by natural numbers and the operations on them we know since we were children. Variables are therefore supposed to range on natural numbers,[3] and "0" is the name of number zero. The intended reading of the one-place functor $Succ(x)$ is "the (immediate) successor of x". Therefore, $Succ(0)$ is 1, that is, the (immediate) successor of zero in the series of natural numbers; $Succ(Succ(0))$ is 2, that is, the successor of 1; etc. $+$ and $\times$, of course, are read as addition and multiplication. Therefore, (PA1) claims that zero is the successor of no

[2]Wittgenstein's remarks had as their background system the one of Russell and Whitehead's *Principia mathematica* (with slight modifications). However, this is a minor point, and sticking to PA allows us to follow a standard way of presenting Gödel's Theorems.

[3]Or, at least, these are our bona fide intuitions when we formulate the theory. The existence of non-standard models shows that things are not so straightforward. I will come to this in a subsequent note.

number; (PA2) claims that if x and y have the same successor, they are the same number. (PA3)–(PA6) represent recursive equations characterizing addition and multiplication. (PA7) is the schematic formulation of the (mathematical) Induction Principle, which claims that if some $\alpha[x]$ holds for the zero and for the successor of a given number x for which it holds, then $\alpha[x]$ holds for all numbers.

Now the Gödelisation procedure allows one to associate a natural number to each symbol, formula and sequence of formulas of PA, so that one can always effectively move back and forth between an expression of the language of PA and the number to which it has been paired (its Gödel number). A k-ary relation (whose extension consists in a set of ordered k-ples) R can be said to be representable in PA iff there is a formula $\alpha[x_1, \ldots, x_k]$, such that, for any ordered k-ple of numbers $\langle n_1, \ldots, n_k\rangle$, we have that:

(a) If $\langle n_1, \ldots, n_k\rangle \in R$, then $\vdash_{PA} \alpha[x_1/\mathbf{n}_1, \ldots, x_k/\mathbf{n}_k]$
(b) If $\langle n_1, \ldots, n_k\rangle \notin R$, then $\vdash_{PA} \neg\alpha[x_1/\mathbf{n}_1, \ldots, x_k/\mathbf{n}_k]$,

where $\mathbf{n}$ is the numeral of number n, see Gödel (1931, p. 607). Now, PA is the typical case of a sufficiently strong theory, that is, it is capable of representing the (primitive) recursive functions. Recursive functions-relations have the role of codifying the syntax of the theory. Metalinguistic claims on PA are mirrored within the officially arithmetic language of PA. As is usually claimed, PA "can talk about" (Boolos et al. 2002, p. 187) some of its syntactic properties. In particular, the property of being a theorem of the theory can be (weakly) represented within the theory itself. The arithmetic predicate no. 45 in Gödel's paper corresponds to something like:

$$Prf(x, y)$$

whose reading via arithmetisation is: "x is (the Gödel number of) a proof of the formula (whose Gödel number is) y". *Prf* is a recursive relation that holds between those pairs of numbers which are, respectively, the Gödel number of a sequence of formulas of PA, and the Gödel number of a formula of PA, such that the former is a proof of the latter. Predicate no. 46 is defined by means of no. 45, thus:

$$Th(y) =_{df} \exists x\ Prf(x, y);$$

therefore, it holds of those numbers which are the Gödel numbers of formulae of PA for which there is a proof in PA.[4] The fundamental condition to prove Gödel's First Theorem concerns provability within PA:

$$\text{(P)}\ \vdash_{PA} \alpha \Rightarrow \vdash_{PA} Th(\ulcorner\alpha\urcorner),^5$$

that is, if α is a theorem of PA, then the formula mirroring this fact within PA is itself a theorem of PA. Now, before the Fixed Point Lemma was employed to obtain

[4] *Th* is not recursive, but semirecursive (see Gödel 1931, p. 606); however, this is of no importance here.

[5] $\ulcorner\alpha\urcorner$ is the numeral of the Gödel number of α.

fully formalised Liar sentences, Gödel used it to build a sentence (be it γ) attributing to itself not falsity, but non-theoremhood:

$$(\mathrm{FP}_\gamma)\gamma \leftrightarrow \neg Th(\ulcorner \gamma \urcorner).$$

γ is a purely arithmetic sentence, but its informal reading via Gödelisation is: "I am not a theorem". Given the definition above, it is equivalent to $\neg\exists x \; Prf(x, \ulcorner \gamma \urcorner)$, that is, "I am not provable".

We have to assume, then, that PA is both consistent and ω-consistent (a system is called ω-consistent iff for no formula $\alpha[x]$ of its language it is possible to prove both $\neg\alpha[x/\mathbf{n}]$ for each natural n, and $\exists x \; \alpha[x]$). Gödel demonstrated that:

1. If PA is consistent, then $\nvdash_{PA} \gamma$;
2. If PA is ω-consistent, then $\nvdash_{PA} \neg\gamma$.

As for (1): if γ were a theorem of PA then, given (P), also $Th(\ulcorner \gamma \urcorner)$ would be. Hence, given (FP_γ), the provability of $\neg\gamma$ would follow. We would have, then, $\vdash_{PA} \gamma$ and $\vdash_{PA} \neg\gamma$, against the assumption that PA is consistent. As for (2): since the proof relation of PA is (primitive) recursive, we have that for each n either $\vdash_{PA}$ $Prf(\mathbf{n}, \ulcorner \gamma \urcorner)$, or $\vdash_{PA} \neg Prf(\mathbf{n}, \ulcorner \gamma \urcorner)$. The former case is ruled out by the fact that, as (1) claims, γ is not provable—therefore, for each n it is not the case that n is the code of a proof of γ in PA. Hence, the latter case holds. It follows, given the assumption that PA is ω-consistent, that $\exists x \; Prf(x, \ulcorner \gamma \urcorner)$ is not a theorem. But $\exists x \; Prf(x, \ulcorner \gamma \urcorner)$ is nothing but $\neg\gamma$ (see e.g., Smullyan 1992, Chap. V; Boolos et al. 2002, pp. 225–227). The conjunction of (1) and (2) gives us Gödel's First Incompleteness Theorem. This tells us that Peano arithmetic includes a sentence, γ (its own Gödel sentence), which is undecidable within PA, that is, not provable and not refutable.[6]

[6]One of the consequences of Gödel's First Theorem is that (first-order) PA is not, as model theorists say, categorical. This means that from Gödel's results follows the existence of non-standard models of PA, structurally different from $\mathbb{N}$. In particular, there is no way to constrain the variables of the theory so that they range exclusively on ordinary natural numbers. In 1957, Goodstein had already claimed that "Wittgenstein with remarkable insight said in the early thirties that Gödel's results showed that the notion of a finite cardinal could not be expressed in an axiomatic system and that formal number variables must necessarily take values other than natural numbers" (Goodstein 1957, p. 551). More recently, Floyd and Putnam have credited the "notorious paragraph" 8 of the Appendix 1 to Part I of Wittgenstein's *Bemerkungen* with a "philosophical claim of great interest" precisely on the role of non-standard models and ω-inconsistency. The claim is to the effect that "if one assumes (and, a fortiori, if one actually finds out) that $\neg P$ [where P is assumed to be the Gödel sentence of the relevant system] is provable in Russell's system one should (or, as Wittgenstein actually writes, one 'will now presumably') give up the 'translation' of P by the English sentence 'P is not provable'" (Floyd and Putnam 2000, p. 625). The point is that if a theory proves $\neg P$ (which may be obtained simply by adding it as an axiom), then it is ω-inconsistent, but consistent. Being consistent, it is supposed to have a model. However, being ω-inconsistent, its model has to be structurally different from the standard model of arithmetics, $\mathbb{N}$. It is a non-standard model, and the "translation" of P as "P is not provable" becomes untenable in this context.

So far, semantics and truth have not poked their nose in the proof[7] which, assuming only the consistency (and ω-consistency) of the system, counts as what logicians usually call a standard "syntactic" one. However, the exposition of the First Theorem usually goes hand in hand with the following short story, which Wittgenstein would probably have labelled as the "prose": since γ claims (via arithmetisation) to be not provable, and we have just proved that it is not provable, then γ just is what it claims to be; hence, it is true. However, this simple reasoning cannot be performed within the theory: the truth predicate for PA, were it expressible within PA, under the usual conditions would originate the Liar paradox; whereas the provability predicate is expressible. Gödel himself pointed at the analogies between his undecidable sentence and such paradoxes as Richard's, or the Liar, see Gödel (1931, p. 598). However, it seems clear that, whereas the Liar sentence, "This sentence is false", produces an antinomy, with the Gödel sentence, metamathematically read as "This sentence is not provable", no contradiction is forthcoming. Or so the usual story goes.[8]

It is often concluded, then, that Gödel's First Theorem establishes a fundamental gap between provability and truth (if a formal system has to be correct, i.e., it must capture only arithmetical truths, then it cannot capture them all). Precisely because of this, it has been taken by some as a keystone of mathematical realism, on the basis of an interpretation encouraged by Gödel himself. Gödel's realist stance emerged, as is well known, only several years after his 1931 paper, mainly in *What is Cantor's Continuum Problem?* (Gödel 1947). Nevertheless, he declared that his mathematical Platonism had been the heuristic key for the discovery of incompleteness, see Feferman (1983). And this is how the Gödelian results have become "one of the great moving forces behind the modern resurgence of Platonism" (Shanker 1988, p. 171).

Now, it is precisely this semantic outcome of Gödel's proof that Wittgenstein challenged as the "prose" (as opposed to the real "proof"). However, this should not be understood as the thesis that the First Theorem shows only the fact that γ is not a theorem of PA, whereas the further semantic conclusion that (if the system is consistent, then) γ is also true would be a "metaphysical claim" (Floyd and Putnam 2000, p. 632). As a matter of fact, quite legitimate and respectable semantic versions of Gödel's result are available, see e.g., Smullyan (1992, Chaps. 3 and 4). This is a minor point with respect to our discussion, though, because two other and quite different aspects of the semantic prose were unacceptable to Wittgenstein: (1) the idea that sentence γ, which is syntactically undecidable within PA, can nevertheless—as is usually said, "with a wave of hands" (Priest 1979, p. 222)—be shown to be true (to be sure, under the hypothesis of the consistency of PA) on the

[7] An anonymous referee has appropriately pointed out to me.

[8] In Kleene's words: "Gödel's sentence 'I am unprovable' is not paradoxical. We escape paradox because (whatever Hilbert may have hoped) there is no *a priori* reason why every true sentence must be provable [...]. The sentence $A_p(\mathbf{p})$, which says 'I am unprovable', is simply unprovable and true" (Kleene 1976, p. 54).

basis of a *meta*theoretic argument conducted "outside" the formal system PA; and (2) the consequent, aforementioned discrepancy between provability in any system capable of expressing elementary arithmetic, and arithmetical truth.

1. As for the first point: the semantic prose is sometimes to the effect that γ is proved by means of an informal or "intuitively correct" argument. One may say, with a little more precision, that it is provable within a theory that can deal with the semantics via the notion of truth (for the language of PA), which is not definable, given Tarski's theorem, within PA. If one asks "how is $[\gamma]$'s truth established? The answer is: by a metamathematical *proof* of $[\gamma]$" (Routley 1979, p. 325), that is, by means of a *detour* through the metatheory. This was stated by Gödel in the opening paragraphs of his paper, where he declared that "the proposition that is undecidable *in the system PM* still was decided by metamathematical considerations" (Gödel 1931, p. 599).

It was probably this claim that initially perplexed Wittgenstein, for in the *Philosophical Remarks* he had already observed:

> What is a proof of provability? It's different from the proof of proposition.
>
> And is a proof of provability perhaps the proof that a proposition makes sense? But then, such a proof would have to rest on *entirely different* principles from those on which the proof of the proposition rests. There cannot be a hierarchy of proofs!
>
> On the other hand there can't in any fundamental sense be such a thing as meta-mathematics. Everything must be of one type (or, what comes to the same thing, not of a type). [...]
>
> Thus, it isn't enough to say that p is provable, what we must say is: provable according to a particular system.
>
> Further, the proposition doesn't assert that p is provable in the system S, but in *its own* system, the system of p. That p belongs to the system S cannot be asserted, but must show itself.
>
> You can't say p belongs to the system S; you can't ask which system p belongs to; you can't search for the system of p. Understanding p means understanding its system. If p appears to go over from one system into another, then p has, in reality, changed its sense. (Wittgenstein 1953, p. 180).

Within this framework, it is not possible that the very same sentence (say, γ), turns out to be expressible, but undecidable, in a formal system (say, PA), and demonstrably true (under the aforementioned consistency hypothesis) in a different system (the meta-system). If, as Wittgenstein maintained, the proof establishes the very meaning of the proved sentence, then it is not possible for *the same* sentence (that is, for a sentence with the same meaning) to be undecidable in a formal system, but decided in a different system (the meta-system).

2. As for the second point: following this general doctrine, Wittgenstein had to reject both the idea that a formal system can be syntactically incomplete, and the Platonic consequence that no formal system proving only arithmetical truths can prove all arithmetical truths. If proofs establish the meaning of mathematical sentences, then there cannot be incomplete systems, just as there cannot be incomplete meanings:

The edifice of rules must be *complete*, if we are to work with a concept at all – *we cannot make any discoveries in syntax*. – For, only the group of rules *defines* the sense of our signs, and any alteration (e.g., supplementation) of the rules means an alteration of the sense. [...]

Mathematics cannot be incomplete; any more than a *sense* can be incomplete. (Wittgenstein 1953, pp. 182, 188).

One may object that Wittgenstein here is collapsing different levels again: he is confusing a theory with what the theory describes. According to the Platonic interpretation of the incompleteness results, it is not arithmetic, in the sense of the "realm of natural numbers", which is incomplete. If we are Platonists, as Gödel certainly was, we will take the "realm of numbers" as perfectly complete, with its properties distributed in a maximal and consistent way among numbers. It is just that this realm cannot be fully captured by any formal system. Formalised arithmetic is incomplete; not the arithmetic reality (say, the standard model $\mathbb{N}$), which the theory was supposed to describe.

However, Wittgenstein intentionally opposed precisely this referential picture of mathematics, according to which the meaning of mathematical sentences consists in their referring to, and describing, an independently existing reality—the picture of "arithmetic as the natural history (mineralogy) of numbers", of which "our whole thinking is penetrated" (Wittgenstein 1953, p. 116e). According to him, the meaning of a mathematical sentence is determined by the rules that govern its use in the calculus and in particular by its own proof (which is why an incompleteness in the theory would become *eo ipso* an incompleteness of meaning):

A psychological disadvantage of proofs that construct *propositions* is that they easily make us forget that the sense of the result is not to be read off from this by itself, but from the *proof*. [...] I am trying to say something like this: even if the proved mathematical proposition seems to point to a reality outside itself, still it is only the expression of acceptance of a new measure (of reality). (Wittgenstein 1953, pp. 76e–77e).

Consequently, also the Platonic separation between provability and truth has to go. The remarks on the First Incompleteness Theorem in the *Bemerkungen* are resolute on this point:

5. Are there true propositions in Russell's system, which cannot be proved in his system? – What is called a true proposition in Russell's system, then?

6. For what does a proposition's '*being true*' mean? '*p*' *is true* = *p*. (that is the answer). (Wittgenstein 1953, p. 50e)

Here Wittgenstein seems to be identifying (mathematical) truth with assertability (see Rodych 1999, pp. 178–179). Therefore, he concludes:

If, then, we ask in this sense: "Under what circumstances is a proposition asserted in Russell's game" the answer is: at the end of one of his proofs [i.e., as a theorem], or as a 'fundamental law' (Pp.) [i.e., as an axiom – and, of course, axioms are theorems]. There is no other way in this system of employing asserted propositions in Russell's symbolism.

7. "But may there not be true propositions which are written in this symbolism, but are not provable in Russell's system?" – "True propositions", hence propositions which are true in *another* system, i.e., can rightly be asserted in another game. [...] [A] proposition which cannot be proved in Russell's system is "true" or "false" in a different sense from a proposition of *Principia mathematica*. (Wittgenstein 1953, p. 50e)

In the end, "'True in Russell's system' means, as was said: proved in Russell's system; and 'false in Russell's system' means: the opposite has been proved in Russell's system" (Wittgenstein 1953, p. 51e).[9] By identifying truth and provability, and by rejecting the very idea of metamathematics, Wittgenstein was opposing some established results of contemporary logic—or, better, of contemporary *classical* mathematics and *classical* logic (whereas his position has often been connected, e.g., by Dummett (1959, pp. 504–505), Bernays (1959, p. 519) and Kielkopf (1970), and others, to a strong mathematical constructivism and to the so-called "strict finitism"). This speaks against Wittgenstein's own claim, according to which "it is my task, not to attack Russell's logic from within, but from without", and "my task is not to talk about (e.g.) Gödel's proof, but to pass it by" (Wittgenstein 1953, p. 174e). As I hinted at, however, it is possible to introduce a single argument that, by reinterpreting Gödel's results in the light of Wittgenstein's general standpoint, gives to the latter an unexpected plausibility precisely from the point of view of modern *non*-classical mathematical logic. Let's have a look.

14.4 Paraconsistency to the Rescue

My strategy exploits an argument proposed by Richard Routley and Graham Priest's various influential essays (Routley 1979; Priest 1979, 1984, 1987). It has not been developed having Wittgenstein in mind,[10] but it allows us to interpret Gödel's proof precisely as a paradoxical derivation. The core idea is to see what happens when one tries to apply the First Incompleteness Theorem to the theory that captures *our intuitive, or naïve, notion of proof.*

By "naïve notion of proof" Routley and Priest apparently mean the one underlying ordinary mathematical activity: "proof, as understood by mathematicians (not logicians), is that process of deductive argumentation by which we establish certain mathematical claims to be true" (Priest 1987, p. 40). Since Hilbert, formal logicians have learned to treat proofs as purely syntactic objects: sequences of strings of symbols, manipulated via transformation rules, etc. However, *proving* something, for a working mathematician, amounts to establishing that some sentence is *true*.

Now, when we want to settle the question whether some mathematical sentence is true or false, we try to deduce it, or its negation, from other mathematical sentences which are already known to be true. The process cannot go backwards *in infinitum*, though. We should therefore reach, eventually, mathematical sentences known to be true without having to be proved—e.g., because they are "self-evident". However,

---

[9]That at the core of Wittgenstein's rejection of the Platonistic "prose" associated to Gödel's proof is his identification of truth with provability, has been argued in detail by Rodych and Shanker in various essays (see Rodych 1999, 2003; Shanker 1988).

[10]In particular, Priest may disagree with the picture of Wittgenstein's attitude towards Gödel proposed here (see Priest 2004).

this is not important (nor is it important to establish *which* are the primal truths; concerning arithmetic, they may be, for instance, principles such as those of Peano, that is, claims according to which every number has a successor, etc.).

Given this characterisation, it is clear that the naïve-intuitive theory Routley and Priest link to the naïve-intuitive notion of proof is rather informal. However, "it is accepted by mathematicians that informal mathematics could be formalised if there were ever a point to doing so, and the belief seems quite legitimate" (Priest 1987, p. 41). Admittedly, this is a step the so-called second Wittgenstein, who disliked formalisations, may have questioned:

> The curse of the invasion of mathematics by mathematical logic is that now any proposition can be represented in a mathematical symbolism, and this makes us feel obliged to understand it. Although of course this method of writing is nothing but the translation of vague ordinary prose. (Wittgenstein 1953, p. 155e)

However, we may reasonably assume that, when Wittgenstein made such claims, he was not questioning formalisation itself, but the overwhelming importance attributed to it by philosophers and logicians looking for the "ideal language". On the contrary, we are now assuming precisely that formalisation is nothing but the "translation of vague ordinary prose": one may regiment the fragment of English in which the naïve theory is expressed, and turn it into a formal language. Then, the primal truths may be written down in the (now) formalised language and taken, say, as axioms; and proofs may be expressed as formal arguments. Priest also claims that, after having been so translated, the naïve theory would certainly be sufficiently strong in the sense explained above, i.e., capable of representing all the (primitive) recursive functions.

Is the naïve notion of proof decidable? This is much less straightforward, and it is likely that the crux of the argument lies here. To assume that the proof relation of naïve arithmetic is decidable challenges the standard perspective, taken as established precisely by Gödel's results. I will come back to this point, though, after exposing the paraconsistent argument, which goes as follows.

Let T be the formalisation of our naïve, intuitive mathematical theory. Assuming that T, just like PA, is sufficiently strong, *if* T is consistent, then Gödel's First Theorem applies: so there is a sentence ϕ which is not a theorem of T, but which can be established as true via a naïve proof, and therefore *is* a theorem of T. Of course, anything that is naïvely-intuitively provable is provable within the naïve-intuitive theory. So "assuming its consistency, it would, therefore, seem to be both complete and incomplete in the relevant sense" (Priest 1984, p. 165). Now we have no way to avoid a paradox: either we accept this one, i.e., $\vdash_T \phi$ and $\nvdash_T \phi$ (which is quite close to Wittgenstein's remark, quoted at the beginning of this paper: "let us suppose I prove the unprovability (in Russell's system) of P; then by this proof I have proved P. Now if this proof were one in Russell's system—I should in this case have proved at once that it belonged and did not belong to Russell's system"); or we have to admit that our naïve mathematical theory, with its naïve notion of proof, is such that the Gödel sentence ϕ for the (formalisation of the) naïve theory can be proved within T itself, together with its negation—so one of the inconsistencies hosted by T is to the effect that $\vdash_T \phi$ and $\vdash_T \neg\phi$.

The philosophical point is that "This sentence is not provable" now has its "provable" understood as meaning "demonstrably true", and, as Wittgenstein conjectured, Gödel's proof becomes the derivation of a real paradox:

> In fact, in this context the Gödel sentence becomes a recognisably paradoxical sentence. In informal terms, the paradox is this. Consider the sentence "This sentence is not provably true". Suppose the sentence is false. Then it is provably true, and hence true. By *reductio* it is true. Moreover, we have just proved this. Hence it is provably true. And since it is true, it is not provably true. Contradiction. This paradox is not the only one forthcoming in the theory. For, as the theory can prove its own soundness, it must be capable of giving its own semantics. In particular, [every instance of] the T-scheme for the language of the theory is provable in the theory. Hence [. . .] the semantic paradoxes will all be provable in the theory. Gödel's "paradox" is just a special case of this (Priest 1987, pp. 46–47; see also Priest 1984, p. 172).

Therefore, Anderson's comment on Wittgenstein, according to which "the conclusion to draw would not be that P at once 'belonged and did not belong' to Russell's system, but rather that Russell's system was inconsistent" (Anderson 1958, p. 458), is really of little importance: either horn of the dilemma makes us end up in a contradiction; and, as we shall see very soon, both contradictions (i.e., a system proving both its Gödel sentence and its negation, and a system both proving and not proving something) are expected in a thoroughly paraconsistent framework, as is shown in (Priest 1987, pp. 239–243).

I claimed that the "semantic prose" on the First Theorem attacked by Wittgenstein has it that the truth of the Gödel sentence is established in the metatheory (under the assumption that the theory is consistent): it can be proved in a metatheoretic context in which we can deal with the semantics of the object theory, i.e., with the truth predicate for (the language of) the object theory. However, T, formalizing as it does our naïve notion of proof, should absorb the metatheory within the theory. After all, as Wittgenstein might have added, mathematicians use ordinary English, and ordinary English may well be (and, according to many philosophers of language, actually is) semantically closed. As Routley has stressed, "everyday arithmetic as presented within a natural language like English appears, unlike say first-order Peano arithmetic, appropriately closed". And "is provable in arithmetic" and "is arithmetically true" are "English, and in a good sense arithmetical, predicates" (Routley 1979, p. 326). So T is semantically closed in the Tarskian sense, and inconsistent. The reasoning behind the proof of the truth of the Gödel sentence is now performed *within* the formal system itself—which is what we should expect in a Wittgensteinian framework that collapses, in the aforementioned sense, the distinction between theory and metatheory. There is no metasystem in which one establishes that (if the object system is consistent, then) the Gödel sentence is true: there are no metasystems. Consequently, one cannot "get out" of a system and solve, in its metasystem, problems that were meaningfully expressible but undecidable within the system.

Now back to the key assumption that the naïve notion of proof is effectively decidable (thus, given Church's Thesis, recursive). The first thing to notice in this respect is that this may well have been Wittgenstein's assumption, too. As we have already hinted at, Wittgenstein believed that the naïve (i.e., the working mathematician's) notion of proof had to be decidable, for lack of decidability meant to him simply lack of mathematical meaning: Wittgenstein believed that everything had to be decidable in mathematics, so the argument coheres with Wittgenstein's position on this point, too. But Routley and Priest also have positive arguments for the view. That the naïve notion of proof is decidable means that we can in principle effectively recognise a naïve proof when we see one. Now, Priest stresses, "it is part of the very notion of proof that a proof should be effectively recognizable as such" (Priest 1987, p. 41)— for the point of a naïve proof is that it is a way of settling the issue whether a given mathematical claim is true or not. As Alonzo Church claims:

> Consider the situation which arises if the notion of proof is non-effective. There is then no certain means by which, when a sequence of formulas has been put forward as a proof, the auditor may determine whether it is in fact a proof. Therefore he may fairly demand a proof, in any given case, that the sequence of formulas put forward is a proof; and until the supplementary proof is provided, he may refuse to be convinced that the alleged theorem is proved. This supplementary proof ought to be regarded, it seems, as part of the whole proof of the theorem... (Church 1956, p. 53)

Besides, by acknowledging that the naïve proof relation is decidable we can explain how we *learn* arithmetic—that is, via an effective procedure:

> We appear to obtain our grasp of arithmetic by learning a set of basic and effective procedures for counting, adding, etc.; in other words, by knowledge encoded in a decidable set of axioms. If this is right, then arithmetic truth would seem to be just what is determined by these procedures. It must therefore be axiomatic. If it is not, the situation is very puzzling. The only real alternative seems to be Platonism, together with the possession of some kind of sixth sense, "mathematical intuition". (Priest 1994, p. 343)

This point, too, meets some Wittgensteinian concerns on teaching and learning mathematical calculi as a public, social phenomenon. Perhaps the most amazing fact about mathematics as a discipline is the unanimity (generally speaking) of mathematicians on what counts as a proof. As Wittgenstein remarked, the whole "language game" of mathematical proofs would be rendered impossible by lack of consensus among mathematicians. If the notion of arithmetic proof were not effectively recognizable, then the process whereby mathematics is learnt, and the general agreement of working mathematicians on what counts as a mathematical proof, would turn out to be a mystery (of course, this is but a particular case of a famous, more general argument to the effect that *language* can only be learnt recursively, and so the grammar of a learnable language must be generated by a decidable set of rules), on which see, famously, Davidson (1984, Chap. 1). On the contrary, as Routley claims, if the truths of mathematics are effective or effectively enumerable we can understand "how one generation of mathematicians learns what counts as true from the previous generation, namely they learn certain basic mathematical truths and how to prove others by making deductions" (Routley 1979, p. 327).

Of course, one can speak against the decidability of the naïve notion of proof on the basis of Gödel's results themselves. But one may argue that, in the context, this would beg the question against paraconsistentists—and against Wittgenstein, too. Both Wittgenstein and the paraconsistentists, on one side, and the followers of the standard view on the other, agree on the following thesis: the decidability of the notion of proof and its consistency are incompatible. But to infer from this that the naïve notion of proof is not decidable invokes the indispensability of consistency, which is exactly what Wittgenstein and the paraconsistent argument call into question. Contrary to what Bernays claimed, the discussion in the *Bemerkungen* does not "suffer from the defect that Gödel's quite explicit premise of the consistency of the considered formal system is ignored" (Bernays 1959). Bernays' charge just begs the question against Wittgenstein, for, as Victor Rodych has forcefully argued, the consistency of the relevant system is precisely what is called into question by Wittgenstein's reasoning (see Rodych 2002, pp. 384–385).

14.5 Paraconsistent Arithmetic

One may wonder how can Wittgenstein's position be made more palatable from a logical point of view by referring it to an inconsistent theory. It is easy to see how audacious the argument itself is: by turning Gödel's proof into a paradox, it places inconsistencies at the very core of (the theory which, supposedly, captures) our mathematical practice. This is not so straightforward, though, if one does not believe, unlike Wittgenstein's early commentators, that contradictions immediately make formal systems uninteresting. Here comes into play the aspect of Wittgenstein's philosophy of mathematics, mentioned at the beginning of this paper, which my interpretation can recapture: his attitude towards contradictions.

That Wittgenstein did not consider the surfacing of contradictions within formal systems as a terrible crisis is well known and testified, for instance, by his discussions with Turing on this point, as reported in the *Lectures on the Foundations of Mathematics*. It is true that Wittgenstein did not comment directly on Gödel's Second Incompleteness Theorem. However, he often commented on the role and the importance of consistency proofs; and his position was clear-cut—he considered this kind of proof as a symptom of "the superstitious fear and awe of mathematicians in face of contradiction" (Wittgenstein 1953, p. 53e)

> And if they now demand a proof of consistency, because otherwise they would be in danger of falling into the bog at every step – what are they demanding? Well, they are demanding a kind of *order*. But was there *no* order before? – Well, they are asking for an order which appeases them now. – But are they like small children, that merely have to be lulled asleep? (Wittgenstein 1953, p. 101e)

After interpreting Gödel's proof as a paradox closely related to the Liar, Wittgenstein asks, rhetorically: "but there is a contradiction here!—Well, then there is a contradiction here. Does it do any harm here?"; "'perhaps', Wittgenstein might say, 'all calculi that admit such sentence-constructions are syntactically inconsistent'" (Rodych 1999, p. 190), but he believed that a calculus within which one can derive a contradiction does not thereby become useless:

> Can we say: "Contradiction is harmless if it can be sealed off"? But what prevents us from sealing it off? [...]
>
> Let us imagine having been taught Frege's calculus, contradiction and all. But the contradiction is not presented as a disease. It is, rather, an accepted part of the calculus, and we calculate with it. [...]
>
> For might we not possibly have wanted to produce a contradiction? Have said – with pride in a mathematical discovery: "Look, this is how we produce a contradiction"? [...]
>
> My aim is to alter the attitude to contradiction and to consistency proofs. (Not to show that this proof shows something unimportant. How could that be so?). (Wittgenstein 1953, pp. 104e–106e)

Because of these insights, Wittgenstein has been considered a precursor of paraconsistent logics. He anticipated the intuition that an inconsistent calculus does not thereby become trivial and uninteresting; on this point, see Marconi (1984):

> "Contradiction destroys the calculus" – what gives it this special position? With a little imagination, I believe, it can certainly be demolished. [...]
>
> And suppose the contradiction [i.e., Russell's paradox] had been discovered but we were not excited about it, and had settled e.g., that no conclusions were to be drawn from it. (Wittgenstein 1953, p. 170e)

Now, if we adopt a paraconsistent logic the theory T mentioned above, which is claimed to capture our naïve-intuitive notion of proof, is not just an argumentative trick anymore. It is possible to provide a respectable logical framework for Wittgenstein's idea according to which Gödel's proof is paradoxical, and nevertheless the derivation of such paradoxes does not render the relevant system(s) useless. Inconsistent arithmetics, i.e., non-classical arithmetics based on a paraconsistent logic, are nowadays a reality. What is more important, the theoretical features of such theories match precisely with some of the aforementioned Wittgensteinian intuitions. Let us see some examples.

First, paraconsistent arithmetics do not fulfil precisely the consistency requisite. This suggests that such theories could emancipate themselves from Gödel's Theorems, and from other limitative results afflicting their consistent cousins based upon a more traditional (classical, or intuitionistic) logic. To be sure, consistency proofs are not at issue, since we are dealing with inconsistent theories. What the theory may hopefully prove, though, is its own non-triviality, which in these contexts is more often called absolute consistency.

Paraconsistent authors have begun to show that this is the case since the 1970s, by building inconsistent but non-trivial theories, whose non-triviality proof can be represented within the very theories and that, in this sense, circumvent Gödel's Second Theorem. Their inconsistency allows them to escape also from Gödel's First Theorem, and from Church's undecidability result: they are, that is, demonstrably complete and decidable.[11] They therefore fulfil precisely Wittgenstein's request, according to which there should not be mathematical problems that canbe

[11] For a quick review, see Bremer (2005, Chap. 13).

meaningfully formulated within the system, but which the rules of the system cannot decide. Hence, the decidability of paraconsistent arithmetics harmonises with an opinion Wittgenstein maintained throughout his philosophical career.

Besides, the perspective of inconsistent arithmetics is (typically, though not necessarily) involved in a form of strict *finitism*. The underlying intuition would be that there is a finite (albeit hardly imaginable and unknown to us) number of things in the world. Although we cannot specify the number, we know that it must be "a number larger than the number of combinations of fundamental particles in the cosmos, larger than any number that could be sensibly specified in a lifetime" (Priest 1994, p. 338) (which should explain why our intuitions on it are rather unconfident); and this largest number is an inconsistent number.[12]

We can get into the details by considering a simple case of inconsistent arithmetic. Suppose n is our largest-inconsistent number. Let N be the theory of $\mathbb{N}$, that is, the set of arithmetic sentences true in the standard model $\mathbb{N}$; and let M_n be the set of sentences true in the paraconsistent model with the inconsistent number n. We may take as the underlying logic of M_n some mainstream paraconsistent logic, such as LP (Priest's logic of paradox), or FDE (Belnap and Dunn's First Degree Entailment). Now, according to Priest (1994) such a theory as M_n has the following enjoyable properties: it is, of course, inconsistent (including, among other things, both its own Gödel sentence and its negation), but provably non-trivial—and its non-triviality proof can be formalised within it. It fully contains N, that is, it includes all the sentences true in the standard model. Finally, M_n includes its own truth predicate. Therefore, the inconsistent arithmetic avoids Gödel's First Incompleteness Theorem; it also avoids the Second Theorem, in the sense that its non-triviality can be established within the theory; and Tarski's Theorem, too—including its own predicate is not a problem for an inconsistent theory.[13]

This is more than enough to get interested in the paraconsistent model of M_n. How is it like? The model can be obtained by applying to $\mathbb{N}$ an appropriate filter that reduces its cardinality. Meyer and Mortensen have initially developed the technique, and some of their main results are summarised in Meyer and Mortensen (1984) that appeared in the *Journal of Symbolic Logic*, in which different finite models are considered. The filter works as follows: let D be the domain of a given model $\mathbb{M}$, and $\approx$ an equivalence relation defined on D, which is also a congruence with respect to the denotations of the function symbols of the language. Given the objects $o_1, \ldots, o_n$ belonging to D, $|o_1|, \ldots, |o_n|$ are the corresponding equivalence classes under $\approx$. Now, let $\mathbb{M}^{\approx}$ be the new model, called the collapsed model, whose domain is $D^{\approx} = \{|o| \mid o \in D\}$. The role of $\mathbb{M}^{\approx}$ is to provide substitutes for the initial objects, and particularly to identify the members of D in each equivalence class, thereby

[12]Such a strict finitism is not unavoidably tied to inconsistency, nonetheless: van Bendegem (1994, 1999) has exploited the properties of paraconsistent arithmetical models to argue for a greatest number, which is not an inconsistent one.

[13]For a detailed account of these facts, see Priest (1994, pp. 337–338) and Priest (1987, pp. 234–237).

producing a composite object that "inherits" the properties of its components: the predicates that were true of the initial objects now apply to the substitute. Now, by induction over the complexity of formulas it is possible to prove the following lemma, called the Collapsing Lemma:

(CL) Given any formula α which has the truth value v in $\mathbb{M}$, α has the truth value v also in $\mathbb{M}^{\approx}$.[14]

Therefore, if the original model satisfied some set of formulas, the collapsed model also satisfies it: when the initial model $\mathbb{M}$ is collapsed into $\mathbb{M}^{\approx}$, no sentence loses a truth value—it can only gain them. Of course, when we begin with the model of a standard theory, the only values around are true and false. But in the collapsed model it may be the case that a formula, which was initially true only, or false only, becomes both true and false (and this, of course, is not a problem within such paraconsistent logics as LP or FDE). This happens when the collapsing filter produces an inconsistent object: for instance, it may identify in an equivalence class two initial objects, one of which had, whereas the other did not have, the very same property. The procedure works even if among the relevant sentences we have formulas that seem to put constraints on cardinality, such as $\exists xy(x \neq y)$, precisely because they can become paradoxical.

In the particular case of M_n, the trick consists in choosing for $\mathbb{N}$ a filter that (a) given a number $x < n$, puts x and nothing else in the corresponding equivalence class, so that $|x|$ inherits all and only the properties of x; and (b) puts every number $y \geq n$ in a single equivalence class. Consequently, all the true/false equations involving any number smaller than n in the standard model are now true only/false only of the substitute. Because of this, the initial segment in the succession (which is sometimes called the *tail*) behaves as usual. Roughly, "up to n" things work like in ordinary arithmetic. Nevertheless, anything that could be truly/falsely claimed of anything bigger than n is now true/false of the inconsistent number. Many things concerning it are therefore paradoxical now (both true and false), and "of course, n is [now] an inconsistent object [...]. In particular, in the model $\mathbf{n} = \mathbf{n} + 1$ is true even though it is also false" (Priest 1994, p. 338), so n is the successor of itself.

Priest has declared that (CL) is "the ultimate downwards Löwenheim-Skolem Theorem" (Priest 1994, p. 339), which is easy to understand. The downward half of the Löwenheim-Skolem Theorem claims that any first-order theory, with a model with an infinite domain has a model with a denumerably infinite domain, too.[15] The filter and the Collapsing Lemma allow us to "shrink" even more, since one can reduce a model with a denumerably infinite domain into one of any smaller size. We can have a collapsed model, $\mathbb{M}^{\approx}$, whose domain, $D^{\approx}$, has cardinality k (smaller than that of the initial model), by choosing an appropriate equivalence relation

[14]See Priest (1994, pp. 346–347); the result was anticipated in Dunn (1979).

[15]One of the consequences of the downward Theorem is the so-called Skolem paradox. Since set theory can be expressed in a first-order language, it has a model whose domain has the cardinality of the set of natural numbers. However, within set theory we can prove the existence of sets whose cardinality is more than denumerable.

that produces precisely k equivalence classes. Bremer has therefore suggested the following Paraconsistent Löwenheim-Skolem Theorem: "Any mathematical theory presented in first order logic has a *finite* paraconsistent model" (Bremer 2005, p. 155).

Now this strong finitism also meets a persistent tendency in Wittgenstein's philosophy of mathematics. Wittgenstein always showed a suspicious attitude towards Cantor's paradise and the non-denumerable infinities which, in Cantor's Platonistic view, were to be discovered by the diagonal argument. Of course, strict finitism and the insistence on the decidability of any meaningful mathematical question go hand in hand. As Rodych has remarked, the intermediate Wittgenstein's view is dominated by "his finitism and his [...] view of mathematical meaningfulness as algorithmic decidability", according to which "[only] finite logical sums and products (containing only decidable arithmetic predicates) *are* meaningful because they are *algorithmically decidable*". But this tendency remains also in the later phase: "as in the middle period, the later Wittgenstein seems to maintain that an expression is a meaningful proposition only *within* a given calculus, and *iff* we knowingly have in hand an applicable and effective DP [decision procedure] by means of which we can decide it" (Rodych 1999, pp.174–176).

14.6 Conclusion: The Costs and Benefits of Making Wittgenstein Plausible

The cost of accepting paraconsistent arithmetics is clear: we have to revise some well-established acquisitions of classical mathematical logic. As I claimed before, by subscribing to such a way of bringing up-to-date Wittgenstein's philosophy of mathematics one will not be allowed to claim—as many commentators did—that such a philosophy does not require any logico-mathematical revisionism, being directed only against the foundational demands of philosophers.

On the other hand, Wittgenstein might have found the situation produced by paraconsistent arithmetics quite plausible. Surprising and (in a broad sense) paradoxical innovations in the history of mathematics— this "motley of techniques of proof" (Wittgenstein 1953, p. 84e)—led to the invention of new kinds of numbers: from Hyppasus' irrational numbers refuting Pythagorism, to infinitesimals, Cantor's transfinite numbers, and all that. The early reception of such new entities among mathematicians has always been controversial, from the Pythagoreans condemning and expelling Hyppasus, to Kronecker making Cantor's life impossible. A process of rethinking mathematics in order to come to grips with the new domain has usually followed. And, as we have seen, such an audacious rethinking in a paraconsistent framework may nowadays vindicate some of Wittgenstein's "outrageous claims", which were dismissed too swiftly by commentators who dogmatically took the logic of Russell and Frege as the One True Logic.

Acknowledgements The non-technical parts of this work draw on a paper published in *Philosophia Mathematica*, 17: 208–219, with the title "The Gödel Paradox and Wittgenstein's Reasons". I am grateful to Oxford University Press and to the Editors of *Philosophia Mathematica* for permission to reuse that material. I am also grateful to an anonymous referee for helpful comments on this expanded version.

References

Ambrose, A., and M. Lazerowitz (eds.). 1972. *Ludwig wittgenstein: Philosophy and language*. London: Allen and Unwin.

Anderson, A.R. 1958. Mathematics and the 'Language Game'. In *Philosophy of mathematics. Selected readings*, ed. P. Benacerraf, H. Putnam, 481–490. Prentice Hall: Englewood Cliffs.

Benacerraf, P., and H. Putnam (eds.). 1964. *Philosophy of mathematics. Selected readings*. Prentice Hall: Englewood Cliffs.

Bernays, P. 1959. Comments on ludwig Wittgenstein's remarks on the foundations of mathematics. In *Philosophy of mathematics. Selected readings*, ed. P. Benacerraf, H. Putnam, 481–490. Prentice Hall: Englewood Cliffs.

Boolos, G.S., J.P. Burgess, and R. Jeffrey. 2002. *Computability and logic*, 4th ed. Cambridge: Cambridge University Press.

Bremer, M. 2005. *An Introduction to paraconsistent logics*, Frankfurt: Peter Lang.

Church, A. 1956. *Introduction to mathematical logic*, Princeton: Princeton University Press.

Davidson, D. 1984. *Inquiries into truth and interpretation*. Oxford: Clarendon Press.

Dawson, W. 1984. The reception of Gödel's incompleteness theorems. *Philosophy of science association* **2**, 74–95. Reprinted in Shanker (1988).

Dummett, M. 1959. Wittgenstein's philosophy of mathematics. In *Philosophy of mathematics. Selected readings*, ed. P. Benacerraf, H. Putnam, 481–490. Prentice Hall: Englewood Cliffs.

Dummett, M. 1978. *Truth and other enigmas*. London: Duckworth.

Dunn, J.M. 1979. A theorem in 3-valued model theory with connections to number theory, type theory and relevant logic. *Studia Logica* 38: 149–169.

Feferman, S. 1983. Kurt Gödel: Conviction and caution. In *Gödel's theorem in focus*, ed. S.G. Shanker, 96–114. London: Croom Helm.

Floyd, J. 2001. Prose versus proof: Wittgenstein on Gödel, tarski and truth. *Philosophia mathematica* 9: 280–307.

Floyd, J., and H. Putnam. 2000. A note on wittgenstein's 'Notorious Paragraph' about the Gödel theorem. *Journal of Philosophy* 97: 624–632.

Gödel, K. (1931). Über formal unentscheidbare Sätze der Principia mathematica und verwandter systeme I. *Monatshefte für Mathematik und Physik* 38: 173–198, translated as: On formally undecidable propositions of principia mathematica and related systems I. In *From Frege to Gödel. A source book in mathematical logic*, ed. J. van Heijenoort, 596–617. Harvard: Harvard University Press.

Gödel, K. 1944. Russell's mathematical logic. In *The philosophy of bertrand russell*, ed. P.A. Schlipp, 125–153. Evanston: Northwestern University Press.

Gödel, K. 1947. What is cantor's continuum problem? *American Mathematical Monthly* 54: 515–525.

Goodstein, R. 1957. Critical notice of remarks on the foundations of mathematics. *Mind* 66: 549–553.

Goodstein, R. 1972. Wittgenstein's philosophy of mathematics. In *Ludwig Wittgenstein: Philosophy and Language*, ed. A. Ambrose and M. Lazerowitz, 271–286. London: Allen and Unwin.

Helmer, O. 1937. Perelman versus Gödel. *Mind* 46: 58–60.

Hintikka, J. 1999. Ludwig Wittgenstein: Half truths and one-and-a-half truths. In *Selected papers*, vol. I, ed. J. Hintikka. Dordrecht: Kluwer.

Holbel, M., and B. Weiss (eds.). 2004. *Wittgenstein's lasting significance*. London/New York: Routledge.
Kielkopf, C. 1970. *Strict finitism: An examination of ludwig Wittgenstein's remarks on the foundations of mathematics*. The Hague: Mouton.
Kleene, S. 1976. The work of Kurt Gödel. *Journal of Symbolic Logic* 41: 761–778. Reprinted in Shanker (1988), 48–73.
Kreisel, G. 1958. Review of Wittgenstein's 'remarks on the foundations of mathematics'. *British Journal for the Philosophy of Science* 9: 135–158.
Marconi, D. 1984. Wittgenstein on contradiction and the philosophy of paraconsistent logics. *History of Philosophy Quarterly* 1: 333–352.
Meyer, R.K., and C. Mortensen. 1984. Inconsistent models for relevant arithmetic. *Journal of Symbolic Logic* 49: 917–929.
Perelman, C. (1936). L'Antinomie de M. Gödel. *Académie Royale de Belgique, Bulletin de la Classe des Sciences* 5(22): 730–733.
Priest, G. 1979. The logic of paradox. *Journal of Philosophical Logic* 8: 219–241.
Priest, G. 1984. Logic of paradox revisited. *Journal of Philosophical Logic* 13: 153–179.
Priest, G. 1987. *In contradiction: A study of the transconsistent*. Dordrecht: Martinus Nijhoff; 2nd expanded edition (2006) Oxford: Oxford University Press.
Priest, G. 1994. Is arithmetic consistent? *Mind* 103: 337–349.
Priest, G. 1995. *Beyond the limits of thought*. Cambridge: Cambridge University Press.
Priest, G. 2004. Wittgenstein's remarks on Gödel's theorem. In *Wittgenstein's lasting significance*, ed. M. Holbel, and B. Weiss. London/New York: Routledge.
Rodych, V. 1999. Wittgenstein's inversion of Gödel's theorem. *Erkenntnis* 51: 173–206.
Rodych, V. 2002. Wittgenstein on Gödel: The newly published remarks. *Erkenntnis* 56: 379–397.
Rodych, V. 2003. Misunderstanding Gödel: New arguments about Wittgenstein and new remarks by Wittgenstein. *Dialectica*, 57: 279–313.
Routley, R. 1979. Dialectical logic, semantics and metamathematics. *Erkenntnis* 14: 301–331.
Shanker, S.G. (ed.). 1988. *Gödel's theorem in focus*. London: Croom Helm.
Smullyan, R. 1992. *Gödel's incompleteness theorems*. Oxford: Oxford University Press.
Tarski, A. 1956. *Logic, semantics, metamathematics. Papers from 1923 to 1938*. Oxford: Oxford University Press.
van Bendegem, J.-P. 1994. Strict finitism as a viable alternative in the foundations of mathematics. *Logique et Analyse* 137: 23–40.
van Bendegem, J.-P. 1999. Why the largest number imaginable is still a finite number. *Logique et Analyse* 165–166: 107–126.
van Heijenoort, J. (ed.). 1967. *From Frege to Gödel. A source book in mathematical logic*. Harvard: Harvard University Press.
Wittgenstein, L. 1921. Logisch-philosophische Abhandlung. *Annalen der Naturphilosophie*, vol. 14. London: Routledge and Kegan Paul; Revised edition 1922. *Tractatus logico-philosophicus*, Reprinted New York: Barnes and Noble.
Wittgenstein, L. 1953. *Philosophische Untersuchungen*. Oxford: Basil Blackwell.
Wittgenstein, L. 1956. *Bemerkungen über die Grundlagen der mathematik*. Oxford: Basil Blackwell.
Wittgenstein, L. 1964. *Philosophische bemerkungen*. Oxford: Basil Blackwell.
Wittgenstein, L. 1974. *Philosophische grammatik*. Oxford: Basil Blackwell.
Wittgenstein, L. 1976. *Lectures on the foundations of mathematics*. Ithaca: Cornell University Press.

Chapter 15
Pluralism and "Bad" Mathematical Theories: Challenging our Prejudices

Michèle Friend

15.1 Introduction

In the philosophy of *logic*, we have fairly well worked out pluralist philosophies (Beall and Restall 2006); we also have good discussions and exchanges over variations of the position. In contrast, in the philosophy of *mathematics*, there is no well worked out position called "pluralism". When the attitude of pluralism is mentioned, it is listed as an attitude amongst others.[1] In this paper, pluralism is defended and developed as a philosophy of mathematics. A pluralist in the philosophy of mathematics is someone who places pluralism as the chief virtue in her philosophy of mathematics. She brings the attitude to bear on mathematical theories, including different foundations of mathematics, and on different philosophies of mathematics.

As a philosophy of mathematics, pluralism is founded on the conviction that we do not have the necessary evidence to think that mathematics is one unified body of truths, or is reducible to one or two theories (foundations). As pluralists, we might *hope* that mathematics will one day turn out to be so unified, but, and this is important, such hope is simply a subjective private feeling, and is not supported on present evidence. So, similarly, a pluralist might hope that there are several irreducible foundations in mathematics, and that there will never, nor can ever, even in principle, be a way of unifying these. Again this is a hope and a private conviction. The pluralist, as (public) philosopher is simply agnostic. Once this agnostic attitude is in place, then the pluralist is free to take an interest in mathematics as a series of theories, where each contains truths-relative-to-a-meta-theory. Or the pluralist might

[1]Examples of attitudes held by various philosophies of mathematics are: ontological parsimony, respect for the phenomenology of mathematics, respect for the views of mathematicians, simplicity etc. Some of the attitudes might be judged as a virtue or as a vice, depending on the philosophy.

M. Friend (✉)
Department of Philosophy, George Washington University, Washington D.C., USA
e-mail: michele@gwu.edu

K. Tanaka et al. (eds.), *Paraconsistency: Logic and Applications*, Logic, Epistemology, and the Unity of Science 26, DOI 10.1007/978-94-007-4438-7_15,

think of mathematics as a process, as opposed to concentrating on mathematics as a unified body of truths. The pluralist is also interested in "bad" mathematics, and how these parts inform the "good" parts. "Bad" mathematics include: intensional theories, intentional theories,[2] not yet completely formally represented theories, paraconsistent mathematics and trivial mathematics. Because of the last two, the pluralist philosopher underpins her philosophy with a paraconsistent logic.

In this paper I give two motivations for pluralism, and then discuss the position itself. I shall begin with three negative arguments why the present day foundationalist philosophies of mathematics are inadequate. The critique mainly comes from the naturalist insight that philosophers should respect mathematical compass and practice and reports by mathematicians concerning their philosophical views (Maddy 2007). In this way, naturalism is a motivation for pluralism. I then discuss the second motivation, which comes from structuralism: that there is no absolute notion of truth in mathematics. Rather, there is a perfectly robust notion of 'truth in a structure' (Shapiro 1997). As a philosophy, pluralism pushes both naturalist and structuralist insights beyond the present developments of these positions. This is what we do in the first two thirds of the paper.

The first of the three negative arguments for pluralism is that many foundationalist philosophies make a slide from a description of mathematics (in terms of a global foundation) to a norm for success for future mathematics, and this slide is illegitimate. The second argument is that the proposed foundations are artificial, so the foundations are a mis-description of mathematics in the first place. The mis-description is twofold, since, in addition to its being a mis-description of contemporary mathematics, whatever founding theory one has, 'it' will grow—by adding new axioms. Some new axioms are independent, and therefore there are no grounds for faith in there being one absolute true foundation for mathematics. The third negative argument is simply that many mathematicians today are pluralist, and that, following the naturalist, philosophers should respect and accommodate this interesting and philosophically challenging fact. Not only are many mathematicians pluralist, but they are highly aware of the context surrounding a theorem or proof, or the limitations of those theorems. In this respect, they are broadly structuralist. For this reason, pluralists are also motivated by structuralism. The structuralism I shall discuss is Shapiro's structuralism (since it, too, is anti-foundationalist). But I want to push it further than Shapiro does, to include more 'structures' than what can be recognised as a 'structure' by model theory. These are the "bad" parts of mathematics. Once we take seriously the idea that we want to do philosophical work concerning such outlandish parts of mathematics, we need to articulate the pluralist position. This will be done in the last third of the paper. We need to fend from the suspicion that pluralism degenerates into history or sociology of mathematics, and that there is no room left for philosophy. We also need to fend from the related suspicion that pluralism is really a rampant relativism. To fend from this last suspicion, we deploy a paraconsistent logic. This allows the pluralist to cope

[2]We shan't say much about these in this paper.

with situations when, (1) no decision can be made as to the truth of a well-formed formula in a theory (so we need a logic with truth-value gaps), and when, (2) a theory contains a well-formed formula which is both true and false (a logic with truth-value gluts).[3] By way of drawing some final conclusions about the limitations of pluralism we shall visit the semantics of the logic underlying the pluralist philosophy.

15.2 Setting up the Negative Arguments

15.2.1 Monism and Dualism, Revisionism and Foundationalism

Pluralism is anti-foundationalist. In this respect, pluralism takes issue with most of the standard traditional philosophies of mathematics. Let us study the enemy camp. In developing a foundationalist philosophy of mathematics, a philosopher seeks to give a unified philosophy of all of mathematics, where the unity is made plain by reduction to a foundation. These foundationalist philosophies of mathematics are monist and usually revisionist of mathematics. Alternatively, standard philosophies of mathematics split mathematics into two parts, the good part and the more suspect part; these are the dualists. Examples of monist philosophies are: platonism/realism, intuitionism and Whitehead and Russell's logicism. Examples of dualist philosophies are: Fregean logicism, Hilbertian formalism and Cantorian realism. In the development of both monist and dualist philosophies,[4] it was presumed that to give a *philosophy of mathematics at all*, one had to give a foundation for mathematics. Only sporadically has this assumption been challenged, and the most recent challenge has come from structuralism. Interestingly, anti-foundationalism can be well motivated by some insights from Maddy's naturalism, although, she does not consider herself to be pluralist in the sense developed here.

Since the terms "monism" and "dualism" are unusual in this context, let us briefly survey to what extent the monist and dualist foundationalist assumption has been made by more traditional philosophies of mathematics. Plato is an obvious example of a platonist monist, and so is Gödel.[5] Plato founded mathematics, especially geometry on the Forms. Gödel founds set theory on ZF set theory and the underlying truths to which it, and its extensions are responsible.

---

[3]Strictly speaking it is a relevant logic, which has truth-value gaps, and a paraconsistent logic has truth-value gluts. I need both, and the logic will also be more general than either a relevant or a paraconsistent logic. For lack of a better name, I still call it a paraconsistent logic, since one of the more interesting features is the presence of truth-value gluts.

[4]There were other philosophies of mathematics around too, such as conventionalism. However, these four have become the canon.

[5]I assume familiarity with the basic ideas of these positions, so I shall not discuss them further. For a good sketch of Gödel's platonism see Maddy (1997, pp. 89–94).

A more nuanced example of a monist is Brouwer, who wants to found mathematics on a firm epistemological footing, where mathematics obeys our mathematical intuitions. Brouwer's philosophy does not neatly fit as a "monist" philosophy since he does not propose a formal foundational theory. Nevertheless, his attitude is still monist, since he founds mathematics in intuition. Brouwer resists giving a formal representation of intuition because he distrusts formal representations as flawed.

> Used as a means of communicating thought to others, language is bound to remain defective [*mutatis mutandis* for formal representations of mathematics], given the essential privacy of thought and the nature of the "sign," the arbitrary association of the thought with a sound or visual object. (van Stigt 1998, p. 7)

Instead, for Brouwer, mathematics is really carried out in the mind.

> Intuitionistic mathematics is a mental activity and for it every language, including the formalistic one, is only a tool for communication. It is in principle impossible to set up a system of formulas which would be equivalent to intuitionistic mathematics, for the possibilities of thought cannot be reduced to a finite number of rules set up in advance. (Mancosu 1998, p. 311; quoted from Heyting 1930)

For Brouwer, the founding intuition is not thought to be subjective and personal.[6] For this reason, he does not accept rival but equally legitimate intuitions. *Our participation in intuition* is personal and subjective but mathematical intuition *itself* is not, since Brouwer is drawing on a Kantian notion of intuition.[7] It is this "objective" intuition which founds mathematics for Brouwer, and since it is objective and unique, this makes Brouwer a monist.

An example of logicist monism can be found in *Principia Mathematica*, (Whitehead and Russell, 1997). They sought to give a foundation for most of mathematics in a logical type theory. The type theory was not proposed as "the new way of doing mathematics" or as "the new notation to adopt". Whitehead and Russell understood that it was impractical for mathematicians to carry out all of their work in type theory, they were not fanatics; but they thought it important to develop a logical foundation in which, *in principle*, mathematics could be carried out. This was enough to make the philosophical point that mathematics is really only logic.[8]

---

[6]'Intuition' here is meant in the Kantian sense of a shared intuition in which we all participate. Degrees of participation might differ from one person to the next, but the mathematical intuition itself does not.

[7]Of course, philosophically, this is quite problematic, especially when Brouwer takes it upon himself to not recognise some mathematics (as accepted as such by others) as legitimate. He is not making the simple minded argument that "if I can't understand it, then it is not legitimate mathematics". However, he does end up restricting mathematics to what we would call today "effective" mathematics. This is difficult to defend in light of the problems with the Kantian or Brouwerian notion of intuition.

[8]Other examples of monists are: Curry, Dummett, Tennant and Hellman. Most of twentieth century philosophy of mathematics is monist, and this is partly because of an underlying idea that in order to be counted as a philosophy of mathematics at all, one has to give a unified philosophy of the whole of mathematics, and to do this, there has to be one founding discipline.

Amongst the standard philosophies of mathematics, we also find dualists.[9] These include Frege, Hilbert and Cantor. Frege tried to prove that analysis, arithmetic and logic are analytic, whereas geometry is synthetic. We have two systems of mathematics working in parallel: the arithmetic/analysis system which is founded in logic and the geometrical theories[10] which are somewhat corrupted, since they require Kantian-type spatial intuition. Hilbert too was a dualist, but like Brouwer, his case is a little nuanced. Hilbert distinguished between real (finitistic) and ideal mathematics. Unlike Frege, Hilbert did not think that the two realms were necessarily irreconcilable. The hope behind his programme was to show that even ideal mathematics was real[11] (see Giaquinto 2002, pp. 142–151). So, arguably, while his programme was set up to ultimately vindicate monism, he recognised that mathematics in his day was (hopefully only temporarily) dualist. Cantor too divided mathematics into two realms: the mathematical realm of the finite and infinite, and the more metaphysical realm of "absolute infinite multiplicities" (see Giaquinto 2002, pp. 42–43). The realm of the finite and the infinite includes multiplicities on which we can perform mathematical operations. In contrast, "absolute infinite multiplicities" cannot be operated upon. They can just be thought up and wondered at. A close modern analogy to "absolute infinite multiplicities" is proper classes. We cannot operate on these, but we can think them up, and say some limited things about them.[12,13] From this brief survey, we can understand how the terms "monist" and "dualist" are being contrasted to pluralism and we can appreciate that many standard philosophies of mathematics are either monist or dualist.

[9]I am not aware of other writers referring to Frege et. al. as dualists per se, but it is implicit in their writings. I hope that the terminology is helpful.

[10]Frege did not accept all of the geometrical systems as on a par. He thought of Euclidean geometry as "the true geometry".

[11]Note that there is some controversy over what Hilbert's programme really is: to reduce the whole realm of ideal mathematics to the formal finitistic part (in which case Gödel showed that this is impossible), or to investigate and maximally extend the scope of the finitistic realm.

[12]Examples of ways to 'think up' a proper class is to think of a totality or think of the complement class to a class which has been 'constructed' in appropriate (set theoretic) ways. More precise examples are: the proper class of all models which do not satisfy the Peano axioms or all of the ordinals.

[13]Kant is an interesting case. If logic is to be included in mathematics (a little anachronistically), then Kant too is a dualist because logic is analytic, whereas arithmetic and geometry are synthetic. He is not the same as the other dualists because he does not think that analytic truths are better. In fact, he needs synthetic mathematics in order to explain how it is possible for us to do metaphysics rigorously and to allow for the possibility of experience. So he does not show a preference for one of the mathematical foundations, in the way that Frege, Hilbert or Cantor did. Nevertheless, Kant does hold arguments for analytic truths and for synthetic truths to different standards.

15.2.2 The Foundationalist Argument

Let us now give the monist argument which is properly revisionist of mathematics. We shall then point out how this differs from the dualist case. The arguments will be important since we shall refer back to them when we run the negative arguments against foundationalism. To fix an example, consider ZFC (Zermelo-Fraenkel set theory with the axiom of choice) as a foundation for mathematics. Presenting a 'foundation' has two parts: a technical part and a philosophical part (which in turn has two parts). The technical result is achieved through a reduction of all (or most) existing successful mathematics to the foundation. We show, for most areas of successful mathematics, that they can be translated into the language of ZFC, and the theorems or results of the area of mathematics can be generated in ZFC too. In other words, we show that the language and proof apparatus of the reduced area of mathematics is strictly redundant with respect to ZFC. This is not enough to convince mathematicians to cease to work in the language of the reduced theories and to use the proof apparatus of the reducing area. For, the original language was designed to suit that area, and might be much more workable, less awkward, more suggestive etc. Nevertheless, the technical part is achieved since all the philosopher needs to know is that *in principle* it is possible (if a little awkward) to do all the work of the reduced discipline in the reducing discipline.

After we have the technical result, we make two philosophical moves. The first philosophical move is to state something to the effect that mathematics is 'essentially' the reducing discipline.[14] That is, we have captured almost all of mathematics in the reducing discipline, and therefore, all other languages, symbols, supposed ontology of reduced disciplines is strictly (philosophically/conceptually) redundant. Therefore, all we *strictly need* for most of mathematics is the apparatus of the reducing discipline. We have unified mathematics into one foundation. This is quite a philosophical coup!

The second philosophical move is to introduce a normative element; the essence of mathematics becomes a norm for *success* in mathematics. When we make this slide, we judge future "successful" mathematics against the backdrop of the essence capturing theory. If a proposed area of study does not fit into the founding discipline, then it is not "properly" or not "really" mathematics. At the very least it is not successful mathematics.

Dualists will run a similar argument, but it will include an added complication. The technical result will split mathematics into two, the best part and the suspect part. For an example, let us consider Fregean logicism. The two parts are the arithmetic part, and the geometrical part. Similarly, there will be two "essences" in mathematics, in the Fregean case, they will be analytic and synthetic, respectively. "Success" is relative to the different parts, there will be different norms, according

[14]Many philosophers are leery of using the term 'essence', so euphemisms are used instead. Feel free to replace 'essence' with your favourite substitute.

to which part of mathematics one is operating in. For example, a proof in the analytic-arithmetic part has to be able to be turned into a gapless proof. In contrast, a proof in geometry may invoke intuitive gaps (which draw on our spatio-temporal intuitions). The normativity in the dualist philosophy surfaces either when we favour one part of mathematics over another, or when we refuse to consider the purported mathematics which lie outside these two parts (for example, modal operators are seldom included in the language of founding mathematical theories, and modality is often not considered to be part of mathematics). Outlying parts of "mathematics" are either not recognised as mathematics or simply not discussed.

15.2.3 Naturalism

For my negative arguments against foundationalist philosophies of mathematics I shall capitalise on the naturalist insight that the philosopher is not there to set norms for success in mathematics on purely philosophical grounds. Thus, the naturalist rejects the normative move of both the monist and the dualist. Rather, the naturalist philosopher has two responsibilities towards the mathematician. One is to take seriously what the mathematician, himself, says about mathematics, on a philosophical level.[15] The other is to give a philosophy of mathematics which looks at all of mathematics, as it is practiced, and not just some philosophically convenient proper sub-part of mathematics. We now have enough information to run the negative arguments against formalism.

15.3 The Negative Arguments from Naturalism

15.3.1 Argument One: The Foundationalist's Slide from Description to Prescription is Illegitimate

The first argument broadly concerns the role of the philosopher vis-à-vis the mathematician. Implicitly, or explicitly, and to different degrees, foundationalist philosophies endorse the general idea that once the philosopher has developed a philosophy of mathematics, that philosophy should determine the limitations, and the future development, of mathematics. We saw this in the normative phase of the argument for the foundationalist philosophies.

[15] I take this insight from Maddy's development of naturalism. This departs from more Quinean naturalist philosophies who take their cues from scientists, and not mathematicians. I do not go as far as Maddy, in taking second place to mathematics. Instead, I follow Colyvan (2001) in thinking that philosophy of mathematics and mathematics sometimes influence each other. So the philosopher of mathematics is also allowed his input!

Illustrating the slide from description to setting the norm for success, Vopěnka complains that set theory is so powerful, that it comes to delimit our interests in mathematics, so while set theory was proposed as a reducing discipline; once this was done, set theory became a *norm for success* in mathematics: if a proposed mathematical topic is not interesting in set theory, then it should not be interesting for mathematicians.

> Set theory opened the way to the study of an immense number of various structures and to an unprecedented growth of knowledge about them. This caused a scattering of mathematics. [It is interesting that Vopěnka does not say "unifying"!] *Moreover, most results of this kind derive their sense only from the existence of the respective structure in Cantor set theory. Mathematics based on Cantor set theory changed to mathematics* [only being recognised in terms] *of Cantor set theory*. (Vopěnka 1979, p. 9)

In other words, Cantorian set theory became the standard by which proposed mathematics was judged to be "good" mathematics.[16] Today, ZFC has replaced Cantor's set theory as a point of reference. Under the ZFC norm for success, much of category theory is not mathematics, nor is the ramified type theory, nor, ironically, is all of Cantorian set theory. What is important here is that from Vopěnka's analysis it becomes plain that in keeping with a monist attitude we slide from description to norm for success, and that the norm precludes some potential further developments in mathematics, just because they are not recognised as *bona fide* mathematics. Unfortunately for the monist, history has not born out her slide from description to norm setting. Alternatives to ZFC have been developed. Some have been proved to be equi-consistent to ZFC, (and this is much weaker than using ZFC as the norm for success). Other areas of mathematics have not been shown to be equi-consistent, but might be so in the future. Other areas might never be, and might in principle never be able to be (such as, for example, a paraconsistent set theory, where a proposed proof of equi-*consistency* would be unrecognisable to the monist who wanted ZFC to set a norm for success in mathematics). Excluding mathematical developments, by means of norm setting, runs against the naturalist insight that philosophers should (if they are naturalists) want to observe not only what mathematicians say about their subject, but also observe their behaviour.

How do mathematicians think of set theory? It is true that a lot of present day mathematicians take it as a good verification of their work that it can be done in first-order set theory. But this does not mitigate against my point, since there is a difference between using first-order set theory as one, amongst other, means of verification, and counting set theory as the *only* means of verification. There is a further complication which should be addressed. When Vopěnka makes his anti-foundationalist complaint, the naturalist should observe that it was mathematicians,

[16] Vopěnka is highly revisionary of mathematics too. But he proposes a different founding theory. This does not interest me here. What is important is that we should realise that the platonist or realist proposal to found mathematics in set theory is, arguably, normative of mathematics.

not philosophers,[17] who set said norm. It seems then, that, as naturalists, respecting the practice of mathematics, we should give a philosophy that advocates ZFC or Cantorian set theory, as a foundation. According to the naturalist attitude, if mathematicians are setting norms in this way, then the philosophers should take the norm setting seriously. But, even as naturalists, we can be more careful. As we saw with the quotation from Vopěnka, not all mathematicians agree to follow the norm, trivially, since he is an example of a mathematician who does not. So then what are the *philosophers* to do about the rival norms internally set in mathematics? The pluralist observes that the Cantorian set theory norm was temporary. This is why Vopěnka's alarm is illegitimate. Contemporaneous with, and subsequent to, when Vopěnka's was writing the quoted passages, many developments in set theory have taken place. A number of higher-cardinal axioms have been proposed as extensions of ZFC set theory, and the prevailing attitude (I think) is that, in the light of the rival foundations, pluralism has succeeded set theoretic monism.

At this point, the monist might well dig in her philosophical heels. For, the monist starts with the presumption that the reducing discipline is the correct foundation. A philosophical consequence of this use of "correctness" is that the reducing discipline is taken as an a priori norm for success for the whole of mathematics.[18] This runs directly counter to observation of practice in mathematics, and thus to the naturalist. Here the naturalist and the monist part company.

The dualist does not fare much better. For, he proposes a foundation for some part of mathematics, and this part will suffer from the same criticisms. The part of mathematics for which we provide a proper foundation: second-order logic for the Fregean logicist, finististic (real) mathematics for the Hilbertian, all of mathematics save "absolutely infinite magnitudes" for the Cantorian; is good mathematics, the rest is suspect. With Hilbert, we then engage in a project of trying to widen, or determine the scope of and the limitations of the "good" part of mathematics, to minimise the "suspect" part. The naturalist observer of mathematics will disagree with this nuanced normative attitude. He will observe that mathematicians work in both "good" and "suspect" areas of mathematics, and do not always agree that the "suspect" part of mathematics really is suspect. Take, for example, all of the work on higher cardinal axioms. A Hilbertian would find less value in this work than in the "proper" engagement in the Hilbertian programme of reducing the existing bad part of mathematics to the good part, since, for Hilbert, mathematicians

[17]The distinction is, of course, somewhat artificial, and if we do not accept it, then we rephrase the structure of the foundationalist philosophy appropriately. Many mathematicians are also philosophers, and the same person can play both roles. I follow Colyvan (2001) in not recognizing a clear distinction between philosophy and mathematics, either in terms of persons or in terms of roles. Despite my agreement with Colyvan, it will be useful for the arguments here to adopt this artificial distinction.

[18]An example would be any attempts at intensional logics not counting as part of mathematics, just because "mathematics" i.e., set theory, is extensional, and cannot recognize intensional differences.

should not be *extending* the suspect part![19] Sporting my naturalist hat, I am not sure that the mathematician working on the higher-cardinals would agree! The very notions of "good" and "suspect" mathematics are not happily applied to the practice of mathematicians. So, here, the naturalist parts company with the dualist. In rejecting the normativity of monism and dualism. In this sense, pluralism is anti-foundationalist.[20]

15.3.2 Second Negative Argument Against the Foundationalist: The Argument from Mis-description

As was mentioned in the argument of the foundationalist, the foundationalist begins with the technical result that most of mathematics can be reduced to the founding discipline. This is a twofold mis-description. First, the reduction is sometimes artificial, and therefore not successful. The second mis-description concerns future growth of the foundation. Whatever the founding theory is, 'it' grows. So there is no fixed foundation.[21]

Vopěnka signals the reduction mis-description: "Some (mathematical) disciplines pursued in pre-set-mathematics [mathematics before the development of Cantorian set theory] had to be considerably violated in order to include them in set theory" (Vopěnka 1979, p. 9). Vopěnka cites the calculus as an example of such a violation. There are many other examples. Try proving that $7+92=99$ in Frege's logic (without using the inconsistency generated from Basic Law V) or in type theory. It is possible to "do calculus" in set theory, but it is so awkward that no one does it. Why? Because the proofs are too long or not explanatory, so we lose sight of what we are trying to do, and much of the proof is very mechanical, and should be skipped, since going through all of the mechanical steps is not informative, and certainly not "doing mathematics". In this way, the reductions do not give the "essence" of what mathematics is about, how it is practiced, what is interesting about it. Nevertheless, the reducing discipline does give some philosophical insights. For example, we might learn, with Frege, that arithmetic is really analytic, *pace* Kant.

---

[19]'Bad', of course, is an over-simplification, especially in light of Hilbert's famously stating that he was not willing to be expelled from the paradise Cantor had introduced. Nevertheless, there is a tension in Hilbert's attitudes towards the finitistic and the ideal.

[20]More precisely, it is anti-foundationalist at the level of discourse where the foundational philosophies of mathematics do their work.

[21]Brouwer agrees with this, so in this respect, he too, parts company with the monist. The issue about where Brouwer fits in my account is quite subtle. Where Brouwer and I part company is in his emphasis on intuition. I think that mathematical intuition is interesting, but I disagree with Brouwer that "mathematics (all and only) takes place in the mind". I save this issue for another paper.

Apart from the artificiality of the reduction, there is the second problem of instability of the foundation. Even lovely, all-encompassing, mathematical theories grow. New axioms and techniques are suggested and tried. If we endorse the naturalist attitude, then we can observe (rather than resist) co-variance between "founding theories" and "essences": as the founding theory changes, so the essence changes.[22] For the pluralist, this observation makes a mockery of foundationalism as essentialism. This is not a logically necessary argument, since we could insist on fixing the foundation by reference to one formal theory, and resist new extensions. Rather, the argument is inductive: based on the history of mathematics. Foundational theories spawn new developments or additions to the mother theory, and foster the development of rival theories. No sooner had Whitehead and Russell introduced their simple type theory, than they developed the ramified type theory. Other type theories have sprung up since, some more successful (studied more), than others. After Cantor developed his naïve set theory, rivals were forthcoming: Zermelo-Fraenkel set theory, Gödel-Bernays set theory. Moreover, additions were made to these, with the axiom of choice, the development of class theory, higher cardinal axioms were added etc. Category theory too has seen development.

Looking more closely, we *extend* the foundational theory with new axioms which make a new theory (assuming we individuate theories by the language, plus axioms, plus inference rules). For example, we can extend ZF set theory with the axiom of choice, which gives us ZFC. Moreover, as is well documented, there is considerable dispute over the *admissibility* of new axioms which extend ZF set theory. Admissibility (classically) requires at least consistency with the original theory, but some proposed axioms are independent of the original theory; and therefore we can add the independent axiom, or an axiom which is inconsistent with it. The problem now is to arbitrate between the alternative proposed extensions, since some pairs of new axioms will lead to contradictions. To arbitrate, we have to modify our original notion of the *essence* of mathematics—since it no longer rests in the founding theory. It is strictly broader since we think it can accommodate extensions to the founding theory. Moreover, we have to do this in such a way as to accommodate only one subset of the proposed additional axioms. Witness the debates about $V = L$. If we choose $V = L$, then we preclude a number of other axioms. If we choose $V \neq L$ as a new axiom, then we preclude other proposed axioms. Accommodation, in the face of choices which preclude other additions, is no easy task, since the founding theory itself cannot arbitrate. Here, "choice" relies on some underlying sense of "the" theory—not individuated by a language, set of

[22]There is an important distinction I am glossing over, but it will be addressed in the next section. The distinction is between the presentation of the formal theory, and whatever it is that the formal theory is trying to capture. Here, I am assuming that the essentialist believes that he has in hand a formal theory which captures the essence of mathematics. The technical result is completed. If we draw apart the formal theory and what it is the theory is supposed to capture (which is intentionally different from an intended interpretation), then we might say that the formal theory imperfectly captures the essence, so the formal theory is allowed to 'grow' as and when we discover new aspects to the essence.

axioms and rules of inference—but by some vague intuitions which, one hopes, will become explicit through discovery and formal representation. But these vague intuitions are not good philosophical justifications for foundationalism, since these sorts of intuition vary from one mathematician to the next. We might dress up the intuitions by introducing considerations of fruitfulness, simplicity, elegance etc. But these considerations alone will not do, since, remember, we are providing a foundation, not for generating "lots" of mathematics,[23] or aesthetically pleasing mathematics, but for correct mathematics. If the foundationalist does go ahead, and opts for one extension over another, to fix the "essence" of mathematics, then he shows a weakness in the original monist argument for his first chosen foundation. For, arguing for one extension over another, is a covert admission that he did not have the full essence properly captured in the first place.

15.3.3 The Third Negative Argument Against the Foundationalist: De dicto and de re many Mathematicians Are Pluralist

De dicto many mathematicians are anti-foundationalist. Or, more mildly, they view foundations with suspicion.

> Many working mathematicians (though by no means all) are suspicious of logicians' [and philosophers'] apparent attempt to take over their subject by stressing its foundations. ... [Moreover,] I have been persuaded by Edwin Coleman that foundationalism in mathematics should be regarded with considerable suspicion; or at least that proper 'foundations', ... would be much more complex and semiotical than twentieth century mathematical logic has attempted. In which case it would be arguable whether 'foundations' is an appropriate term. (Mortensen 1995, p. 4)

In conversation, Ali Enayat, Joe Mourad, Jennifer Chubb, István Németi, Hajnal Andréka, Russell Miller and many other working mathematicians have all declared themselves to be pluralist, in some sense of "pluralist". I think that pluralism "is in the air", but it has not been worked out as a whole philosophical position, only as part of other positions.

Moreover, many mathematicians are not only *de dicto* pluralist, many are *de re* pluralist. That is, their behaviour at conferences and in their written work, displays an open-mindedness and acceptance of alternative foundational theories. More than this, in their proofs and methodology, mathematicians will often avail themselves of whatever hypotheses are useful and can support the desired result. According

[23]We should be careful about the accolade 'fruitful'. It pre-supposes quantifying over mathematical results. For, adding any axiom will add an effectively enumerable number of new theorems, so axioms are equally fruitful. Alternatively, we might count only "important" new results, but how these are determined/chosen is again a problem; at least at any given time, since we might later discover that a theorem or result is important only many years later.

to Thurston, for mathematicians, the "*reliability* (of proof) does not need to come from mathematicians *formally* checking formal arguments (so working within one foundation): it comes from mathematicians *thinking carefully and critically* about mathematical ideas." These ideas are not restricted to the ideas found in one foundation. The choice of which method or result to use in a proof is pragmatic, and there is a sense in which said method or result is considered to be trustworthy because it is "quite good at producing reliable theorems that can be solidly backed up." (Thurston 1994, p. 171). Real "mathematical" proofs are non-deductive derivations of plausible hypotheses from problems, in some sense of "plausible", where a problem is an open question; a hypothesis is any means that can be used to solve a problem (Cellucci 2008, p. 2). And, more important, a hypothesis is said to be plausible if and only if it is compatible with existing data—which includes any mathematical results and notions available at the time of inquiry (Goethe and Friend 2010).

This runs directly against the picture drawn by the monist philosophies of mathematics, but maybe the dualists are more accommodating. We might think that, as good dualists, mathematicians avail themselves of the better part of mathematics, when they can, and use the more suspect part with an uneasy conscience, such as when a constructive mathematician knows very well that there is a constructively unacceptable proof for a result, but, nevertheless, believes that the result is true, and works on giving a constructively acceptable proof of the same result. Of course this happens, and there are *bona fide* dualists amongst mathematicians. But the monist or the dualist stories are not the only stories to be told, and many mathematicians completely disregard the advice of the dualists. There is no "bad conscience". In other words, for many mathematicians, the purported distinction between good and suspect mathematics completely dissolves. In the light of the *de dicto* and *de re* observations, the naturalist aspect of pluralism makes the pluralist anti-foundationalist.

We have some *prima facie* evidence for pluralism from the claims and behaviour of mathematicians. However, this is simply an observation about the state of play in mathematics today. As philosophers we have to decide whether or not to take the observations seriously, or to think of them as a temporary glitch. We might excuse the observations on the grounds that the working mathematician is simply "not a very good philosopher of mathematics and has not thought through the implications of her pluralism,"[24] or is engaged in "cognitive dissonance" or treats mathematical theories as tools and therefore her pluralism is due to a lack of philosophical thought about the matter. This might well be true in some cases. But, as philosophers, we should say more.

---

[24]"... what the mathematician says [about the philosophy of mathematics] is no more reliable as a guide to the interpretation of their work than what artists say about their work, or musicians [about theirs]." (Potter 2004, p. 4). Even if we do not quite have such a strong point of view, it remains that mathematicians express very different philosophical attitudes. At the risk of being repetitive, my personal observation is that most mathematicians are pluralists.

Pluralism motivated by naturalism does not prevent a philosopher or mathematician from working within the strictures of a philosophy, but we want to distinguish between being wedded to a theory for technical reasons, historical reasons or reasons of personal taste, on the one hand, and being wedded to a philosophical or mathematical theory for foundationalist reasons. For the pluralist, the normative force of foundationalist philosophies is confined to a class of theories. So, it is qualified, and not absolute. Speaking pluralistically: the normativity of a philosophical position stays internal to that philosophy, and is limited to the scope of the foundation. This should not upset traditional philosophers of mathematics too much: they have, after all, a whole class of theories at their disposal. Moreover this might be a proper class. So they have quite a large play-ground. But, at the end of the day, the pluralist asks them to admit to the parameters of the play-ground. In other words, according to the pluralist, what the foundationalist may not do is claim to give a philosophy for "all" of mathematics. There is perfectly legitimate and interesting mathematics outside the foundation.[25]

For the pluralist, there is no absolute mathematical truth, only truth within a theory. There is not one essence of mathematics characterised or represented, by a particular mathematical founding theory. Pluralism in mathematics is a challenge for philosophers, since it is foreign to the more traditional foundational philosophies of mathematics. However, it is not an insurmountable challenge. Shapiro, for one, considers himself to be a pluralist, and supplies a philosophy "without foundationalism".

15.4 The Second Motivation for Pluralism

15.4.1 Shapiro's Structuralism

Shapiro's structuralism is interesting to us because it is neither monist, nor dualist, it self-avowedly anti-foundationalist and pluralist, and advocates a "mathematics-first" attitude which is close to the naturalist insight we adopted in the previous section to make our negative arguments in favour of pluralism.[26] I think that his is the closest position extant to the pluralism supported here. Shapiro's structuralism is a philosophy of mathematics where the notion of "truth" is always qualified by that of "in-a-structure". He uses the highly expressive language of second-order logic to capture important mathematical concepts, such as "is Dedekind

[25]This is a comment about the state of play today. It might turn out one day that we have a unified foundation, which encompasses all of mathematics.

[26]The title of one of Shapiro's books is: *Foundations Without Foundationalism*. There is a sort of foundation, based on second-order logic and model theory. I am calling this an "organisational perspective" to distinguish it from a unifying revisionist foundation.

infinite"[27] and model theory to pick out structures (which, for Shapiro, are what mathematicians are interested in). For Shapiro, model theory is not a foundation, but an organisational perspective allowing for the clear individuation of mathematical theories, and for the comparison of various theories/structures from the point of view of chosen further meta-structures.[28] There is no ultimate structure, on pain of paradox. There is no absolute perspective. No structure is ultimately favoured over others (since model theory does not have one global structure). Model theory is not an axiomatised theory, and therefore has the flexibility to grow, without jeopardising stability. What is admitted as a structure will, undoubtedly, change over time, since model theory is a developing theory.

There are two types of individuation of theory taking place side-by side. We can either individuate theories in terms of the language of the theory, the proof theory and axioms, which is, roughly, how the model theorist thinks of a theory. A structure is just a set of objects together with some structure imposing relations which bear on the objects. Equally, we can individuate theories, in terms of an underlying idea which is not necessarily known to be fully captured by the formal representation of the theory. For example, if we think of model theory, then the formal representation is not yet fully achieved. In Shapiro's structuralism: model theory itself should be individuated in the latter way, since it is a growing theory; whereas particular structures should be individuated in the former way.[29] This is a very pluralist way of speaking. For the pluralist, the model theorist is able to "see" a lot of mathematics, make sense of it, organise it within the strictures of his model theory and make contributions and offer insights. He individuates structures up to isomorphism and recognises all concepts expressible in a second-order language.

The pluralist will now detect a limitation. Shapiro's structuralist can see quite a lot of mathematics, but not all of mathematics. As a result, Shapiro's pluralism is

---

[27]The definition of Dedekind infinite is that: a set is Dedekind infinite iff it has a proper sub-set with which it can be placed into one-to-one correspondence. The natural numbers are Dedekind infinite, as are the integers, the rationals, the reals and so on. In contrast, finite sets have no proper sub-set which can be placed into one-to-one correspondence with them. To capture the notion of Dedekind infinite, we need the expressive power of second-order logic. See Shapiro (1991, p. 100). The formula for set X being infinite is: $INF(X) : \exists f[\forall x \forall y(fx = fy \rightarrow x = y) \& \forall x(Xx \rightarrow Xfx) \& \exists y(Xy \& \forall x(Xx \rightarrow fx \neq y))]$. This is read: There is a function which is such that if (two) of its values are identical, then the (two) arguments are equal. Moreover, the function operates on a proper subset of the set X.

[28]As previously mentioned, the title of Shapiro's first book on structuralism is: *Foundations Without Foundationalism The Case for Second-Order Logic*. Note the "Without Foundationalism". Foundationalism is identified with the monist or the dualist. Shapiro is anti-foundationalist in the sense that all mathematical theories which he recognizes are on a par. Insofar as he has a foundation, Shapiro's "foundation" is model theory. Model theory allows him to individuate mathematical theories (as structures). The model theory does not favour one structure as against another.

[29]This could be turned into a criticism of Shapiro's structuralism. It is inspired by Potter and Sullivan (1997, pp. 135–152). The criticism, adapted from the Potter and Sullivan paper is that Shapiro makes different ontological and metaphysical claims concerning individual models, on the one hand, and model theory itself, on the other. So there is a double standard.

restricted to what is recognized through the lens of model theory and what can be expressed in second-order logic, and this lens then sets a norm for what is to count as successful mathematics. In the spirit of friendly banter, we might say that Shapiro, too, is guilty of some of the sins of foundationalism.

15.4.2 Moving Beyond Shapiro's Structuralism

Where Shapiro and I part company is over the very important issues of what is to count as success in mathematics and what is of interest to the philosopher. The pluralist takes her cues from observed mathematical practice. So she can recognise parts of mathematics, which some mathematicians count as successful, but which are not recognised by model theorists (Friend 2006, 73). These include: intensional logics,[30] mathematical theories which are still in a stage of development and paraconsistent mathematics. Unsuccessful mathematics are trivial mathematics. Shapiro and I agree that these are unsuccessful. The difference between us, on this issue, is that the pluralist thinks that trivial theories are interesting and useful in mathematics. I shall call all of the above "bad" mathematics, since they are all implicitly rejected by Shapiro and each is rejected by most other philosophies of mathematics. The philosophical move of rejecting "bad" mathematics offends against the naturalist insight, runs the risk of instability or of begging the question. The pluralist philosophy developed here is more stable than Shapiro's pluralism and the more traditional philosophies as well[31] since the pluralist has the flexibility to adapt to changes over time in what counts as successful mathematics. Odd theories suddenly find an application; some obscure result proves useful to a more central mathematical concern. Theories which were viewed as highly suspect come to be accepted in more main-stream mathematics, such as the study of non-standard arithmetic. More important, there are revolutions in mathematics, such as the discovery on non-Euclidean geometries, or of the incompleteness results. These changes radically alter our conception of mathematics.

---

[30]Model theory is extensionalist, and only individuates structures and objects in those structures "up to isomorphism", only recognizing certain properties (predicates, relations, functions) as "counting" for mathematics. But we find, in mathematical practice, that considerations, not recognized by model theory, are also pertinent to mathematics.

[31]One might think that I am being somewhat unfair, and ignoring a lot of philosophical activity. For example one might point out that Russell was much aggrieved by the paradoxes, and theorised a lot about them. And Russell's investigation into the paradoxes shaped his philosophy and formal system. Moreover, some very important philosophical work has been done in looking very closely at Frege's trivial theory—such as the work of Dummett, Wright and Heck. I appropriate such activity, and call it pluralist. What is anti-pluralist is any accompanying norm setting revisionism. So, we should be carful about our interpretation of the intension behind the excellent work cited above; we might say that these philosophers engage in pluralist work despite themselves.

To remedy the instability, we can be more circumspect by qualifying our measure of success by means of a temporal index, then any "rejection" is made relative, stable and harmless. For example, we might say that we are giving a philosophy of early nineteenth century mathematics. Provided we are historically accurate, the pluralist will not object. But the pluralist is more ambitious than this.

The rejection of bad mathematical theories might also beg the question, as when the reductionist foundationalist philosopher re-trenches and says that whatever fails to conform to her conception of what counts as successful mathematics is, by definition, unsuccessful. That is, she sets an a priori norm for success in mathematics. But the force of such an argument is limited, for it begs the question against the naturalist perspective. My diagnosis is that there is an inevitable tension between the naturalist attitude and the desire to give a traditional philosophical account of successful mathematics.

15.4.3 Prejudices: Optimal Versus Maximal Pluralism

To distinguish "success", from the "rest" of mathematics, while remaining pluralist, let us distinguish between an *optimal* pluralist philosophy and a *maximal* pluralist philosophy. The optimal pluralist gives norms for philosophically well motivated theories, i.e., for "successful" mathematics. Shapiro is an example of an optimal pluralist. There might be several competing norms. They might include: consistency, constructive considerations, definitions of validity, search for a robust ontology etc. In contrast, the maximal pluralist is maximally descriptivist: tries to philosophically account for the whole corpus of mathematical activity. The maximal pluralist is loath to set a norm for success in mathematics, and he will *accommodate, account for, or study* "bad" theories (without slipping into triviality; see the next section). Under the maximalist attitude, the pluralist can, of course, *observe* norms, but does not arbitrate between competing norms (unlike the optimal pluralist). We had a taste of this earlier. For this reason, the pluralist has to entertain, what were traditionally thought of as "bad" mathematical theories.

Since, in this paper I am advocating maximal pluralism, let us give the motivations for considering "bad" mathematics at all. Beginning with paraconsistent mathematics: there is now a corpus of literature on paraconsistent logics and paraconsistent mathematics. These are taken seriously by some mathematicians.[32] A philosophy of mathematics which does not treat of these is incomplete, and

[32]There is plenty of sociological evidence for this. Witness publications by "major" publishers, both as books and in journals; numbers, sizes, and sections newly contained in conferences. One telling example is the history of the world congress on paraconsistent logic.

violates the naturalist attitude.[33] The second sort of important (and potentially successful) mathematics, for the pluralist, is nascent theories. These are ignored by other philosophies, so count as "bad" mathematics for everyone else. These are theories which are still in progress. All theories go though a stage of "construction", "becoming" or (more platonistically) "coming to be known" or "coming to be formally represented". Depending on how we individuate theories in mathematics, we might even say, with the Gödelian optimist,[34] that set theory is nascent! More specifically: if we do not individuate theories in the standard way in terms of a language, a set of axioms and rules of inference, but rather in terms of "some theory to be discovered" or as a "construction of the mind", or as having an "informative semantics",[35] then many mathematical theories are nascent. A philosophy of mathematics which did not accommodate nascent theories would be found lacking by the pluralist.

Trivial theories are the most controversial of the "bad" theories. A trivial mathematical theory is one where every well formed formula in the language of the theory is true. They are distinguished from each other by their language.[36] For a trivial mathematical theory two factors have to be in place. The underlying logic of the theory has to be classical (has to allow *ex falso quodlibet* inferences) and there has to be a contradiction derivable from the axioms using the rules of inference of the theory. Historically, there are three (to my knowledge) mathematical theories which had a profound impact on mathematics or logic, and were found to be trivial. These are: Cantor's naïve set theory, Frege's formal theory of logic and the first version of Church's formal theory of mathematical logic. All three had profound repercussions on subsequent mathematics. None led to the collapse of "all

---

[33]Shapiro's pluralist structuralism cannot recognize paraconsistent logics and mathematics, since they cannot have a structure, since the logic Shapiro uses is classical second-order logic, and only consistent theories have a model—in a classical theory.

[34]The Gödelian optimist thinks that in the end, given an open problem, we shall discover a technique to make an absolute decision about that problem. Tennant has several good discussions about the Gödelian optimist in Tennant (1997).

[35]For the distinction between an "informative" and a merely "technical" semantics see Priest (2006, p. 181). A semantics is uninformative if it is developed simply for technical reasons, to prove consistency (in a classical setting). In contrast, a semantics can be informative in two ways. Either it is informative in the sense of being the intended interpretation. That is, the semantics is developed with complete reference to some logical or mathematical meaning. In these cases the syntax is developed after, or conceptually comes after, and is developed soundly—to be in harmony with the semantics. The more subtle case of an informative semantics is found when we developed a semantics for technical reasons, so we know that the syntax is consistent. But then we find an application, or an interpretation, that suits the syntax. This semantics is informative *post facto*.

[36]A trivial theory is got by espousing a classical logic (i.e., which allows *ex falso quodlibet* inferences) that contains a contradiction. We then have the result that every sentence written in the language is provable. If the languages of (what we suppose are) two trivial theories are the same, then the theories are the same. However, if (what we suppose to be) two trivial theories have different languages, then they can be distinguished from each other. Some sentences will be true in one, but not recognizable in the other. I thank Priest for pressing me on this point at the Logica conference 2005.

of mathematics". None led even to the collapse of "that part of mathematics infected by the theory".[37] The important proofs contained in the above trivial theories do not proceed as *ex falso quodlibet* inferences, which is one of the reasons *why* the theories are considered to be important despite their being trivial. The good trivial theories are studied and trawled for good ideas and insights. Contrast these trivial theories to Prior's "Tonk" theory (Prior 1960, pp. 8–39). This is not an interesting theory because we see immediately that it is inconsistent. Not all inconsistent theories are uninteresting as witnessed by the work of Dummett, Wright and Heck on Frege's trivial theory. In general, after spotting an inconsistency, mathematicians try to fix the theory with *minimal* changes. To ignore the mathematical and philosophical influence of such theories, again, would be to provide a philosophy of mathematics which is lacking in scope.

Having considered various "bad" theories, and finding that they should not be dismissed as not part of the mathematical corpus by the naturalist, it follows that we should try to adopt a maximal pluralist attitude, and not only an optimal pluralist attitude. The virtues of maximal pluralism are greater inclusion and stability. We should turn to the criticisms of maximal pluralism before giving a more detailed account of the maximal pluralist view.[38]

15.4.4 Critique of Maximal Pluralism from Trivialism

The criticism runs: if the maximal pluralist is so loath to set norms or arbitrate between existing norms, then everything goes, and the position is actually trivialist.[39] For any theory, we can find a meta-theory or an attitude that endorses the theory, so there is no real philosophical judgment, there are just relative judgments or descriptions. No one wants a trivial philosophy. We might end up with a trivial philosophy because we took seriously some trivial mathematical theories. The language of these theories is a proper sub-part of the philosophy, so the triviality spreads through the philosophy. Oddly, for the trivialist, but it comes as a relief to

---

[37]For example, we did not stop doing arithmetic when Russell discovered paradox in Frege's reduction of arithmetic to logic. This is also evidence against trivialism.

[38]I should like to thank Norma B. Goethe and Göran Sundholm for sustaining some of these criticisms against me in conversation. Note that they were much more delicate and kind in their tone than what is reported in the imagined quotation!

[39]Trivialism is the position that every grammatical, categorically correct, sentence is true. A sentence is categorically correct if it makes no "category mistakes": where we confuse what type of object we are talking about. For example, it makes no sense to talk of water dreaming, angry chairs, kilograms travelling etc., unless, of course, we are in a fantastical/super-natural setting or using a metaphor. Trivialism is the dual of scepticism: where every grammatical, categorically correct sentence is subject to doubt. However, unlike the sceptic, the trivialist position does not "implode" since its own very trivialism is true. It is an entirely robust and stable position. However, it is highly uninteresting to maintain it.

the maximal pluralist that this has not in fact occurred.[40] In practice we observe a clean break between trivial mathematical theories and philosophical positions which entertain them. The infection does not spread to the philosophy.

A trivial philosophy of mathematics holds that every well-formed mathematical formula, in any language of mathematics is true (and its negation is true), and any philosophical sentence about mathematics is also true.[41] Anything goes, and all judgments are as good or as correct as the next.[42] The notion of judgment is so degenerate that we might say it is absent, since it is not discriminating. Trivialism is pretty hopeless *as a philosophy*, although it is very easy to defend/maintain verbally! The main criticism against it which is pertinent to this project is well expounded in Priest (2006, pp. 68–69), and it is an argument from meaning. It is not clear that a trivialist can mean anything by his utterance or written statement, since there is no recognisable *judgment* attending sentences. They are all true, so there is no distinction between true and false, since they are also all false. So there can be no meaningful intentionality,[43] since there is no distinction between a belief, a known fact, a subject of fear, desire or what have you. Since there is neither judgment nor intentionality attending the use of language, the philosophy of mathematics being presented is degenerate. According to someone who is not a trivialist, the trivialist theory renders[44] mathematics and philosophy meaningless. Provided that we hold that some wffs are false (and not true), we do not have a trivial theory. An example of a wff which the maximal pluralist holds false (and not true) is: $\vdash_{PA} 2+9 = 34$.[45] We read this: "in Peano Arithmetic, two plus nine equals thirty-four". This is enough to distinguish the maximal pluralist from the trivialist about arithmetic, the whole of mathematics or philosophy. Note that we have not *defeated* the trivialist. Rather, we have simply shown that maximal pluralism is distinct from trivialism which is

[40] I know of no discussion of trivialism which has degenerated into trivialism, except in moments of jest.

[41] We might come to this position by supposing, say, that ZF contains a contradiction. More precisely, we need a theory which is considered to be foundational to mathematics, we need for it to be a classical theory: allowing *ex falso quodlibet* inferences, and we need to be able to derive a contradiction from the axioms using the rules of inference.

[42] For a good discussion of trivialism see Priest (2006, pp. 56–71).

[43] There might, of course be reported or avowed intentionality, such as when the trivialist reports: "I believe that snow is white". He will equally assent to: "I believe that snow is any colour but white."

[44] The trivialist will "hold", in the sense of assert, any position. This is not the point. Trivialism arises from the idea that mathematics is classical and there is a contradiction in mathematics, and therefore (under our old classical reasoning) all of mathematics is true, we then get to the meaninglessness of any particular mathematical statement, and wallow in our degenerate theory. There is a sequence to the reasoning which gets us to the degenerate position. Once there, reasoning, as such, is impossible.

[45] The trivialist will, of course, agree that "$\vdash_{PA} 2 + 9 = 34$ is false", since the trivialist will agree to everything. The maximal pluralist will disagree that " "$\vdash_{PA} 2 + 9 = 34$" is true". (I think that) this is all we need is to distinguish the positions.

enough to fend from the criticism that maximal pluralism is a trivial philosophy. The differences will be fleshed out when we look more closely at the paraconsistent logic underpinning pluralism.

15.4.5 Critique from "Disdain for Sociology"

A less technical critique comes from a "disdain for sociology". An imagined interlocutor might object: "Michèle, if you give up on giving a philosophical account of successful mathematics, then you let in all sorts of abominations: trivial theories, crankish scribbles, numerology... Moreover with your moral-high-ground pluralism you are loath to judge, rate and order rubbish-posing-as-mathematics as quite inferior to very good and fruitful mathematics. What sort of a philosophy are you hoping to give here? It might be stable, but it will also be empty/uninteresting. Have you lost all philosophical ambition? Have you turned Wittgensteinian (later, and only under some interpretations)? Are you not left with only doing sociology, history or historiography of mathematics since your naturalist attitude only allows description?" There are a number of complaints included in the imagined quotation. The interlocutor accuses the maximal pluralist of philosophical, or logical, degeneration in the sense that whatever philosophy there was initially threatens to "degenerate" to the rank of sociology. The pluralist has a response. There is, in fact, lots of philosophical work to be done under a pluralist banner. The best way to see this is to look at the philosophical position of pluralism, and pay special attention to the sort of judgments which pluralists make. Even at the level of description, there is good description, and mis-description. Aside from the descriptions, there are *bona fide* value judgments, and these are modeled by the semantics of the logic.

15.5 Maximal Pluralism

15.5.1 The View

There are three levels of philosophical activity and three corresponding levels of mathematical activity. The first level of both mathematics and philosophy concerns particular results in mathematics. Examples on the mathematical side are particular theorems, lemmas, definitions and proofs. On the philosophical side we have discussions concerning particular results. For example, we might discuss: theorems, definitions or the completeness of a theory, a compactness result or a proof in a theory. These might include discussions about limitative results, since these results are given within a particular mathematical theory. Thus, we might include the proof of equi-consistency of two theories at this first level. At the second level, we have full mathematical theories, or theories which are being developed. Examples of fully developed theories are: Euclidean geometry, first-order arithmetic, modal logic S4, Zermelo-Fraenkel set theory and Topos theory. The larger of these mathematical theories are theories *within which* we make mathematical comparisons between other (smaller)[46] theories. For example we might show the reduction of one theory to another, we might give an equi-consistency proof between two smaller theories, we might show embeddings, and so on. The larger whole theories are often thought of, by philosophers, as foundational and are often accompanied by a philosophy. For this reason, on the philosophical side, at this level, we have the more traditional philosophies of mathematics, such as: set theoretic realism, Maddy's set theoretic naturalism, the constructive philosophies, logicism and so on. In fact, just about every philosophical position in the philosophy of mathematics is found at this level. *At the third level, we have pluralism as a philosophy which is pluralist towards the activity which takes place at the first and second level.*

The logic accompanying pluralism is paraconsistent.[47] The paraconsistent logic provides the "space of reason"[48] within which mathematical activity can take place. The view is that the forces of mathematical history: particular inspirations, insights, the publication of articles, the disseminating of mathematical information, conspire to trace complex paths which merge and split within the paraconsistent space of reason. As a philosophy, pluralism favours the pluralist attitude over other philosophical virtues. The pluralist attitude combines anti-foundationalism while maintaining an interest in foundations—as good mathematical theories in their own right, and as accompanied by philosophies of mathematics which affect the development of mathematics. A pluralist who took an interest in "foundations" would have plenty to say about axioms which are independent of a foundational mathematical theory, such as the higher-cardinal axioms. The pluralist observes the bifurcations of set theory with the addition of different sets of axioms. The pluralist will not feel any need to favour one extension over another. Note that this demurring is not due to "lack of knowledge", but, rather, to an acceptance that in the present state of play in mathematics, there simply is no definitive mathematical way to arbitrate between theories. There is no unique absolute perspective.[49] Since some pairs of mathematical theory contradict each other, pluralism requires a paraconsistent logic at the third level.

---

[46]The terms "smaller" and "larger" refer to the expressive power of a theory. Roughly, the more theories can be reduced to, or embedded in a theory, the more expressive power the theory has.

[47]There are actually different versions of pluralism, varying with choice of underlying logic, but to simplify, here, I give only one logic, which in this case is paraconsistent.

[48]This is not a term I like, but it is useful in this context.

[49]It might be instructive to compare this attitude to Gödelian optimism, which is the thought that in the end, given an open problem, we shall discover a technique to make an absolute decision about that problem. Tennant has several good discussions about the Gödelian optimist in Tennant (1997). In contrast, here we have the agnostic, who demurs. This character is either a pessimist (the demurring is then based on an inductive argument, and the pessimism might be reversed in a particular instance), or the character is a principled agnostic. It is the principled agnostic position which is explored in this paper.

The logic helps us to individuate and reason over theories, foundations and philosophies of mathematics. The virtue of paraconsistent logic is that in deploying it, we can cope with contradictions within and between theories, and this is very important. Byers remarks: "No description of mathematics would be complete without a discussion of its *subtle* relationship to the contradictory (my emphasis)" (Byers 2007, p. 81). Our only hope of engaging in a subtle discussion is through the use of a paraconsistent logic, since the more traditional philosophies are anything but subtle in this respect! The same author remarks later:

> Moreover, paradox has great value. Thus paradox should be seen as a generating force within the domain of mathematical practice. ... Where do that power and dynamism come from? Well, they come from ambiguity, contradiction and paradox. These things are therefore of great value. They need to be unravelled, explored, developed, and not excised. (Byers 2007, p. 112)[50]

Ambiguity, paradox and contradiction need to be unravelled if one wants to give an account of the practice and development of mathematics. This is partly a psychological task, but it is also philosophical, since it raises epistemological questions largely ignored by traditional philosophies of mathematics. For, if Byers is correct, then it is through awareness of, and confrontation with: ambiguity, paradox and contradiction that we develop mathematics. They are epistemological tools, not strict limitations or parameters on reasoning or on the corpus of mathematics.

As a step towards developing this epistemological sophistication, the logic the maximal pluralist uses is a little different from regular paraconsistent logics in that one type of variable ranges over whole mathematical theories (sets of wffs)[51] and another type of variable will range over classes of mathematical theory.[52] For, it is with these units that we meet conflict and contradiction. With careful use of the Routley/Priest Characterisation Principle,[53] which we shall explore in the next section, the maximal pluralist can compare "bad" theories to each other: cordoning off the trivial theories, and comparing mutually contradictory theories. We can even compare trivial theories to each other. To distinguish between different trivial theories, we look to the differences in characterisation between, say, Frege's theory and Cantorian set theory. The difference lies in the vocabulary and languages of the theories. In both cases we have, what are sometimes called "consistent contradictions".[54] In contrast, if we compare classical Euclidean geometry to

---

[50]Note that Byers makes no mention of paraconsistent or relevant logics. I therefore point out that he, himself, is not advocating a paraconsistent point of view or anything of the sort. Nevertheless, the quotations, and in many other places in the book, I found support for the position advocated in this paper. I do not know what Byers' reaction would be to the mention of paraconsistent logics.

[51]It is not very different, for, we could imagine a very long conjunction of wffs, each conjunct of which is put in normal form and arranged in some ordering.

[52]To preserve the pluralism, we allow all symbols of mathematics to be included in the language. The language is growing, not fixed.

[53]The principle is: *A theory is just whatever it is characterized to be.*

[54]The notion of "consistent contradiction" was introduced to me by Marcelo Coniglio in the presentation of Coniglio and Carnielli (2008). "Consistent contradictions" are explosive. Anything

projective geometry, we find that, together they give us a "normal contradiction"; one we can resolve by keeping the theories separate. This sort of observation, made by a pluralist, is enough to parry the accusation, made by the imaginary interlocutor who accused the pluralist of precluding judgment of theories (letting in abominations). There is plenty of properly philosophical work to be done for the pluralist. However, we should be aware that the use of a paraconsist logic brings with it its own philosophical stamp concerning the "truth" of a mathematical theory.

15.5.2 *The Semantics of the Logic Underlying Pluralism: Judgment Values*

The semantics of the logic concerns the first and second level of analysis.[55] We use the semantics to make sense of the following judgments, where we have a notion of truth indexed to a philosophy X. The units 'y' being judged are whole mathematical theories (sets of wffs) or individual wffs. Philosophies and mathematical theories are individuated by the *characterisation principle*. The original characterisation principle, as developed by Routley is:

> *An object* is characterised by its properties.

Which object it is depends therefore only on its properties, so an object just is (is individuated by) its characterisation (see Priest 2003, p. 4). We use a more general version of the principle: namely, that

> *A mathematical theory* just is its wffs.

Note that the set of wffs might not be closed (under certain operations) since we might not have decided which are the admissible operations for generating new wffs. The truth of a mathematical theory is indexed to a philosophy at the second level. A philosophy, X, is not as easy to characterise, since developing a philosophy is not as rigorous and formal a task as developing a mathematical theory. So, we shall have to be more circumspect. A philosophy of mathematics (at the second level) just is the set of sentences used to characterise the philosophy. The easiest way to do this is to refer to a definition, if there is one, or a chapter in a book. Thus we have the philosophy of whatnot as characterised by whoever in chapter whatever of some book. More broadly, how we determine philosophies might be no

---

can be derived logically from them. These are contrasted to "normal contradictions" from which not everything follows, only a very few things follow. With normal contradictions, we have a very controlled explosion. The use of the word "normal" refers to the fact that we encounter what, at first appear to be, contradictions quite frequently in "real life", but we deal with these quite well.

[55] I prefer the term "judgment values" to "semantics", since "semantics" comes with too many connotations about giving truth-values, interpretations and domains of interpretation. "Judgment values" are part of the semantics, in a broad sense of "semantics".

easy matter, whence the subtleties, intricacies and delights philosophy. Therefore, the characterisation is not always fixed, but we can usually fix it temporarily for the sake of coming up with some judgments.

The value judgments are as follows. In all of the clauses, fallibility and revisability are understood. That is, words like "success", "recognise", "determine", "wrong" make the judgments revisable.

1. "y is a true mathematical theory given philosophy X" gets the value judgment (T), iff X recognises y as a successful mathematical theory.

   An example of such a judgment is Euclidean geometry, given Zermelo-Fraenkel set theoretic realism is true. Projective geometry is also true if we are Zermelo-Fraenkel set theoretic realists. Intuitionist ordinal arithmetic is also true if we are Zermelo-Fraenkel set theoretic realists, as is any theory which can be reduced to Zermelo-Fraenkel set theory.
2. "We do not yet know if y is true given philosophy X", or "the philosophy X is neutral with respect to the mathematical theory y" gets the value judgment (U) (for "unknown" or truth-value gap), iff philosophy X is not able to determine whether or not mathematics y is true.[56]

   Examples show up if we choose, say, a Gödelian optimist philosophy and consider some of the set theories made by taking Zermelo-Fraenkel set theory as a base and adding some of the higher cardinal axioms. Another example of 2 shows up if we ask whether particular modal logics are true given Hellman's structuralism, as presented in Hellman (1989) since he does not come clean on either his metaphysical views concerning modality, nor on the formal theory which best represents mathematical possibility.
3. "y is false, given philosophy X" gets the value judgment (F), X recognises y as false, incorrect, ill conceived or wrong in some sense.

   Examples of number 3 are: if we start with Martin Löf's constructive type theory, then any mathematical theory which ineluctably contains the full classical law of excluded middle will be false. The law of excluded middle is eliminable in theories which allow only a finite domain, or which can be interpreted by the ordinals. Otherwise the law of excluded middle is ineluctable, and the theory is false or meaningless or misguided, when indexed to Martin-Löf's constructivism. Other examples can be found if we consider "bad" theories. Shapiro's structuralism considers paraconsistent mathematical theories to be false, since they are not recognised by model theory (since, in classical model theory, models demonstrate consistency).

[56]There is an ambiguity between our *not knowing* that philosophy X endorses y, and in principle, philosophy X is neutral with respect to y. This ambiguity runs through all of the judgments. I leave it in place with the counsel to make it clear when deploying judgments whether one means them in the epistemic or the ontological sense. Of course in a constructive vein, the distinction does not arise.

4. "y has contradictions, but these are true (and false), given philosophy X" gets the value judgment (*Ŧ*) (Truth value glut, T favoured), iff Philosophy X recognises, entertains and tolerates the mathematical theory y which contains contradictions.

   Examples of 4 are: relevant or paraconsistent theories with their accompanying philosophies.
5. "y has contradictions but these are false (and true), given philosophy X" gets the value judgment (*F̵*) (truth-value glut, F favoured) iff the philosopher works with a trivial theory, despite its being trivial.

   Examples of 5 are trivial theories. These are true and false: studied and trawled, but badly flawed, since they enjoy "consistent contradictions". They are not rejected altogether. So, for example, Dummett's work on Frege's inconsistent formal system gives us plenty of examples of sentences which would be judged as *F̵* because they are embedded in a trivial formal theory. The theory gets this judgment because there are quite acceptable things Frege writes within his own formal theory.[57]

The notions of truth and falsity of a whole mathematical theory is not classically bivalent. Nor should it be, for the pluralist. The five sorts of judgment are a bit unusual and call into question some of our philosophical prejudices. In particular, we are replacing the more traditional "truth-value" notion with the more nuanced notion of value judgment. The motivation for this is the simple thought that, for the pluralist, it does not make sense to say of a mathematical theory that it is true as such. It is more interesting, and informative, to say that it is consistent, or, it is a theory to which most of mathematics can be reduced, or, that it can be used to analyse another theory, and then reveal interesting problems. The choice of value judgments forces us to be explicit about our perspective, philosophy X. This mechanism in the judgment turns what used to be a normative claim into a descriptive claim. The force of this turn is to be more subtle, more precise, and then allows us to move on to other judgments. Note also that we have two sorts of judgment value glut: the one where contradiction is handled and constrained already in the theory, and the one where we have a trivial theory. As far as I know, this is original to the semantics for this pluralist theory, and is not a distinction found in other paraconsistent logics. The draw to be more nuanced in our value judgments is echoed in Byers.

> Mathematics is so commonly identified with its formal structure that it seems peculiar to assert that an idea [in mathematics] is neither true nor false. What I [William Byers] mean by this is similar to what David Bohm means when he says "theories are insights which are neither true nor false, but, rather, clear in certain domains, and unclear when extended

[57]The definition of the logical connectives and operators has not yet been set. There are several possibilities. Consider, for example conjunction. We can define this as true when: both conjuncts are true, when both conjuncts are favoured (truth by itself is truth favoured) or when neither conjunct is false. Negation also merits careful consideration in the face of value gaps and gluts. In face of such choices, we can make particular choices, so conjunction is one thing, or we could even introduce several conjunctions, several negations, several conditionals—defined in terms of the other connectives and so on. Presumably, the syntax would then be designed to make, at least a sound system.

> beyond those domains" (Bohm 1980, p. 4). Classifying ideas as true or false is just not the best way of thinking about them. Ideas may be fecund; they may be deep; they may be subtle; they may be trivial. These are the kinds of attributes we should ascribe to ideas. Prematurely characterising an idea as true or false rigidifies the mathematical environment. Even a "false" idea can be valuable. For example, Goro Shimura once said of his late colleague Yutaka Taniyama, "He was gifted with the special capability of making many mistakes, mostly in the right direction. I envied him for this and tried in vain to imitate him, but found it quite difficult to make good mistakes" (Singh 1997, p. 174). A mistake is "good" precisely because it carries within it a legitimate mathematical idea. (Byers 2007, pp. 256–257)

It is too easy for philosophers of mathematics to restrict their task to giving an account of the "realm of mathematical truths". This conception of the task of the philosophy of mathematics is a relic of Platonism which is rejected by the pluralist.

The pluralist is not alone in his rejection. There have been notable exceptions to this conception of the philosophy of mathematics: Brouwer was interested in mathematics as living in the mind. For Brouwer, there is a supervenience relationship between the psychology of learning about mathematics and the content of mathematics. In some ways this was explored further by Husserl. Husserl was interested in the phenomenology of mathematics, in our interaction with whatever it is that mathematicians treat as objects of study, in studying the phenomenology of these objects, since they are presented to us as objective and rigid, see Tieszen (2005). Husserl was not interested in the traditional metaphysical question concerning the ontology of mathematics: whether the objects of mathematics are independent of us. Instead, Husserl "bracketed" the traditional/metaphysical question of ontology and focused on what it is for us to encounter or experience such an object. These philosophies of mathematics break with our inherited Platonistic tradition, and, those of us who were raised in that tradition typically have difficulty recognising the importance of their place in the philosophy of mathematics, since we cannot easily classify them as "realist or anti-realist" etc. So, we cannot easily compare them to the more standard philosophies of mathematics. But this is not enough to claim that pluralism is correct. We have yet to answer the traditional philosopher's concerns.

To traditional philosophers of mathematics, discussion of "judgments" and "insights" can sound rather suspect. After all, "indexed truth" is not far removed from "subjective truth": "true for me" but "false for you", since this is a type of indexing—indexing to a person at a time. Prompted by such worries, our opponent interlocutor of the previous critique might step in and ask if we are not really interested in the psychology of mathematics more than the philosophy of mathematics. The answer is that we are interested in the psychology, but only insofar as it can inform the philosophy. For example, it would be quite significant if we were to discover that there is a neuro-psychological block to our conceiving a consistent contradiction.[58] Or, there might be a neuro-psychological basis for

[58] I'm afraid I only have an anecdote. A student of mine, Thom Genarro gave a talk on some recent findings in psychology which he reckoned had some impact on the philosophy of mathematics. One such finding was that at the very primitive level, our brains are so constructed as to preclude

our holding some mathematical theorems as ineluctable. For example, we might discover that when faced with said theorem, a very primitive part of the brain is active, as opposed to a more esoteric theorem, where several complex areas of the brain are activated, and therefore, tenuously, we might think that esoteric mathematics cannot be grasped by everyone, but that they are easier to accept in the sense of there being several alternative neurological pathways which serve the function of allowing us to grasp the concepts. For pluralists, this line of enquiry is legitimate, if highly tenuous. Pluralists are able to acknowledge the insightful work done by neuro-scientists, or before them, Brouwer and Husserl, because we are able to acknowledge that psychological findings might partly explain the paths our mathematical investigations have taken. If we replace the traditional absolute truth values with judgment values, then we have more subtle tools to use for our philosophical work of comparing and evaluating philosophical theories and their fit with formal mathematical theories.

There is other philosophical work to be done, which is quite untainted by psychology or neuro-science. Take a paraconsistent logic "off the shelf" and put the value-judgments to work. The pluralist blocks *ex falso quodlibet* inferences in order to entertain pair-wise inconsistent theories, and whole trivial theories. Our new judgment values can accommodate nascent theories, where the "U" judgment will be useful. Contradictory theories are ones where the $\not{T}$ and $\not{F}$ judgments become important. Add indexes to the turnstile symbol to indicate the context of derivation (so essentially these turnstiles will pick out a class of formal theories). We can now prove things about combining mathematical theories. When the pluralist puts the value-judgments to work, he can give a subtle and precise interpretation of what is going on in mathematics. Working on the logic which best accommodates maximal pluralism is a philosophical and logical task. Giving philosophies for mathematical theories or classes of theories is another philosophical task (at the second level). Bringing philosophical insights to bear on particular results in mathematics is a philosophical task of the first level. See Corfield (2003). Thus there is plenty of philosophical work to be done; and it is work that cannot be done purely by a sociologist, historian or psychologist. This parries, again, the critique from "disdain for sociology". But what about the third level: the one occupied by the pluralist? Is the pluralist a pluralist about pluralism, or is he a foundationalist?

15.5.3 Conclusion: The Third (Paraconsistent) Level

The third level is pluralist *about* the levels below, but what happens at the third level? If we wanted to be optimally pluralist, and, say, we were constructivist, we

---

our conceiving a contradiction. Some of the audience, including myself shot our hands up at this point. Either paraconsistent logicians are some sort of ubermenschen since they have overcome this primitive block, or the experiments which indirectly "show" this pre-suppose that we cannot conceive of a contradiction. I'll let you decide which is the more likely disjunct.

would choose a constructive logic at this level. We would then insist on a criterion for success in mathematics, namely, that the theory be constructive. What falls under which value judgments would then change. In this sense, pluralism has a perspective which is informed by our logic. Within these constraints, we could then be pluralist about different constructive mathematical theories—all of those which fit within the constraints. Similarly, if we are ultimately convinced by classical logics and convinced that contradictions can only be supported in the short term, in the sense that we have to adapt when faced with contradictory information, then we would probably favour a classical adaptive logic.[59] Optimal pluralism is not pluralist towards itself, since an optimal pluralist cannot tolerate competing optimal pluralist philosophies.

In this paper, I have presented maximal pluralism, and maximal pluralism is pluralist towards itself. The maximal pluralist favours some form of paraconsistent logic. The paraconsistent logic brings with it its own to philosophical stamp. In this case, the stamp is (at least) dialetheist, since there are true contradictory theories (which are also false). Truth is favoured. But it is not only dialetheist, since there are also false contradictory theories (which are also true). These are the trivial theories. Falsity is favoured. Thus, the logic might be more than dialetheist. When making judgments about the first and second levels, we have to hold the logic and philosophy of the third level fixed. However, pluralism can also be pluralist concerning the third level by moving up to a fourth level. When we judge pluralism, as a philosophy, we are occupying a fourth level of analysis. This is the level we have occupied in this section. Occupying this fourth level, we can say that, for example, pluralism accommodates mathematics as it is practiced today. So, maximal pluralism is pluralist about pluralism at the third level. In this sense, we might say that pluralism enjoys "closure" of its theory, as Priest uses the term, see Priest (2002). But we transcend pluralism too. Pluralists are well aware that philosophy and mathematics are historically (conceptually) situated. That is, philosophy and mathematics are developing with reference to each other, what looks like a good philosophy of present day mathematics, might not look so good of future mathematics. More important dialetheically, as pluralists concerning the third level, we take seriously the possibility that there could be significant improvement to the logic or class theory, in which case we would revise the underlying logic.

To sum up: if we take a paraconsistent logic to underpin our pluralism, and we think that mathematics contains dialetheias. We think that pluralism commits us to recognising some contradictions as true and to being pluralists about our own pluralism. This is because we take seriously the idea that there might be rival paraconsistent foundations, each meriting its own take on the lower levels of analysis. After all, the term "paraconsistent" is adopted by quite different philosophical and logical traditions. Each has its own motivations, and will bring its own philosophical stamp to bear on its version of pluralism. Maximal pluralism is a thoroughgoing pluralism.

---

[59]See Batens' http://logica.rug.ac.be/adlog/al.html for an introduction to adaptive logics.

References

Batens, D. 1997. Inconsistencies and beyond. A logical-philosophical discussion. *Revue International de Philosophie* 200: 259–273.

Beall, J.C., and G. Restall. 2006. *Logical pluralism*. Oxford: Clarendon Press.

Bohm, D. 1980. *Wholeness and the implicate order*. London: Routledge.

Byers, W. 2007. *How mathematicians think: Using ambiguity, contradiction and paradox to create mathematics*. Princeton: Princeton University Press.

Cellucci, C. 2008. Why proof? What is a proof? In *Deduction, computation, experiment. Exploring the effectiveness of proof*, ed. G. Corsi and R. Lupacchini, 1–27. Berlin: Springer.

Colyvan, M. 2001. *The indispensability of mathematics*. Oxford: Oxford University Press.

Coniglio, M.E., and W. Carnielli. 2008. *On discourses addressed by infidel logicians*. Presented at Fourth World Congress of Paraconsistency in Melbourne, Australia.

Corfield, D. 2003. *Towards a philosophy of real mathematics*. Cambridge: Cambridge University Press.

Friend, M. 2006. *Meinongian Structuralism*. The Logica Yearbook 2005. Marta Bílkova and Ondřej Tomala (eds.). Filosofia, Prague, pp. 71–84.

Giaquinto, M. 2002. *The search for certainty: A philosophical account of the foundations of mathematics*. Oxford: Clarendon Press.

Goethe, N.B., and M. Friend. 2010. Confronting ideals of proofs with the ways of proving of the research mathematician: Philosophy, history and practice-based mathematics. *Studia Logica* 96: 273–288.

Hellman, G. 1989. *Mathematics without numbers*. Oxford: Oxford University Press.

Heyting, A. 1930. Die formalen Regeln der intuitionistischen Logik. *Sitzungsberichte der Preussischen Akademie der Wissenschaften* 1930: 42–56. Translated and quoted in Paulo Mancosu (1998); *From brouwer to Hilbert: The debate on the foundations of mathematics in the 1920s*. New York: Oxford University Press.

Körner, S. 1962. *The philosophy of mathematics: An introductory essay*. New York: Harper and Row.

Maddy, P. 1997. *Naturalism in mathematics*. Oxford: Clarendon Press.

Maddy, P. 2007. *Second philosophy*. Oxford: Oxford University Press.

Mancosu, P. 1998. *From Brouwer to Hilbert: The debate on the foundations of mathematics in the 1920s*. New York: Oxford University Press.

Mortensen, C. 1995. *Inconsistent mathematics and its applications*. Dordrecht: Kluwer.

Potter, M. 2004. *Set theory and its philosophy*. Oxford: Oxford University Press.

Potter, M., and P. Sullivan. 1997. Hale on caesar. *Philosophia Mathematica* 5: 135–152.

Priest, G. 2002. *Beyond the limits of thought*. Oxford: Clarendon Press.

Priest, G. 2003. Meinongianism and the philosophy of mathematics. *Philosophia Mathematica* vol. XI Series III(1): 3–15.

Priest, G. 2006. *Doubt truth to be a liar*. Oxford: Oxford University Press.

Prior, A. 1960. The roundabout inference ticket. *Analysis* 21: 8–39.

Shapiro, S. 1991. *Foundations without foundationalism: The case for second-order logic*, Oxford logic guides, vol. 17. Oxford: Clarendon Press.

Shapiro, S. 1997. *Philosophy of mathematics: Structure and ontology*. Oxford: Oxford University Press.

Singh, S. 1997. *Fermat's enigma: The epic quest to solve the world's greatest mathematical problem*. New York: Penguin Books.

Tennant, N. 1997. *The taming of the true*. Oxford: Clarendon Press.

Thurston, W.P. 1994. On proofs and progress in mathematics. *Bulletin of the American Mathematical Society* 30(2): 161–177.

Tieszen, R. 2005. *Phenomenology, logic and the philosophy of mathematics*. Cambridge: Cambridge University Press.

van Stigt, W.P. 1998. Brouwer's intuitionist programme. In *From Brouwer to Hilbert: The debate on the foundations of mathematics in the 1920s*, ed. Paulo Mancosu. New York: Oxford University Press.

Vopěnka, P. 1979. *Mathematics in the alternative set theory*. Leipzig: Teubner-texte zur Mathematik.

Whitehead, A.N., and B. Russell. 1997. *Principia mathematica to *56*. Cambridge, Cambridge University Press.

Chapter 16
Arithmetic Starred

Chris Mortensen

16.1 Introduction

An interesting and unstudied "theory" is the Routley star * of classical Peano Arithmetic, that is **PA***. To recall for the reader, if S is any set of sentences, then S* is defined as $\{A{:}\neg A \notin S\}$. The star operation was first defined in Routley and Routley (1972), and plays a key role in the semantics of paraconsistent logics, particularly relevant logics. Now PA* has a number of interesting properties.

1. It is closed under deducibility in classical logic (in a sense to be made clearer later).
2. It is inconsistent (if PA itself is consistent): by Gödel's incompleteness theorem neither the Gödel sentence G nor its negation $\neg G$ are in PA, so both are in PA*.
3. It is non-trivial (PA if consistent contains many sentences consistently, such as $0 = 0$, $\neg 0 = 1$ while lacking their negations, so that those negations do not appear in PA*).

How can this be? A classical theory which is inconsistent but non-trivial? The answer, as we will see, lies in a careful dissection of what a theory should be.

16.2 Semitheories and Star-Invariance

It was necessary for the semantics of first degree entailment that deducibility be star-invariant. But this raises the general question of what properties are star-invariant. It is convenient to separate single-premiss deducibility from multi-premiss deducibility. This calls for a distinction between *semitheories* and *theories*. The former are closed on some form of single premiss deductions.

C. Mortensen (✉)
Department of Philosophy, The University of Adelaide, Adelaide, Australia
e-mail: chris.mortensen@adelaide.edu.au

K. Tanaka et al. (eds.), *Paraconsistency: Logic and Applications*, Logic, Epistemology, and the Unity of Science 26, DOI 10.1007/978-94-007-4438-7_16,

The other thing to take note of, is controversy about choice of logic. In particular we have to distinguish classical theories, closed under the deducibility relation of classical logic, and theories of other logics L, closed under different deducibility $\vdash_L$. We start with:

4. T is an *L-semitheory* iff ((A $\in T$ and A$\vdash_L$B) implies B $\in T$).

The important property following from this definition is:

5. If L obeys rule contraposition in the form: (if A$\vdash_L$B then ¬B$\vdash_L$ ¬A), then L-semitheoryhood is *-invariant.

Proof. Let T be a semitheory, and suppose $A \in T^*$ and $A \vdash_L B$. From the latter, assuming rule contraposition, $\neg B \vdash_L \neg A$. From the former, $\neg A \notin T$. Hence $\neg B \notin T$, that is $B \in T^*$.

We note that rule contraposition is a minimal constraint on L. A related alternative definition is:

6. T is an L-semitheory iff ((A $\in T$ and $\vdash_L$A→B) implies B $\in T$).

This is equivalent to (4), given only the deduction theorem for →, which holds for many logics with implication operators. Hence semitheoryhood under definition (6) is also *-invariant.

Henceforth we drop the "$L-$" prefix for semitheories where the logic L is understood. Now apparently similar to (4) and (6), but in fact quite different, is:

7. T is *closed under modus ponens* for → iff ((A $\in T$ and A→B $\in T$) implies B $\in T$).

This is a property which is independent of the logic L. It is desirable property for the semitheories and theories of a logic to have: it can be argued that a theory generally contains contingent or non-logical conditionals from which consequents can be detached given the antecedents, for example "If it rains, the match is cancelled; it rains, so the match is cancelled". In support, see Meyer et al. (1979).

Closure under modus ponens in sense (7) is neither necessary nor sufficient for semitheoryhood in sense (4). However, for any *regular* logic (all theorems belong to all theories), closure under modus ponens implies semitheoryhood in sense (6). Classical logic is regular; however the relevant logic R and all sublogics are non-regular. Also, we must take note of the operator →, which is some sort of implication, but can have different properties, such as those of relevant implication or instead classical ⊃.

The important thing to note here is that being closed under modus ponens is not generally ∗-invariant. We note a weaker result, under a minimal assumption about the logic:

8. If T is complete and regular and T^* is closed under contraposition, then being closed under modus ponens is ∗-invariant.

Proof. Let T be complete and closed under modus ponens, and let A and $A \to B$ be in T^*. Since T^* is closed under contraposition, $\neg B \to \neg A$ is in T^*. Since T is complete, $T^* \subseteq T$ (see (13) below). But since $A \in T^*$, $\neg A \notin T$. Hence since T is closed under modus ponens, $\neg B \notin T$. Hence $B \in T^*$ as required.

But there is no obvious way to strengthen (8) to general $*$-invariance of modus ponens, for example in the way that (5) was proved. This fact will serve to explain the puzzle posed at the end of the last section. But first, we need to take account of multi-premiss arguments.

16.3 Conjunction

The Routleys realised that being a semitheory does not in general imply being closed under conjunctions. So they defined:

9. A *theory* is a semitheory which is also closed under conjunctions.

It is clear that a theory in this sense is closed under multi-premiss deductions: if $A_1, \ldots A_n \in T$ and $A_1, \ldots A_n \vdash B$ then $B \in T$. Of course a theory may not have conjunction in its language, when it would be necessary to use the more general definition if such closure is desired.

Now being a theory is not generally $*$-invariant. Instead we have the following. A semitheory is *prime* iff for any disjunction in the semitheory, at least one disjunct is too. Then, given weak properties for the background logic (De Morgan and double negation), the Routleys showed:

10. If T is a theory, then T^* is a prime semitheory.

It follows that being closed under conjunctions is not generally $*$-invariant, nor is being closed under multi-premiss deductions. However, they also showed, under the same assumptions:

11. If T is a prime theory then so is T^*, and
12. If T is consistent then T^* is complete and $T \subseteq T^*$
13. If T is complete then T^* is consistent and $T^* \subseteq T$
14. $T = T^{**}$. This is clearly equivalent to Double Negation.
15. T is inconsistent iff T^* is incomplete.
16. If T is consistent and complete, then $T = T^*$, so that T^* is consistent and complete also.

That is, (11) and (16) say: being a prime theory, and being consistent and complete, are both $*$-invariant.

16.4 PA* Revisited

We now have the ingredients to clarify the puzzle we started with. PA is a theory of classical logic. But assuming consistency it is not prime, since $G \vee \neg G \in PA$, but $G \notin PA$ and $\neg G \notin PA$, where G is the Gödel sentence. Hence PA* is a prime semitheory but not a prime theory. That is, the rule of conjunction fails for PA*. In particular, both G and $\neg G$ are in PA*, but their conjunction is not. To put it otherwise, since $G \vee \neg G$ is consistently in PA, it is also consistently in PA*. Thus $\neg(G \vee \neg G)$ is not in PA*. But, given De Morgan and Double Negation, neither is the equivalent $G \& \neg G$ in PA^*. Hence the deduction of triviality from the contradictory conjunction is blocked: PA* is "non-adjunctive".

Non-adjunctiveness is not the whole story however. PA*, being a regular classical semitheory, is closed under classical deducibility and classical theorems. These include three versions of *Ex Contradictione Quodlibet*: (a) $G, \neg G \vdash$ any, (b) $G \vdash (\neg G \supset$ any), and (c) $\vdash G \supset (\neg G \supset$ any). These might seem to imply the triviality of PA*, which we are assured does not obtain. However, PA* is not closed under multi-premiss deductions, so (a) does not hold. For both (b) and (c), since $G \in PA^*$ which is a semitheory, we can detach $\neg G \supset$ any. But then, we note that PA is not complete, so by the observations on (8) above, we have no reason to think that PA* is closed under modus ponens (though it does satisfy the other conditions for (8)). Hence, we are unable to detach every wff using $\neg G$. Indeed, stronger than that, we know that modus ponens *must* fail. For if it did not, we would have triviality, which we know we do not have.

16.5 Sundry Additional Extensions of PA

In this section, we survey some additional extensions of PA built using star and complementation. We first note, for future reference, that:

17. PA^* is co-r.e.

To see this, recall that $PA^* = \{X : \neg X \notin PA\}$. Hence its complement $\overline{PA^*} = \{X : \neg X \in PA\}$. Hence the complement of PA^* can be enumerated by enumerating the formulae of PA and choosing just those beginning with a negation. We return to this complement presently.

First, however, consider the complement of PA, that is $\overline{PA}$. Obviously, $\overline{PA}$ is also co-r.e. Now, it is easily shown that:

18. A set of sentences T is a semitheory iff its complement $\overline{T}$ is a *co-semitheory*, that is, if $A \in \overline{T}$ and $B \vdash A$ then $B \in \overline{T}$.

That is to say, since $\overline{PA}$ reverses the deductive order in comparison with PA, $\overline{PA}$ is a co-semitheory. This immediately suggests that:

19. If L obeys rule composition in the same sense as (5), then co-semitheoryhood is *-invariant.

Proof. Let T be a co-semitheory, it has to be shown that T* is a co-semitheory. So let $A \in T^*$ and $B \vdash A$, it has to be shown that $B \in T^*$. Since $A \in T^*$, then $\neg A \notin T$. Since $B \vdash A$, then $\neg A \vdash \neg B$ (rule contraposition). Hence, since T is a semitheory, $\neg B \notin T$, so that $B \in T^*$ as required.

We also note that $\overline{PA}$ is closed under conjunctions. For if even just $A \in \overline{PA}$ then since in classical logic $A\&B \vdash A$, then also $A\&B \in \overline{PA}$; so that closure under conjunctions follows trivially. Of course, $\overline{PA}$ is not closed under &–elimination, for this would require to go from $A\&B \in \overline{PA}$ to $A \in \overline{PA}$ via $A \vdash A\&B$, but the latter does not hold. Dually, $\overline{PA}$ is prime, since if $A \vee B \in \overline{PA}$ then since both $A \vdash A \vee B$ and $B \vdash A \vee B$, then *both* of A, B are in $\overline{PA}$.

Assembling what we know:

20. PA^* is a semitheory, co-r.e., inconsistent, complete, contains both G and $\neg G$, but not $G\&\neg G$, contains its negation, and equivalently contains $G \vee \neg G$, but does not contain its negation. Contains all axioms and no negations of axioms.

In contrast:

21. $\overline{PA}$ is a co-semitheory, co-r.e., inconsistent, complete, contains both G and $\neg G$, and also $G\&\neg G$, does not contain its negation, and equivalently does not contain $G \vee \neg G$, but does contain its negation. Contains no axioms and all negations of axioms.

Combining (20) and (21):

22. $\overline{PA^*}$ is a co-semitheory, r.e., consistent, incomplete, contains neither G nor $\neg G$, but does contain $G\&\neg G$, but not its negation, and equivalently does not contain $G \vee \neg G$, but does contain its negation. Contains no axioms and all negations of axioms.

Also:

23. $(\overline{PA})^*$ is a co-semitheory, r.e. consistent, incomplete, contains neither G nor $\neg G$, but does contain $G\&\neg G$, but not its negation, and equivalently does not contain $G \vee \neg G$, but does contain its negation. Contains no axioms and all negations of axioms.

Further:

24. $(\overline{PA^*})^*$ is a co-semitheory, co-r.e., inconsistent, complete, contains both G and $\neg G$, and their conjunction, but not its negation, and equivalently does not contain $G \vee \neg G$, but does contain its negation. Contains no axioms and all negations of axioms.

And finally:

25. $\overline{((\overline{PA})^*)}$ is a semitheory, co-.r.e., inconsistent, complete, contains both G and $\neg G$, but not their conjunction, contains its negation, and equivalently contains $G \vee \neg G$, but does not contain its negation. Contains all axioms and no negations of axioms.

In the light of these similarities and differences, it is reasonable to ask whether (22) and (23) are the same, that is:

26. $\overline{PA^*} = (\overline{PA})^*$

Proof. $A \in \overline{PA^*}$ iff $A \notin PA^*$ iff $\neg A \in PA$ iff $\neg A \notin \overline{PA}$ iff $A \in (\overline{PA})^*$

As corollaries we have:

27. $PA^* = \overline{((\overline{PA})^*)}$ and $\overline{PA} = (\overline{PA^*})^*$

These follow by barring, and starring, both sides of (26).

16.6 Conclusion

Our main interest has been in the puzzle about PA^*. But we have also seen that there are interesting, well-behaved structures, including $\overline{PA}$ and $(\overline{PA})^*$, in the vicinity.

References

Meyer, R.K., R. Routley, and J.M. Dunn. 1979. Curry's paradox. *Analysis* 39: 124–128.
Routley, R., and V. Routley. 1972. The semantics of first degree entailment. *Noûs* 6: 335–359.

Chapter 17
Notes on Inconsistent Set Theory

Zach Weber

17.1 Introduction

The standard axioms of naive set theory state existence and uniqueness conditions for sets (see Routley 1980; Priest et al. 1989; Brady 2006). The axioms are:

Axiom 17.1 (Abstraction) $x \in \{z : \Phi(z, u)\} \leftrightarrow \Phi(x, u)$.

Axiom 17.2 (Extensionality) $(\forall z)(z \in x \leftrightarrow z \in y) \leftrightarrow x = y$.

The purpose of this paper is to highlight and discuss two ideas that play in to the axiomatic development of a paraconsistent naive set theory, as detailed in Weber (2010b). We will focus on aspects of the theory that can be read right off the axioms, concerning intensional identity and unrestricted set existence. Both relate to inconsistency and are dealt with here as follows.

First, the extensionality axiom says that identity is governed by entailments. As we will define below, $\rightarrow$ is an intensional, relevant implication and so, as with an extensionality axiom formulated using a material conditional, this leads to some distinctive properties for identity. With these new properties in hand I extend some results of Arruda and Batens from da Costa's set theory (from da Costa 2000 in Batens et al. 2000).

Second, the set formation principle is fully unrestricted, so the set being defined may appear in its defining condition. We will explore how this makes modelling recursive phenomena particularly easy and natural, elaborating on ideas from Routley's set theory in Routley (1977).

To begin I lay out a relevant background logic, placing a strong emphasis on the restrictions such a logic must have in order to support an inconsistent set theory. The sections that follow proceed on the understanding that, while highly

Z. Weber (✉)
Department of Philosophy, University of Otago, PO Box 56, Dunedin 9054, New Zealand
e-mail: zach.weber@otago.ac.nz

K. Tanaka et al. (eds.), *Paraconsistency: Logic and Applications*, Logic, Epistemology, and the Unity of Science 26, DOI 10.1007/978-94-007-4438-7_17,

inconsistent, a good deal of control is being exerted on the theory through the weakened logic. The two features of a fully naive theory, identity and self-reference, dovetail throughout.[1]

17.2 Logic

The main purpose of this section is to summarize the known restrictions on a logic for naive set theory; see also Weber (2010a). The subsidiary purpose is to fix the logic used in this paper; the logic may be altered for different results, as long as all the restrictions are observed. Thus not much emphasis is placed on the particular choice here, except to provide exactness.

The language of first order set theory has primitives $\wedge, \neg, \rightarrow, \forall$, $=$ and $\in$, as well as a term-forming operator $\{\cdot : \cdot\}$; variables $x, y, z, \ldots$; names $a, b, c, \ldots$; and formulae $\Phi, \Psi, \Upsilon, \ldots$, built up by standard formation rules. The usual shorthand is used: $\Phi \vee \Psi$ for $\neg(\neg\Phi \wedge \neg\Psi)$; $\Phi \leftrightarrow \Psi$ for $(\Phi \rightarrow \Psi) \wedge (\Psi \rightarrow \Phi)$; $\exists$ is $\neg\forall\neg$. (Taking these as definitions means that e.g, $\Phi \vee \Psi \rightarrow \neg(\neg\Phi \wedge \neg\Psi)$ is no more than an instance of axiom I below.)

17.2.1 Axioms

All instances of the following schemata are theorems:

I	$\Phi \rightarrow \Phi$	
IIa	$\Phi \wedge \Psi \rightarrow \Phi$	
IIb	$\Phi \wedge \Psi \rightarrow \Psi$	
III	$\Phi \wedge (\Psi \vee \Upsilon) \rightarrow (\Phi \wedge \Psi) \vee (\Phi \wedge \Upsilon)$	*(distribution)*
IV	$(\Phi \rightarrow \Psi) \wedge (\Psi \rightarrow \Upsilon) \rightarrow (\Phi \rightarrow \Upsilon)$	*(conjunctive syllogism)*
V	$(\Phi \rightarrow \Psi) \wedge (\Phi \rightarrow \Upsilon) \rightarrow (\Phi \rightarrow \Psi \wedge \Upsilon)$	
VI	$(\Phi \rightarrow \neg\Psi) \rightarrow (\Psi \rightarrow \neg\Phi)$	*(contraposition)*
VII	$\neg\neg\Psi \rightarrow \Psi$	*(double negation elimination)*
VIII	$\Phi \vee \neg\Phi$	*(excluded middle)*

[1]Following a distinction I first saw in Libert (2005), Axiom 17.1 is called *abstraction*, while the formulation in Theorem 17.3 below is called *comprehension*. There is a syntactic difference between abstraction and comprehension, and in weak paraconsistent logics the principles are not equally user-friendly, because the quantifier $\exists$ is sometimes tricky to eliminate. Nevertheless, both formulations capture a core intuition and in informal discussion the names are used interchangeably, without intending to mark an important difference.

IXa $(\Phi \to \Psi) \to [(\Psi \to \Upsilon) \to (\Phi \to \Upsilon)]$
IXb $(\Phi \to \Psi) \to [(\Upsilon \to \Phi) \to (\Upsilon \to \Psi)]$ *(hypothetical syllogisms)*
X $(\forall x)\Phi \to \Phi(a/x)$
XI $(\forall x)(\Phi \to \Psi) \to (\Phi \to (\forall x)\Psi)$
XII $(\forall x)(\Phi \vee \Psi) \to \Phi \vee (\forall x)\Psi$

Axioms XI and XII have the caveat that x does not appear free in Φ. The hypothetical syllogism pair *IXa* and *IXb* are called *suffixing* and *prefixing*, respectively.

17.2.2 Rules

The following rules are valid:

I $\Phi, \Psi \vdash \Phi \wedge \Psi$ *(adjunction)*
II $\Phi, \Phi \to \Psi \vdash \Psi$ *(modus ponens)*
III $\Phi, \neg\Psi \vdash \neg(\Phi \to \Psi)$
IV $\Phi \vdash (\forall x)\Phi$
V $x = y \vdash \Phi(x) \to \Phi(y)$ *(substitution)*

Brady proves that set theory in this logic has a model and is non-trivial (Brady 1989 and 2006, p. 242). If rule *III*, called *counterexample*, is brought up to arrow strength, the resulting logic is *DLQ* from Routley and Meyer (1976); with hypothetical syllogism, Axioms *IXa*, *b*, the logic is called *TLQ*. Non-triviality of naive set theory in these stronger logics is an open problem.

The fact that Brady's universal logic DJQ is not strong enough for some of these results is important. The key non-DJQ principles, excluded middle and counterexample, restores a connection between the intensional $\to$ and the extensional connectives, via the derived rule

$$\Phi \to \Psi \vdash \neg\Phi \vee \Psi$$

More to the point, the axiom does a lot of work. The preponderance of the results discussed below cannot be recovered (as given) using only DJQ. For more on the considerations going in to the choice of this particular logic, see Weber (2010a).

17.2.3 Restrictions

For a logic of naive set theory, *DLQ* is quite strong. For instance, it has a robust negation. But it is very spare, and for good reason. The first phase of paraconsistent set theoretical research has shown that there are several key restrictions to respect, on pain of triviality, which we recite here for ease of reference.

An inference is invalid if it does not preserve truth, and in the context of inconsistent set abstraction one must take extra care. *Disjunctive syllogism*,

$$\Phi, \neg\Phi \vee \Psi \vdash \Psi,$$

is invalid in this context, due to C.I. Lewis' famous argument in (Lewis and Langford 1959, p. 250). Also invalid is *contraction*,

$$\Phi \rightarrow (\Phi \rightarrow \Psi) \vdash \Phi \rightarrow \Psi$$

as shown by Curry (1942). Closely related is *axiom modus ponens* (or pseudo-modus ponens or mp-contraction),

$$\Phi \wedge (\Phi \rightarrow \Psi) \rightarrow \Psi$$

as found in Meyer et al. (1978) and Restall (1994). There is also a trouble with *permutation*,

$$\Phi \rightarrow (\Psi \rightarrow \Upsilon) \vdash \Psi \rightarrow (\Phi \rightarrow \Upsilon)$$

due to the argument in Slaney (1989). Slaney's argument shows that excluding the middle and permutation are not jointly tenable. The cause is, again, a close relative of Curry's paradox.

Since the logic is relevant it does not include weakening, $\Phi \vdash \Psi \rightarrow \Phi$. With weakening, we would have to drop contraposition. Else, we could argue from Λ to $\neg\Psi \rightarrow \Lambda$, then to $\neg\Lambda \rightarrow \Psi$. But if also $\neg\Lambda$, i.e. Λ is a true contradiction, then Ψ follows by *modus ponens*, where Ψ is arbitrary. The improper inference here is just $\Phi \vdash \neg\Phi \rightarrow \Psi$, a form of explosion.

17.2.4 A Case Study

Here is an example of how the weakened logic must be attended to. Consider the two way inference

$$\Phi \wedge \Psi \rightarrow \Upsilon \dashv\vdash \Phi \rightarrow (\Psi \rightarrow \Upsilon). \tag{17.1}$$

In classical logic (and set theory) this is obvious—because, materially, it just says

$$\neg(\Phi \wedge \Psi) \vee \Upsilon \dashv\vdash \neg\Phi \vee (\neg\Psi \vee \Upsilon). \tag{17.2}$$

The two-way derivation (17.2) is valid here, but in an intensional logic, the two sentences in (17.1) certainly do not say the same thing. They must, on pain of triviality, not be inter-derivable. Suppose the inference (1) from left to right. Now, we have as an axiom $\Phi \wedge \Psi \rightarrow \Phi$. So we would infer $\Phi \rightarrow (\Psi \rightarrow \Phi)$ as a valid scheme, which is weakening and so trivializing in this logic. From right to left on (1), because $(\Phi \to \Psi) \to (\Phi \to \Psi)$ is an instance of an axiom, $\Phi \wedge (\Phi \to \Psi) \to \Psi$ would be a valid scheme, which is mp-contraction. Given the relevance logic we are using, both directions of (17.1) must fail.

With this in mind, though, let us look at an example with a subset relation $\subseteq$.

Definition 17.1. $x \subseteq y := (\forall z)(z \in x \to z \in y)$. Then $x \subset y := x \subseteq y \wedge (\exists z)(z \in y \wedge z \notin x)$.

Then consider two ways of understanding transitivity,

$$y \subseteq z \to (x \subseteq y \to x \subseteq z),$$

$$x \subseteq y \wedge y \subseteq z \to x \subseteq z.$$

If subset is understood with arrows, as it is in Definition 17.1, then the first is an instance of hypothetical syllogism (Axioms IX) and the second of conjunctive syllogism (Axiom IV). But we can see that these are almost certainly independent of one another, based on the problems just discussed. So it is required in proofs that we be very clear about which forms we are using. More generally, in any formulation of definitions, for subset, ordinal number, or function, a great deal of thought is required. For example, f could be called a function when $\langle x, y\rangle \in f \wedge \langle x, z\rangle \in f \to y = z$, or when $\langle x, y\rangle \in f \to (\langle x, z\rangle \in f \to y = z)$, but with different results.

Although we do not need the following here, it is worth flagging a useful notion: a *relevant singleton* is written $\{x\}_y := \{z : z = x \wedge z \in y\}$. This is for relevance purposes, to fix $\{x\}_y \subseteq y$ iff $x \in y$.

17.3 Basics

Existential generalization (the contrapositive of axiom X) on the abstraction Axiom 17.1 immediately yields the principle:

Theorem 17.3 (Comprehension). $(\exists y)(\forall x)(x \in y \leftrightarrow \Phi(x, u))$.

Under abstraction, the substitution rule is $x = y \vdash (\forall z)(x \in z \to y \in z)$.

Proposition 17.1. $y = \{z : \Phi(z)\} \leftrightarrow (\forall x)(x \in y \leftrightarrow \Phi(x))$.

Proof. By extensionality, $y = \{z : \Phi(z)\} \leftrightarrow (\forall x)(x \in y \leftrightarrow x \in \{z : \Phi(z)\})$. By abstraction, $(\forall x)(x \in \{z : \Phi(z)\} \leftrightarrow \Phi(x))$.Then by conjunctive syllogism, $(\forall x)(x \in y \leftrightarrow \Phi(x))$. For the converse, we again invoke the abstraction scheme, where $(\forall x)(\Phi(x) \leftrightarrow x \in \{z : \Phi(z)\})$, so by conjunctive syllogism $(\forall x)(x \in y \leftrightarrow x \in \{z : \Phi(z)\})$. And this with the extensionality axiom completes the proof.

Abstraction and extensionality can then be reconnected, as in Frege's axiom:

Theorem 17.4 (Basic Law V). $\{x : \Phi\} = \{x : \Psi\} \leftrightarrow (\forall x)(\Phi \leftrightarrow \Psi)$.

As Routley (1977) points out, Zermelo's axioms now follow instantly as theorems—unsurprisingly, since Zermelo explicitly picked out instances of comprehension. For example, *Aussonderung* is just a weaker comprehension scheme, $(\exists y)(\forall x)(x \in y \leftrightarrow x \in a \wedge \Phi(x))$, while union, intersection and pairing are all as usual; e.g. the last is obtained by abstraction on the condition $x = a \vee x = b$. Given a working theory of functions, Fraenkel's replacement axiom scheme is easily obtainable, too.[2] Of some more interest is a proof of the axiom of infinity, which is an artefact of full comprehension, Proposition 17.3 below.[3]

A universe and an empty set both exist. The *universe* is

$$V = \{x : (\exists y)(x \in y)\},$$

and as one would expect, both $(\forall x)(x \in V)$ and $(\forall x)(x \subseteq V)$ hold. The *empty set* is the complement of V,

$$\emptyset = \{x : (\forall y)(x \in y)\},$$

and both $(\forall x)(x \notin \emptyset)$ and $(\forall x)(\emptyset \subseteq x)$ hold, too. See Dunn (1988) (in Austin 1988) for study of the uniqueness of these sets. For now the main fact to know about the empty set is that it is explosive. For example, to show that the empty set is empty, we argue by cases. Either $x \notin \emptyset$ or $x \in \emptyset$. If the former, stop. So suppose that $x \in \emptyset$. Then $(\forall y)(x \in y)$; then $x \in \{z : z \notin \emptyset\}$ and therefore $x \notin \emptyset$. More generally, $(\forall y)(x \in y) \rightarrow x \in \{z : \Psi\}$ for any Ψ at all. So $x \in \emptyset \rightarrow \Psi$. This property of $\emptyset$ is very useful; see also Slaney (1989).

17.4 Identity

The properties of $\rightarrow$ make identity an equivalence relation,

$$x = x,$$

$$x = y \rightarrow y = x,$$

$$x = y \wedge y = z \rightarrow x = z.$$

With hypothetical syllogism, additionally, $x = y \rightarrow (y = z \rightarrow x = z)$.

---

[2]The first step in securing a set theoretic account of functions is defined ordered pairs and show them to behave according to the law $\langle a, b \rangle = \langle c, d \rangle \dashv\vdash a = c, b = d$. We will be assuming throughout that some approximation of standard mathematical functions is available.

[3]Without full comprehension, one can prove that the set of all sets is *Dedekind infinite* by producing an injection into itself, say by a map $x \mapsto \{x\}$, but, again, functions and cardinality arguments are mostly beyond our scope here.

By the counterexample axiom, $\rightarrow$ retains a connection to material implication, namely that if all Φs are Ψs, then everything is either not Φ or else Ψ. Contrapositively, if some Φs are not Ψs, then not all Φs are Ψs. This leads to the following surprising-and-intuitive result:

Proposition 17.2. *Sets that differ with respect to membership are not identical. In particular,* $(\exists x)(x \in a \wedge x \notin a) \vdash a \neq a$.

Proof. This is by rule III and the axiom of extensionality.

When a set a is such that its membership is inconsistent, some $b \in a$ and $b \notin a$, then a is *inconsistent*. And $(\exists x)(x \neq x)$, since by comprehension we have (at least) Russell's set,

$$R = \{x : x \notin x\}.$$

Excluding the middle, $R \in R \wedge R \notin R$. Since R differs from itself with respect to membership,

$$R \neq R.$$

Let us briefly expand on this theme, by seeing what happens when not only $=$ but parthood is tied to entailment,[4] as given by Definition 17.1.

For any a, we use the name $\mathscr{P}(a)$ for $\{x : x \subseteq a\}$.

The eccentricities of R enrich a result of Arruda and Batens (1982) from the set theory of da Costa (2000). Define by finite recursion (Theorem 17.7 below), $\mathscr{P}^0 = \mathscr{P}$ and $\mathscr{P}^{n+1} = \mathscr{P}\mathscr{P}^n$. Then

Theorem 17.5. $(\forall n)[\mathscr{P}^{n+1}(R) \subset \mathscr{P}^n(R)]$.

Proof. Arruda has found that

$$\ldots \mathscr{P}\mathscr{P}\mathscr{P}(R) \subseteq \mathscr{P}\mathscr{P}(R) \subseteq \mathscr{P}(R) \subseteq R.$$

To see that $\mathscr{P}(R) \subseteq R$, suppose $x \notin R$. Then $x \in x$. So $x \in x \wedge x \notin R$, meaning that $\exists y(y \in x \wedge y \notin R)$, so $x \not\subseteq R$. By contraposition, then, $x \subseteq R \rightarrow x \in R$, ergo $\mathscr{P}(R) \subseteq R$. Now suppose $x \in \mathscr{P}\mathscr{P}(R)$. Then $x \subseteq \mathscr{P}(R)$, so $x \subseteq R$ by transitivity. Therefore $x \in \mathscr{P}(R)$, and thus $\mathscr{P}\mathscr{P}(R) \subseteq \mathscr{P}(R)$.

To strengthen Arruda's finding, we employ contraposition at each arrow. For $\mathscr{P}(R) \subset R$, recall that $R \in R$ and $R \notin R$; this implies $R \not\subseteq R$. And $R \not\subseteq R \wedge R \in R$ gives $\mathscr{P}(R) \subset R$. For $\mathscr{P}\mathscr{P}(R) \subset \mathscr{P}R$, notice that $R \subseteq R$, but $R \in R \wedge R \not\subseteq R$; so by generalizing, $(\exists y)(y \subseteq R \wedge y \not\subseteq \mathscr{P}(R))$, as required. Again the argument may be continued:

$$\ldots \mathscr{P}\mathscr{P}\mathscr{P}(R) \subset \mathscr{P}\mathscr{P}(R) \subset \mathscr{P}(R) \subset R.$$

In general, then, this argument can be carried out for $\mathscr{P}^{n+1}(R)$ and $\mathscr{P}^n(R)$, which gives the full result by $\forall$-introduction.

[4]There is a debate about the right definition of subset—see (Mares, 2004, p. 198), and Beall et al. (2006), for instance using a more restricted implication.

That R 'implodes' in this way can be read as simple structure. With ordinal indices, one could go on to define by recursion (Theorem 17.7 below)

$$R_0 = R,$$
$$R_{\alpha+1} = \mathscr{P}(R_\alpha),$$
$$R_\lambda = \bigcup_{\kappa\in\lambda} R_\kappa,$$

where λ is a limit ordinal.

17.5 Full Comprehension

Since naive set theory formalizes the idea that all predicates determine sets, in the comprehension principle the occurrence of the set being defined in the defining predicate Φ is not ruled out. Following Routley, this is completely unrestricted or full comprehension. Priest and Routley write that

> The naive notion of set is that of the extension of an arbitrary predicate... This is as tight an account as can be expected from any fundamental notion. It was thought to be problematical only because it was assumed (under the ideology of consistency) that 'arbitrary' could not mean arbitrary. However, it does. (Priest et al. 1989, p. 499)

Set theory with a fully unrestricted comprehension principle is covered by the non-triviality proof in e.g. Brady (1989); Brady notes that Chang in 1965 had already noticed that a set theory with unrestricted comprehension can be consistent. In this section we look at some of the work a full comprehension principle can do—from supplying the concept of recursion to justifying a global choice principle.

When unrestricted, the abstraction axiom generates 'circular' or self-referring cases. These are neither necessarily inconsistent nor unique, e.g. cases like

$$x \in J \leftrightarrow x = J,$$
$$x \in K \leftrightarrow x = K,$$

mean that $J = \{J\}$ and $K = \{K\}$. But there is no way to say, absent further postulation, whether or not $J = K$. Compare this to other non-well-founded set theories, like Aczel's (discussed in Exercise 2.4 of Barwise and Moss (1996)). Aczel adds an axiom asserting, in effect, that $J = K$ in cases like these (since his anti-foundation axiom implies that all systems of equations have unique solutions). Here we allow the indeterminacy, in exchange for axiomatic simplicity.

To guarantee, meanwhile, that such instances are valid abstractions—to ensure that every predicate, even groundless ones, determines a set—we have abstraction instances of the form

$$x \in \{z : \Phi(z, u)\} \leftrightarrow \Phi\left[z/x,\, u/\{z : \Phi(z, u)\}\right]$$

where the right-hand-side indicates a simultaneous substitution in Φ of z by x, and u by the term $\{z : \Phi(z,u)\}$. (At first, in Brady and Routley (1989, p. 419), a new quantifier, formation rule, and reflection axiom were added to handle circular predicates; but by Brady (2006, p. 177), the idea is streamlined as above.) Axiom 17.1 in this way includes cases

$$x \in \{z : \Phi(z,u)\} \leftrightarrow \Phi(x, \{z : \Phi(z,u)\}).$$

To start, the axiom makes for some very direct expressions of natural phenomena. For example, (consistent) infinite descents have extreme expressions, like

$$x \in \Delta \leftrightarrow (\exists y)(y \in x \wedge y \in \Delta).$$

The simplest members of Δ could be a pair a, b such that $a \in b$ and $b \in a$. That there are such sets at our disposal might have application to models of inconsistent arithmetic (see Priest 2000), where circular periods occur in the successor relation, if the ordering $<$ on natural numbers is reduced to $\in$.

To take a simpler, and inconsistent, example, Routley identifies the limiting case of diagonal sets,

$$x \in \mathscr{Z} \leftrightarrow x \notin \mathscr{Z}$$

which is a kind of 'ultimate Russell set'. Non-self-identity $\mathscr{Z} \neq \mathscr{Z}$ is by Proposition 17.2, but actually something much stronger follows. By excluded middle, either $x \in \mathscr{Z}$ or not, for every x, from which it follows that $(\forall x)(x \in \mathscr{Z})$ and $(\forall x)(x \notin \mathscr{Z})$.

While this in some sense does make $\mathscr{Z}$ both universal and empty, we do not have $\mathscr{Z} = V$ or $\mathscr{Z} = \emptyset$, since identity is controlled by relevance. Because of relevance, $(\exists y)(x \in y)$ does not entail $x \in \mathscr{Z}$, so $V \subseteq \mathscr{Z}$ does not obtain and a fortiori neither does $\mathscr{Z} = V$. This is actually good news; the alternative is triviality (see Weber 2010a).

The universe is not the only set to have a highly inconsistent '$\mathscr{Z}$'-part. Any non-empty set a, for example, will have a subset $\mathscr{Z}(a) = \{x : x \in a \wedge x \notin \mathscr{Z}(a)\}$. Now, just as with unrestricted $\mathscr{Z}$, we have $(\forall x)(x \notin \mathscr{Z}(a))$. For $x \in a$, though, this is just the property needed to show $x \in \mathscr{Z}(a)$. So every member of a both is and is not a member of $\mathscr{Z}(a)$. This subset of a acts as a reflection of a over which inconsistency can be 'dialled up' as high as we like. Some points to note about this $\mathscr{Z}(a)$ phenomenon:

- Full comprehension is not required to give this result. Instead of $\mathscr{Z}$, just take $\{x : x \in a \wedge R \in R\}$, for R the Russell set. The same arguments go through. This is the inconsistent aspect of the dopplegänger phenomenon (see Weber 2010a).
- While $\mathscr{Z}(a)$ is inconsistent for non-empty a, this does not prove that a is inconsistent. By the $\rightarrow$-logic of parthood, a set can have inconsistent parts and yet be perfectly consistent as a whole. The universe V is only the biggest example.

- There are consequences here for cardinality. For example, one can provide a proof of Cantor's theorem, of the form $|a| < |\mathscr{P}a|$, essentially by appealing to $\mathscr{Z}(a) \in \mathscr{P}a$. In a sense, this is good news, as it confirms an important theorem. On the other hand, consider singletons:

$$\{a\} = \{x : x = a\}$$
$$\mathscr{Z}(\{a\}) = \{x : x = a \wedge x \notin \mathscr{Z}(\{a\})\}$$

It is simple to check that $\mathscr{Z}(\{a\}) \subseteq \{a\}$. In fact, though, by the argument for Theorem 17.5, it is almost as straightforward that $\mathscr{Z}(\{a\}) \subset \{a\}$, a *proper* subset. Now, if a set X is *Dedekind infinite* when there is an injection from X to a proper subset of X, then we just proved that $\{a\}$ is Dedekind infinite for any set a. This strongly suggests that a finer grained notion of cardinality is required than in the classical definitions of infinity.

On this note, we derive a classical axiom of infinity.[5]

Proposition 17.3 (Infinity). *There is a non-empty set i isomorphic to an ω-sequence of Zermelo ordinals,*

$$i = \{i, \{i\}, \{\{i\}\}, \{\{\{i\}\}\}, \ldots\}$$

Proof. Consider $i = \{x : x = i\}$. Since $i \in i$, the set is not empty. Since $i = \{i\}$, by substitution, $\{i\} \in i$.

For the development of Peano arithmetic, we could then define the natural numbers as

$$\omega = \{x : [x = \emptyset \vee (\exists y)(\{y\} = x)] \wedge x \subseteq \omega\}$$

using full comprehension to ensure that numbers are preceded only by other numbers.

We turn then to the theory of ordinal numbers, which includes the natural numbers. In standard set theory, ordinals are understood as the set of all preceding ordinals, ordered by membership. This is plainly recursive and can be captured in a definition: An ordinal is a transitive, well-ordered set of ordinals.

Let $Wo(x)$ mean that x is well-ordered—that there is a linear $\in$-order on x where also every non-empty subset of x has a least member. Let $Conn(x)$ mean that for every $y \in On$, either $x \subseteq y$ or else $y \subseteq x$. The formalism of this is not a concern now; the rendering of (Brady, 2006, p. 310), could do. The matter at hand is full comprehension.

[5]Compare this to Petersen's characterization of the natural numbers, (Petersen, 2000, p. 386).

Proposition 17.4. *There is a set On such that*

$$\begin{aligned} x \in On \leftrightarrow \quad & x \subseteq On \\ & \wedge\ Conn(x) \\ & \wedge\ y \in x \rightarrow y \subseteq x \\ & \wedge\ Wo(x). \end{aligned}$$

For short, $On = \{x : x \textit{ is an ordinal}\}$.

Notice immediately that On is transitive. Since $\alpha \in On \rightarrow \alpha \subseteq On$, the set of all ordinals satisfies one of the key conditions for being an ordinal. With this definition, one can work from $\emptyset \in On$ up to Burali-Forti's paradox that $On \in On$.

Theorem 17.6. *[Burali-Forti 1897]* $On \in On$.

Proof. On is a transitive, well-ordered set of connected ordinals (see Weber 2010b). Checking the definition of 'ordinal' gives the result.

Because $\in$ is irreflexive on ordinals, and because of what we know about identity from the last section (Proposition 17.2), we have some contradictions:

Corollary 17.1. $On \notin On$, *and then* $On \neq On$.

Full comprehension is well suited to modelling recursive processes, as we have been seeing. We would like a transfinite recursion theorem. Barwise and Moss (1996) use the nice example of a function $g : \omega \longrightarrow \omega \times \omega$ defined as $g(n) = \langle n, g(n+1)\rangle$, which delivers a sequence $g(0) = \langle 0, \langle 1, \langle 2, \langle \ldots \rangle\rangle\rangle\rangle$, and with full comprehension, it is easy enough to prove that something like g exists, namely

$$\langle x, y\rangle \in g \leftrightarrow x \in \omega \wedge y = \langle x, g(x+1)\rangle.$$

There is no guarantee that this g is a function, though. Instead, a general form of recursion on the ordinals (and ipso facto the natural numbers) is captured in the next proof. Let $f|x$ be the restriction of f to x, defined as $\{\langle u, v\rangle \in f : u \in x\}$.

Theorem 17.7 (Transfinite Recursion). *Let h be a function from V to V. There is a function f from On to V such that*

$$f(\alpha) = h(f|\alpha).$$

Proof. The set $\langle x, y\rangle \in f \leftrightarrow y = h(f|x)$ exists, and is a function because h is.

Full comprehension, then, is very powerful. It is time to consider one of its most arresting, and earliest, applications, to the axiom of choice.

Routley (1977) produced an argument for the axiom of global choice from full comprehension. He did this by defining a function to be either univocal or empty, since classically an empty set is a function by dint of material implication. The instance of comprehension

$$x \in f \leftrightarrow \exists u \exists v(u \in X \wedge x = \langle u, v\rangle \wedge v \in u) \wedge f \textit{ is a function}.$$

then allows the following proof: Either f is empty or not. Either way, f is a function, because if it is non empty then f is a function by the definition of f, while if it is empty then f is a function by definition of function. So there is a choice function on any X—including the universe, V. This is the axiom of *global choice*.

There is something unsatisfactory about the argument. Full comprehension is not even required here, since a 'function' like

$$\{\langle u, v\rangle : R \in R\},$$

with R the Russell set, supports the same reasoning.[6] (Since $R \notin R$, the set has no members, and so satisfies Routley's criteria to be a function.) But this does not appear to be a function in any mathematical sense, since every ordered pair whatsoever is a member.

Later Routley (Priest et al. 1989, p. 374) reprised the attempt with the comprehension instance

$$x \in f \leftrightarrow \exists u \exists v(u \in X \wedge x = \langle u, v\rangle \wedge v \in u)$$
$$\wedge \forall u \forall y \forall z(\langle u, y\rangle \in f \wedge \langle u, z\rangle \in f \rightarrow y = z),$$

again looking to say that f is a function on X. But the argument is really just a version of Curry's paradox, and is blocked by the failure of contraction, since to show that f is a function leads us to consider $\langle u, v\rangle \in f \rightarrow \forall z(\langle u, v\rangle \in f \wedge \langle u, z\rangle \in f \rightarrow v = z)$. If f is non-empty, then it is a function, but there is no telling whether or not f is empty; we first need to know whether or not it is a function. So this second formulation is a contraction away from choice, but also from proving anything at all.

In a sense, Routley is trying to use a paradox to make choice true. The first attempt uses a paradox of material implication (that when f is empty, $\langle x, y\rangle \in f$ materially implies that y is unique). That idea can be presented in terms of Russell's paradox, or it can be rephrased in terms of the implicational form of Russell's antinomy, Curry's paradox. But none of these are making meaningful use of full comprehension per se, and more seriously, none of these give us reason to think that the axiom of choice is true.

[6] Conrad Asmus pointed this out.

On the other hand, defining the ordinals self-referentially by full comprehension leads to Burali-Forti's paradox, and, as I now outline, this paradox not only delivers an equivalent theorem, but gives us a good mathematical reason to think that what we have proved is true. We derive Cantor's well-ordering principle, by giving an easy way for the universe to be injected into a particular subset of On. Suppose we say that a function $f : a \longrightarrow b$ is *injective*, or one-one, iff $(\forall x)(\forall y)\neg(x \neq y \wedge f(x) = f(y))$.

Theorem 17.8. *The universe can be well-ordered.*

Proof. An injection $f : V \longrightarrow On$ is required. Consider the constant function $f(x) = On$. The range of f is a segment of the ordinals. Because $On \neq On$, we have that $(\forall x)(\forall y)(x = y \vee On \neq On)$, so $(\forall x)(\forall y)(x = y \vee f(x) \neq f(y))$. Therefore f is an injection. Thus

$$\{x_{f(x)} : f(x) \in On\}$$

is a well-order on V.

17.6 Conclusion

Whether a useful choice principle really obtains, and so whether this line fares better than Routley's arguments, remains to be seen. Indeed, most of an elementarily paraconsistent set theory—elementary in the sense that no appeal is made to classical results—remains to be seen. From the point of view of inconsistent mathematics, I only hope to have suggested there is a great deal of the universe of sets still waiting to be explored. Drawing again on one venerable tradition in paraconsistent set theory, I join da Costa in his structural initiative:

> It would be as interesting to study the inconsistent systems as, for instance, the non-Euclidian geometries: we would obtain a better idea of the nature of certain paradoxes, could have a better insight on the connections amongst the various logical principles necessary to obtain determinate results, etc. (da Costa 1974, p. 498)

And drawing again on another tradition, Routley claimed more. In a programmatic polemic, Routley (1977) hypothesized that standard mathematics, beginning with set theory, can be recaptured using a suitable "ultramodal" logic. Later reprinted in his magnum opus, he writes

> There are whole mathematical cities that have been closed off and partially abandoned because of the outbreak of isolated contradictions. They have become like modern restorations of ancient cities, mostly just patched up ruins visited by tourists.
> In order to sustain the ultramodal challenge to classical logic it will have to be shown that even though leading features of classical logic and theories have been rejected, ... by going ultramodal one does not lose great chunks of the modern mathematical megalopolis. ... The strong ultramodal claim—not so far vindicated—is the expectedly brash one: we can do everything you can do, only better, and we can do more. (Routley 1980, p. 927)

References

Arruda, A.I., and D. Batens. 1982. Russell's set and the universal set in paraconsistent set theory. *Logique et Analyse* 25: 121–133.

Austin, D.F. (ed.). 1988. *Philosophical analysis*. Dordrecht: Holland

Barwise, J., and L. Moss. 1996. *Vicious circles*. Stanford: CSLI Publications.

Batens, D., C. Mortensen, G. Priest, and J.-P. van Bendegem. eds. 2000. *Frontiers of paraconsistent logic*. Philadelphia: Research Studies Press.

Beall, J.C., R.T. Brady, A.P. Hazen, G. Priest, and G. Restall. 2006. Relevant restricted quantification. *Journal of Philosophical Logic* 35: 587–598.

Brady, R.T. 1989. The non-triviality of dialectical set theory. In *Paraconsistent logic: Essays on the inconsistent*, ed. G. Priest, R. Routley, and J. Norman, 437–470. Munich: Philosophia Verlag.

Brady, R.T. 2006. *Universal logic*. Stanford: CSLI.

Brady, R.T., and R. Routley. 1989. The non-triviality of extensional dialectical set theory. In *Paraconsistent logic: Essays on the inconsistent*, ed. G. Priest, R. Routley, and J. Norman, 415–436. Munich: Philosophia Verlag.

Curry, H.B. 1942. The inconsistency of certain formal logics. *Journal of Symbolic Logic* 7: 115–117.

da Costa, N.C.A. 1974. On the theory of inconsistent formal systems. *Notre Dame Journal of Formal Logic* 15: 497–510.

da Costa, N.C.A. 2000. Paraconsistent mathematics. In *Frontiers of paraconsistent logic*, ed. D. Batens, C. Mortensen, G. Priest, and J.-P. van Bendegem, 165–180. Philadelphia: Research Studies Press.

Dunn, J.M. 1988. The impossibility of certain higher-order non-classical logics with extensionality. In *Philosophical analysis*, ed. D.F. Austin, 261–280. Dordrecht: Holland

Lewis, C.I., and C.H. Langford. 1959. *Symbolic logic*. New York: Dover.

Libert, T. 2005. Models for paraconsistent set theory. *Journal of Applied Logic* 3: 15–41.

Mares, E. 2004. *Relevant logic*. Cambridge: Cambridge University Press.

Meyer, R.K., R. Routley, and J.M. Dunn. 1978. Curry's paradox. *Analysis* 39: 124–128.

Petersen, U. 2000. Logic without contraction as based on inclusion and unrestricted abstraction. *Studia Logica* 64: 365–403.

Priest, G. 2000. Inconsistent models of arithmetic, ii: The general case. *Journal of Symbolic Logic* 65: 1519–29.

Priest, G., R. Routley, and J. Norman (eds.). 1989. *Paraconsistent logic: Essays on the inconsistent*. Munich: Philosophia Verlag.

Restall, G. 1994. *On logics without contraction*. Ph.D. thesis. The University of Queensland.

Routley, R. 1977. Ultralogic as universal? *Relevance Logic Newsletter* 2: 51–89. Reprinted in Routley (1980).

Routley, R. 1980. *Exploring Meinong's jungle and beyond*. Departmental Monograph, Philosophy Department, RSSS, Australian National University, vol. 3. Canberra: RSSS, Australian National University, Canberra.

Routley, R., and R.K. Meyer. 1976. Dialectical logic, classical logic and the consistency of the world. *Studies in Soviet Thought* 16: 1–25.

Slaney, J.K. 1989. Rwx is not curry-paraconsistent. In *Paraconsistent logic: Essays on the inconsistent*, ed. G. Priest, R. Routley, and J. Norman, 472–480. Munich: Philosophia Verlag.

Weber, Z. 2010a. Extensionality and restriction in naive set theory. *Studia Logica* 94(1): 87–104.

Weber, Z. 2010b. Transfinite numbers in paraconsistent set theory. *Review of Symbolic Logic* 3(1): 71–92.

Chapter 18
Sorting out the Sorites

David Ripley

18.1 Introduction

Supervaluational theories of vagueness have achieved considerable popularity in the past decades, as seen in e.g., Fine (1975) and Lewis (1970). This popularity is only natural; supervaluations let us retain much of the power and simplicity of classical logic, while avoiding the commitment to strict bivalence that strikes many as implausible.

Like many non-classical logics, the supervaluationist system SP has a natural dual, the subvaluationist system SB, explored in e.g., Hyde (1997) and Varzi (2000).[1] As is usual for such dual systems, the classical features of SP (typically viewed as benefits) appear in SB in 'mirror-image' form, and the non-classical features of SP (typically viewed as costs) also appear in SB in 'mirror-image' form. Given this circumstance, it can be difficult to decide which of the two dual systems is better suited for an approach to vagueness.[2]

The present paper starts from a consideration of these two approaches—the supervaluational and the subvaluational—and argues that neither of them is well-positioned to give a sensible logic for vague language. The first section presents the

[1]Although there are many different ways of presenting a supervaluational system, I'll ignore these distinctions here; my remarks should be general enough to apply to them all, or at least all that adopt the so-called 'global' account of consequence. (For discussion, see Varzi 2007.) Similarly for subvaluational systems.

[2]The situation is similar for approaches to the Liar paradox; for discussion, see e.g., Beall and Ripley (2004) and Parsons (1984).

D. Ripley (✉)
Department of Philosophy, University of Connecticut, Storrs, CT, USA

School of Historical and Philosophical Studies, The University of Melbourne, Parkville, VIC, Australia
e-mail: davewripley@gmail.com

K. Tanaka et al. (eds.), *Paraconsistency: Logic and Applications*, Logic, Epistemology, and the Unity of Science 26, DOI 10.1007/978-94-007-4438-7_18,

systems SP and SB and argues against their usefulness. Even if we suppose that the general picture of vague language they are often taken to embody is accurate, we ought not arrive at systems like SP and SB. Instead, such a picture should lead us to truth-functional systems like strong Kleene logic (K_3) or its dual LP. The second section presents these systems, and argues that supervaluationist and subvaluationist understandings of language are better captured there; in particular, that a dialetheic approach to vagueness based on the logic LP is a more sensible approach. The next section goes on to consider the phenomenon of higher-order vagueness within an LP-based approach, and the last section closes with a consideration of the sorites argument itself.

18.2 S'valuations

Subvaluationists and supervaluationists offer identical pictures about how vague language works; they differ solely in their theory of truth. Because their overall theories are so similar, this section will often ignore the distinction between the two; when that's happening, I'll refer to them all as s'valuationists. First I present the shared portion of the s'valuationist view, then go on to lay out the difference between subvaluational and supervaluational theories of truth, and offer some criticism of both the subvaluationist and supervaluationist approaches.

18.2.1 *The Shared Picture*

It's difficult to suppose that there really is a single last noonish second, or a single oldest child, & c.[3] Nonetheless, a classical picture of negation seems to commit us to just that. After all, for each second, the law of excluded middle, $A \vee \neg A$, tells us it's either noonish or not noonish, and the law of non-contradiction, $\neg(A \wedge \neg A)$, tells us it's not both. Now, let's start at noon (since noon is clearly noonish) and move forward second by second. For a time, the seconds are all noonish, but the classical picture seems to commit us to there being a second—just one second!—that tips the scale over to non-noonishness.

Many have found it implausible to think that our language is pinned down that precisely, and some of those who have found this implausible (e.g., Fine 1975) have found refuge in an s'valuational picture. The key idea is this: we keep that sharp borderline (classicality, as noted above, seems to require it), but we allow that there are many different places it might be. The s'valuationists then take vagueness to be something like ambiguity: there are many precise extensions that a vague

[3]It's not un-supposable, though; see e.g., Sorensen (2001) and Williamson (1994) for able defences of such a position.

predicate might have, and (in some sense) it has all of them.[4] It's important that this 'might' not be interpreted epistemically; the idea is not that one of these extensions is the one, and we just don't know which one it is. Rather, the idea is that each potential extension is part of the meaning of the vague predicate. Call these potential extensions 'admissible precisifications'.

The phenomenon of vagueness involves a three-part structure *somehow*; on this just about all theorists are agreed. Examples: for a vague predicate F, epistemicists (e.g., Williamson 1994) consider (1) things that are known to be F, (2) things that are known not to be F, and (3) things not known either to be F or not to be F; while standard fuzzy theorists (e.g., Machina 1972; Smith 2008) consider (1) things that are absolutely F, or F to degree 1, (2) things that are absolutely not F, or F to degree 0, and (3) things that are neither absolutely F nor absolutely not F. S'valuationists also acknowledge this three-part structure: they talk of (1) things that are F on every admissible precisification, (2) things that are not-F on every admissible precisification, and (3) things that are F on some admissible precisifications and not-F on others.

Consider 'noonish'. It has many different precisifications, but there are some precisifications that are admissible and others that aren't. (37a)–(38b) give four sample precisifications; (37a) and (37b) are admissible precisifications for 'noonish', but (38a) and (38b) aren't:[5]

(37) a. $\{x : x$ is between 11:40 and 12:20$\}$
b. $\{x : x$ is between 11:45 and 12:30$\}$

(38) a. $\{x : x$ is between 4:00 and 4:30$\}$
b. $\{x : x$ is between 11:40 and 11:44, or x is between 12:06 and 12:10, or x is between 12:17 and 12:22$\}$

There are at least a couple ways, then, for a precisification to go wrong, to be inadmissible. Like (38a), it might simply be too far from where the vague range is; or like (38b), it might fail to respect what are called *penumbral connections*. 12:22 might not be in every admissible precisification of 'noonish', but if it's in a

[4]NB: S'valuationists differ in the extent to which they take vagueness to be like ambiguity, but they all take it to be like ambiguity in at least this minimal sense. Smith (2008) draws a helpful distinction between supervaluationism and what Smith calls 'plurivaluationism'. Although both of these views have both travelled under the name 'supervaluationism', they are distinct. Supervaluationism makes use of non-bivalent semantic machinery (for example, the machinery in Fine 1975), while plurivaluationism makes do with purely classical models, insisting merely that more than one of these models is the intended one. My discussion of supervaluationism here is restricted to the view Smith calls supervaluationism.

[5]At least in most normal contexts. Vague predicates seem particularly context-sensitive, although they are not the only predicates that have been claimed to be (see e.g., Recanati 2004; Wilson and Sperber 2002). For the purposes of this paper, I'll assume a single fixed (non-wacky) context; these are theories about what happens *within* that context. Some philosophers (e.g., Raffman 1994) have held that taking proper account of context is itself sufficient to dissolve the problems around vagueness. I disagree, but won't address the issue here.

certain admissible precisification, then 12:13 ought to be in that precisification too. After all, 12:13 is more noonish than 12:22 is. Since this is a connection within the penumbra of a single vague predicate, we can follow Fine (1975) in calling it an *internal* penumbral connection.

Admissible precisifications also must respect *external* penumbral connections. The key idea here is that the extensions of vague predicates sometimes depend on each other. Consider the borderline between green and blue. It's sometimes claimed that something's being green rules out its also being blue. If this is so, then no admissible precisification will count a thing as both green and blue, even if some admissible precisifications count it as green and others count it as blue. In order to handle this phenomenon in general, it's crucial that we count precisifications not as precisifying one predicate at a time, but instead as precisifying multiple predicates simultaneously. That way, they can give us the requisite sensitivity to penumbral connections.

18.2.1.1 S'valuational Models

Now that we've got the core of the idea down, let's see how it can be formally modeled.[6] An SV model M is a tuple $\langle D, I, P \rangle$ such that D, the domain, is a set of objects; I is a function from terms in the language to members of D; and P is a set of precisifications: functions from predicates to subsets of D.[7]

We then extend each precisification $p \in P$ to a valuation of the full language. For an atomic sentence Fa, a precisification $p \in P$ assigns Fa the value 1 if $I(a) \in p(F)$, 0 otherwise. This valuation of the atomics is extended to a valuation of the full language (in $\wedge, \vee, \neg, \forall, \exists$) in the familiar classical way. In other words, each precisification is a full classical valuation of the language, using truth values from the set $V_0 = \{1, 0\}$.

Then the model M assigns a value to a sentence simply by collecting into a set the values assigned to the sentence by M's precisifications: $M(A) = \{v \in V_0 : \exists p \in P(p(A) = v)\}$. Thus, models assign values to sentences from the set $V_1 = \{\{1\}, \{0\}, \{1, 0\}\}$. Note that so far, these values are uninterpreted; they work merely as a record of a sentence's values across precisifications. $M(A) = \{1\}$ iff A gets value 1 on every precisification in M; $M(A) = \{0\}$ iff A gets value 0 on every precisification in M; and $M(A) = \{1, 0\}$ iff A gets value 1 on some precisifications in M and 0 on others.

---

[6]There are many ways to build s'valuational models. In particular, one might not want to have to *fully* precisify the language in order to assign truth-values to just a few sentences. Nonetheless, the approach to be presented here will display the logical behaviour of s'valuational approaches, and it's pretty simple to boot. So we can get the picture from this simple approach.

[7]And from propositional variables directly to classical truth-values, if one wants bare propositional variables in the language. Vague propositional variables can be accommodated in this way as well as precise ones.

18.2.2 *Differences in Interpretation and Consequence*

The s'valuationists agree on the interpretation of two of these values: {1} and {0}. If $M(A) = \{1\}$, then A is true on M. If $M(A) = \{0\}$, then A is false on M. But the subvaluationist and the supervaluationist differ over the interpretation of the third value: {1,0}.

The supervaluationist (of the sort I'm interested in here) claims that a sentence must be true on *every* precisification in order to be true *simpliciter*, and false on *every* precisification in order to be false *simpliciter*. Since the value {1,0} records a sentence's taking value 1 on some precisifications and 0 on others, when $M(A) = \{1, 0\}$, A is neither true nor false on M for the supervaluationist.

The subvaluationist, on the other hand, claims that a sentence has only to be true on *some* precisification to be true *simpliciter*, and false on *some* precisification in order to be false *simpliciter*. So, when $M(A) = \{1, 0\}$, the subvaluationist says that A is both true and false on M.

Both define consequence in the usual way (via truth-preservation):

(39) $\Gamma \vDash \Delta$ iff, for every model M, either δ is true on M for some $\delta \in \Delta$, or γ fails to be true on M for some $\gamma \in \Gamma$.[8]

Since the subvaluationist and the supervaluationist differ over which sentences are true on a given model (at least if the model assigns {1,0} anywhere), this one definition results in two different consequence relations; call them $\vDash_{SB}$ (for the subvaluationist) and $\vDash_{SP}$ (for the supervaluationist).

18.2.2.1 $\vDash_{SB}$ and $\vDash_{SP}$

One striking (and much-advertised) feature of these consequence relations is their considerable classicality. For example (where $\vDash_{CL}$ is the classical consequence relation):

(40) $\vDash_{CL} A$ iff $\vDash_{SB} A$ iff $\vDash_{SP} A$

(41) $A \vDash_{CL}$ iff $A \vDash_{SB}$ iff $A \vDash_{SP}$

(40) tells us that these three logics have all the same logical truths, and (41) tells us that they have all the same explosive sentences.[9] What's more, for any classically

[8] Note that this is a multiple-conclusion consequence relation. One can recover a single-conclusion consequence relation from this if one is so inclined, but for present purposes the symmetrical treatment will be more revealing. See e.g., Restall (2005) for details, or Hyde (1997) for application to s'valuations. See also Keefe (2000) for arguments against using multiple-conclusion consequence, and Hyde (2010) for response.

[9] *Explosive* sentences are sentences from which one can derive any conclusions at all, just as logical truths are sentences that can be derived from any premises at all. It's a bit sticky calling them 'logical falsehoods', as may be tempting, since some sentences (in SB at least) can be false without

valid argument, there is a corresponding SB-valid or SP-valid argument:

(42) $A_1, \ldots, A_i \vDash_{CL} B_1, \ldots, B_j$ iff $A_1 \wedge \ldots \wedge A_i \vDash_{SB} B_1, \ldots, B_j$

(43) $A_1, \ldots, A_i \vDash_{CL} B_1, \ldots, B_j$ iff $A_1, \ldots, A_i \vDash_{SP} B_1 \vee \ldots \vee B_j$

Let's reason through these a bit.[10] Suppose $A_1, \ldots, A_i \vDash_{CL} B_1, \ldots, B_j$. Then, since every precisification is classical, every precisification p (in every model) that verifies all the As will also verify one of the Bs. Consider the same argument subvaluationally; one might have all the premises true in some model (because each is true on some precisification or other), without having all the premises true in the *same* precisification; thus, there's no guarantee that any of the Bs will be true on any precisification at all. On the other hand, if one simply conjoins all the premises into one big premise, then if it's true in a model at all it guarantees the truth of all the As on (at least) a single precisification, and so one of the Bs must be true on that precisification, hence true in the model.

Similar reasoning applies in the supervaluational case. If all the As are true in a model, then they're all true on every precisification; nonetheless it might be that none of the Bs is true on every precisification; all the classical validity guarantees is that each precisification has some B or other true on it. But when we disjoin all the Bs into one big conclusion, that disjunction must be true on every precisification, so the argument is SP-valid.

Note that (42) and (43) guarantee that $\vDash_{SB}$ matches $\vDash_{CL}$ on single-premise arguments, and that $\vDash_{SP}$ matches $\vDash_{CL}$ on single-conclusion arguments. It is apparent that there is a close relationship between classical logic, subvaluational logic, and supervaluational logic. What's more, for every difference between SB and CL, there is a dual difference between SP and CL, and vice versa. This duality continues as we turn to the logical behaviour of the connectives:

(44) a. $A, B \nvDash_{SB} A \wedge B$
b. $A, \neg A \nvDash_{SB} A \wedge \neg A$

(45) a. $A \vee B \nvDash_{SP} A, B$
b. $A \vee \neg A \nvDash_{SP} A, \neg A$

(44a) and (44b) are dual to (45a) and (45b). It is often remarked about supervaluations that it's odd to have a disjunction be true when neither of its disjuncts is, but this oddity can't be expressed via $\vDash_{SP}$ in a single-conclusion format.[11] Here, in

---

failing to be true. And I want to shy away from 'contradiction' here too, since I understand by that a sentence of the form $A \wedge \neg A$, and such a sentence will be explosive here but not in the eventual target system.

[10]Here I prove only the LTR directions, but both directions indeed hold; see Hyde (1997) for details.

[11]This, essentially, is Tappenden's 'objection from upper-case letters' (Tappenden 1993). With multiple conclusions, there's no need for upper-case letters; the point can be made in any typeface you like.

a multiple-conclusion format, it becomes apparent that this oddity is an oddity in the supervaluational consequence relation, not just in its semantics. And of course, there is a parallel oddity involving conjunction for the subvaluationist.

Disjunction and conjunction can be seen as underwriting existential and universal quantification, respectively, so it is no surprise that the oddities continue when it comes to quantification. A sample:

(46) a. $Fa, Fb, \forall x(x = a \vee x = b) \not\models_{SB} \forall x(Fx)$

(47) a. $\exists x(Fx), \forall x(x = a \vee x = b) \not\models_{SP} Fa, Fb$

The cause is as before: in the SB case there is no guarantee that the premises are true on the same precisification, so they cannot interact to generate the conclusion; while in the SP case there is no guarantee that the same one of the conclusions is true on every precisification, so it may be that neither is true *simpliciter*. In the supervaluational case, consequences for quantification have often been noted;[12] but of course they have their duals for the subvaluationist.

18.2.2.2 What to Say About Borderline Cases?

This formal picture gives rise to certain commitments about borderline cases. (I assume here, and throughout, that every theorist is committed to all and only those sentences they take to be true.) Assume that 12:23 is a borderline case of 'noonish'. The subvaluationist and the supervaluationist agree in their acceptance of (48a)–(48b), and in their rejection of (49a)–(49b):

(48) a. 12:23 is either noonish or not noonish.
b. It's not the case that 12:23 is both noonish and not noonish.

(49) a. It's not the case that 12:23 is either noonish or not noonish.
b. 12:23 is both noonish and not noonish.

On the other hand, they disagree about such sentences as (50a)–(50b):

(50) a. 12:23 is noonish.
b. 12:23 is not noonish.

The subvaluationist accepts both of these sentences, despite her rejection of their conjunction, (49b). On the other hand, the supervaluationist rejects them both, despite her acceptance of their disjunction, (48a). So the odd behaviour of conjunction and disjunction observed above isn't simply a theoretical possibility; these connectives misbehave every time there's a borderline case of any vague predicate.

[12]For example, consider the sentence 'There is a last noonish second'. It is true for the supervaluationist, but there is no second x such that 'x is the last noonish second' is true for the supervaluationist.

A major challenge for either the subvaluationist or the supervaluationist is to justify their deviant consequence relations, especially their behaviour around conjunction and universal quantification (for the subvaluationist) or disjunction and existential quantification (for the supervaluationist). At least prima facie, one would think that a conjunction is true iff both its conjuncts are, or that a disjunction is true iff one disjunct is, but the s'valuationists must claim that these appearances are deceiving.

The trouble is generated by the lack of truth-functional conjunction and disjunction in these frameworks. Consider the subvaluational case. If A is true, and B is true, we'd like to be able to say that $A \wedge B$ is true. In some cases we can, but in other cases we can't. The value of $A \wedge B$ depends upon more than just the value of A and the value of B; it also matters how those values are related to each other precisification to precisification. It's this extra dependence that allows s'valuational approaches to capture 'penumbral connections', as argued for in Fine (1975). Unfortunately, it gets in the way of sensible conjunctions and disjunctions.

18.3 LP and K_3

This trouble can be fixed as follows: we keep the s'valuationist picture for atomic sentences, but then use familiar truth-functional machinery to assign values to complex sentences. This will help us retain more familiar connectives, and allow us to compute the values of conjunctions and disjunctions without worrying about which particular conjuncts or disjuncts we use.

18.3.1 The Shared Picture

The informal picture, then, is as follows: to evaluate atomic sentences, we consider all the ways in which the vague predicates within them can be precisified. For compound sentences, we simply combine the values of atomic sentences in some sensible way. But what sensible way? Remember, we're going to end up with three possible values for our atomic sentences—{1}, {0}, and {1,0}—so we need sensible three-valued operations to interpret our connectives. Here are some minimal desiderata for conjunction, disjunction, and negation:

(51) Conjunction:

a. $A \wedge B$ is true iff both A and B are true.

b. $A \wedge B$ is false iff either A is false or B is false.

(52) Disjunction:

a. $A \vee B$ is true iff either A is true or B is true.

b. $A \vee B$ is false iff both A and B are false.

(53) Negation:

a. $\neg A$ is true iff A is false.

b. $\neg A$ is false iff A is true.

These desiderata alone rule out the s'valuationist options: as is pointed out in Varzi (2000), SB violates the RTL directions of (51a) and (52b), while SP violates the LTR directions of (51b) and (52a).

As before, we'll have two options for interpreting {1,0}: we can take it to be both true and false, like the subvaluationist, or we can take it to be neither true nor false, like the supervaluationist. Since the above desiderata are phrased in terms of truth and falsity, it might seem that we need to settle this question before we find appropriate operations to interpret our connectives. It turns out, however, that the same set of operations on values will satisfy the above desiderata whichever way we interpret {1,0}.

18.3.1.1 LP/K_3

These are the operations from either strong Kleene logic (which I'll call K_3; see e.g., Kripke 1975) or Priest's Logic of Paradox (which I'll call LP; see e.g., Priest 1979). Consider the following lattice of truth values:

$$\begin{array}{c} \{1\} \\ | \\ \{1,0\} \\ | \\ \{0\} \end{array}$$

Take $\wedge$ to be greatest lower bound, $\vee$ to be least upper bound, and $\neg$ to reverse order (it takes {1} to {0}, {0} to {1}, and {1,0} to itself). Note that these operations satisfy (51a)–(53b); this is so whether {1,0} is interpreted as both true and false or as neither true nor false. For example, consider (52a). Suppose we interpret {1,0} LP-style, as both true and false. Then a disjunction is true (has value {1} or {1,0}) iff one of its disjuncts is: RTL holds because disjunction is an *upper* bound, and LTR holds because disjunction is *least* upper bound. On the other hand, suppose we interpret {1,0} K_3-style, as neither true nor false. Then a disjunction is true (has value {1}) iff one of its disjuncts is: again, RTL because disjunction is an *upper* bound and LTR because it's *least* upper bound. Similar reasoning establishes all of (51a)–(53b). So the LP/K_3 connectives meet our desiderata.

18.3.1.2 Differences in Interpretation and Consequence

There are still, then, two approaches being considered. One, the K_3 approach, interprets sentences that take the value $\{1,0\}$ on a model as neither true nor false on that model. The other, the LP approach, interprets these sentences as both true and false on that model. This section will explore the consequences of such a difference.

As before, consequence for both approaches is defined as in (39) (repeated here as (54)):

(54) $\Gamma \vDash \Delta$ iff, for every model M, either δ is true on M for some $\delta \in \Delta$, or γ fails to be true on M for some $\gamma \in \Gamma$.

And as before, differences in interpretation of the value $\{1,0\}$ result in differences about 'true', and so different consequence relations (written here as $\vDash_{K_3}$ and $\vDash_{\mathrm{LP}}$).

First, we should ensure that the connectives behave appropriately, as indeed they do, in both K_3 and LP:

(55) a. $A, B \vDash_{\mathrm{LP}} A \wedge B$
b. $A \vee B \vDash_{\mathrm{LP}} A, B$

(56) a. $A, B \vDash_{K_3} A \wedge B$
b. $A \vee B \vDash_{K_3} A, B$

As you'd expect given this, so do universal and existential quantification:

(57) a. $Fa, Fb, \forall x(x = a \vee x = b) \vDash_{\mathrm{LP}} \forall x(Fx)$
b. $\exists x(Fx), \forall x(x = a \vee x = b) \vDash_{\mathrm{LP}} Fa, Fb$

(58) a. $Fa, Fb, \forall x(x = a \vee x = b) \vDash_{K_3} \forall x(Fx)$
b. $\exists x(Fx), \forall x(x = a \vee x = b) \vDash_{K_3} Fa, Fb$

Both consequence relations have other affinities with classical consequence, although neither is fully classical:

(59) a. $\vDash_{\mathrm{CL}} A$ iff $\vDash_{\mathrm{LP}} A$
b. $A \vDash_{\mathrm{CL}}$ iff $A \vDash_{K_3}$

(60) a. $A, \neg A \nvDash_{\mathrm{LP}} B$
b. $\neg A, A \vee B \nvDash_{\mathrm{LP}} B$

(61) a. $A \nvDash_{K_3} B, \neg B$
b. $A \nvDash_{K_3} A \wedge B, \neg B$

(59a) tells us that LP and classical logic have all the same logical truths, while (59b) tells us that K_3 and classical logic have all the same explosive sentences. (60) shows us some of the non-classical features of $\vDash_{\mathrm{LP}}$; note that the failure of Explosion in (60a) does not come about in the same way as in SB (by failing adjunction), since adjunction is valid in LP, as recorded in (55a). (60b) points out the much-remarked failure of Disjunctive Syllogism in LP. Dual to these non-classicalities are the non-classicalities of K_3 given in (61).

18.3.2 Vagueness and Ambiguity

As we've seen, one clear reason to think that LP and K_3 are better logics of vagueness than SB and SP is the sheer sensibleness of their conjunction and disjunction, which SB and SP lacked. LP and K_3 thus allow us to give an s'valuation-flavoured picture of vague *predication* that doesn't interfere with a more standard picture of *connectives*. But there's another reason why at least some s'valuationists should prefer the truth-functional approach recommended here, having to do with ambiguity.

As we've seen, the s'valuational picture alleges at least some similarities between vagueness and ambiguity: at a bare minimum, they both involve a one-many relation between a word and its potential extensions. Some s'valuationists (e.g., Keefe 2000) stop there, but others (e.g., Fine 1975 in places, Lewis 1982) go farther, claiming that vagueness is actually a species of ambiguity. For these authors, there is an additional question worth facing: what's ambiguity like?

18.3.2.1 Non-uniform Disambiguation

Here's one key feature of ambiguity: when an ambiguous word occurs twice in the same sentence, it can be disambiguated in different ways across its occurrences. For example, consider the word 'plant', which is ambiguous between (at least) *vegetation* and *factory*. Now, consider the sentence (62):

(62) Jimmy ate a plant, but he didn't eat a plant.

It's clear that (62) has a non-contradictory reading; in fact, it has two, assuming for the moment that 'plant' is only two-ways ambiguous. 'Plant' can take on a different disambiguation at each of its occurrences, even when those occurrences are in the same sentence. If this were not the case, if multiple occurrences of an ambiguous word had to be disambiguated uniformly within a sentence, then the standard method of resolving an apparent contradiction—by finding an ambiguity—couldn't work. But of course this method does work.

Now, suppose we wanted to build formal models for an ambiguous language. They had better take this fact into account. But SB and SP cannot—they precisify whole sentences at once, uniformly. Hence, SB and SP could not work as logics for ambiguous language.[13]

LP and K_3, on the other hand, do not have this bad result. They deal with each occurrence of an ambiguous predicate (each atomic sentence) separately, and combine them truth-functionally. Thus, they avoid the bad consequences faced by s'valuational pictures. In fact, it is LP that seems to be a superior logic of

[13] *Pace* Fine (1975), which, in a footnote, proposes SP as a logic for ambiguous language. As noted above, this would make it impossible to explain how one resolves a contradiction by finding an ambiguity—a very bad result.

ambiguous language. Here's why: typically, for an ambiguous sentence to be true, it's not necessary that *every* disambiguation of it be true; it suffices that *some* disambiguation is.[14]

Since it's clear that LP and K_3 (and LP in particular) are better logics of ambiguity than SB and SP, those s'valuationists who take vagueness to be a species of ambiguity have additional reason to adopt LP and K_3.

18.3.2.2 Asynchronous Precisification

For an s'valuationist who does not take vagueness to be a species of ambiguity, the above argument applies little direct pressure to use LP or K_3, but it raises an interesting issue dividing the truth-functional approaches from the s'valuational approaches: when multiple vague predicates occur in the same sentence, how do the various precisifications of one interact with the various precisifications of the other?

Take the simplest case, where a single vague predicate occurs twice in one sentence. What are the available precisifications of the whole sentence? Suppose a model with n precisifications. On the s'valuational pictures, there will be n precisifications for the whole sentence; while on a truth-functional picture there will be n^2; every possible combination of precisifications of the predicates is available. This can be seen in the LP/K_3 connectives; for example, where $\wedge_0$ is classical conjunction,

(63) $M(A \wedge B) = \{a \wedge_0 b : a \in M(A), b \in M(B)\}$

$M(A)$ and $M(B)$, recall, are sets of classical values. $M(A \wedge B)$ is then obtained by pulling a pair of classical values, one from $M(A)$ and one from $M(B)$, conjoining these values, and repeating for every possible combination, then collecting all the results into a set. In other words, on this picture, every precisification 'sees' every other precisification in a compound sentence formed with $\wedge$; multiple predicates are not precisified in lockstep. The same holds, *mutatis mutandis*, for $\vee$.

18.3.3 What to Say About Borderline Cases?

So much for the logical machinery. What do these approaches say about borderline cases of a vague predicate? Suppose again that 12:23 is a borderline case of 'noonish'. Consider the following list of claims:

(64) a. 12:23 is noonish.

b. 12:23 is not noonish.

---

[14] Lewis (1982) argues for LP, in particular, as a logic of ambiguity, and mentions vagueness as one sort of ambiguity.

c. 12:23 is both noonish and not noonish.
d. 12:23 is neither noonish nor not noonish.
e. It's not the case that 12:23 is both noonish and not noonish.
f. 12:23 is either noonish or not noonish.

All of these are familiar things to claim about borderline cases, although a common aversion to contradictions among philosophers means that some of them, like (64c), are more likely to be heard outside the classroom than in it.[15] All these claims receive the value {1,0} in a model that takes 12:23 to be a borderline case of noonish. The LP partisan, then, will hold all of these to be true, while the K_3 partisan will hold none of them to be true. Which interpretation of {1,0} is more plausible, then? If we are to avoid attributing massive error to ordinary speakers (and ourselves, a great deal of the time), the LP story is far superior. Accordingly, for the remainder of the paper I'll focus in on LP, setting K_3 aside (although much of what follows holds for K_3 as well as LP).

18.4 Higher-Order Vagueness

Some objections to any dialetheic approach to vagueness are considered and ably answered in Hyde and Colyvan (2008). But one objection not considered there might seem to threaten any three-valued approach to vagueness, and in particular the LP approach I've presented: the phenomenon of higher-order vagueness. This section evaluates the LP approach's response to the phenomenon, focusing first on the case of second-order vagueness, and then generalizing the response to take in higher orders as well.

18.4.1 Second-Order Vagueness

So far, we've seen a plausible semantics for vague predicates that depends crucially on the notion of an 'admissible precisification'. A vague atomic sentence is true iff it's true on some admissible precisification, false iff false on some admissible precisification. But which precisifications are admissible? Consider 'noonish'. Is a precisification that draws the line at 12:01 admissible, or is it too early? It has seemed to many theorists (as it seems to me) that 'admissible' in this use is itself vague. Thus, we've run into something of the form of a revenge paradox: theoretical machinery invoked to solve a puzzle works to solve the puzzle, but then the puzzle reappears at the level of the new theoretical machinery.[16]

[15]See Ripley (2011) for evidence that ordinary speakers agree with such claims as (64c) and (64d).

[16]See e.g., Beall (2008) for a discussion of revenge.

It would be poor form to offer an account of the vagueness of 'admissible' that differs from the account offered of the vagueness of 'noonish'. After all, vagueness is vagueness, and similar problems demand similar solutions.[17] So let's see how the account offered above applies to this particular case of vagueness.

What do we do when a precisification is borderline admissible—that is, both admissible and not admissible? We consider various precisifications of 'admissible'. This will kick our models up a level, as it were. Models (call them level-2 models, in distinction from the earlier level-1 models) now determine not sets of precisifications, but sets of *sets of* precisifications. That is, a level-2 model is a tuple $\langle D, I, P_2 \rangle$, where D is again a domain of objects, I is again a function from terms in the language to members of D, and P_2 is a set whose members are sets of precisifications.

Every individual precisification p works as before; it still assigns each atomic sentence A a value $p(A)$ from the set $V_0 = \{1, 0\}$. Every set of precisifications assigns each atomic sentence a value as well: a set P_1 of precisifications assigns to an atomic A the value $P_1(A) = \bigcup_{p \in P_1} \{p(A)\}$. These values come from the set $V_1 = \wp(V_0) - \emptyset = \{\{1\}, \{0\}, \{1, 0\}\}$. That is, sets of precisifications work just like level-1 models, as far as atomics are concerned; they simply collect into a set the values assigned by the individual precisifications. A level-2 model $M = \langle D, I, P_2 \rangle$ assigns to every atomic A a value $M(A) = \bigcup_{P_1 \in P_2} \{P_1(A)\}$. It simply collects into a set the values assigned to A by the sets of precisifications in P_2, so it assigns values from the seven-membered set $V_2 = \wp(V_1) - \emptyset = \{\{\{1\}\}, \{\{0\}\}, \{\{1,0\}\}, \{\{1\}, \{0\}\}, \{\{1\}, \{1,0\}\}, \{\{0\}, \{1,0\}\}, \{\{1\}, \{0\}, \{1,0\}\}\}$.

In applications to vagueness, presumably only five of these values will be needed. Not much hangs on this fact in itself, but it will better show the machinery of the theory if we take a moment to see why it's likely to be so. Let's look at how level-2 models are to be interpreted. Take a model $M = \langle D, I, P_2 \rangle$. Each member P_1 of P_2 is an admissible precisification of 'admissible precisification'. Some precisifications, those that are in any admissible precisification of 'admissible precisification', will be in every such P_1. Others, those that are in no admissible precisification of 'admissible precisification', will be in no such P_1. And still others, those precisifications on the borderline of 'admissible precisification', will be in some of the P_1s but not others.

Now let's turn to 'noonish'. 12:00 is in every admissible precisification of 'noonish', no matter how one precisifies 'admissible precisification'; 12:23 is in some admissible precisifications but not others, no matter how one precisifies 'admissible precisification'[18]; and 20:00 is in no admissible precisifications of 'noonish', no matter how one precisifies 'admissible precisification'. So far so good—and so far, it all could have been captured with a level-1 model.

---

[17]This is sometimes called the 'principle of uniform solution'. For discussion, see e.g., Colyvan (2008) and Priest (2002).

[18]Again, assume a context where this is true.

But there is more structure to map. Some moment between 12:00 and 12:23—let's say 12:10 for concreteness—is in every admissible precisification of 'noonish' on some admissible precisifications of 'admissible precisification', and in some admissible precisifications of 'noonish' but not others on some admissible precisifications of 'admissible precisification'. And some moment between 12:23 and 20:00—let's say 12:34—is in no admissible precisification of 'noonish' on some admissible precisifications of 'admissible precisification', and in some admissible precisifications of 'noonish' but not others on some admissible precisifications of 'admissible precisification'.

Here's a (very toy)[19] model mapping the above structure:

(65) a. D = the set of times from 12:00 to 20:00

b. I = the usual map from time-names to times

c. $P_2 = \{\{\{12{:}00–12{:}38\},\{12{:}00–12{:}15\}\},\{\{12{:}00–12{:}25\}, \{12{:}00–12{:}08\}\}\}$

Call this model M. Now let's apply it to some atomic sentences: $M(N12{:}00) = \{\{1\}\}$, $M(N12{:}10) = \{\{1\}, \{1,0\}\}$, $M(N12{:}23) = \{\{1,0\}\}$, $M(N12{:}34) = \{\{1,0\}, \{0\}\}$, and $M(N20{:}00) = \{\{0\}\}$. One suspects that these five values are all one needs of V_2 for (at least most) vague predicates. In order for a sentence to take the value $\{\{1\}, \{0\}, \{1,0\}\}$ on a model, the model would have to be set up so that, depending on the precisification of 'admissible precisification', the sentence could be in all admissible precisifications *or* some but not others *or* none at all. It seems unlikely that many predicates have admissible precisifications that work like this. For a sentence to take the value $\{\{1\}, \{0\}\}$ on a model, something even weirder would have to happen: the model would have to make it so that, depending on the precisification of 'admissible precisification', the sentence could be either true in all admissible precisifications or false in all of them, but there could be no admissible precisification of 'admissible precisification' that would allow the sentence to be true on some admissible precisifications but not others. This too seems unlikely. So I suspect that only five of the seven members of V_2 are likely to be useful for vague predicates, although (as mentioned above) not much hangs on this.[20]

There is of course the question of interpretation: which of these values counts as true? Again, we should give the same answer here as in the case of first-order vagueness, to avoid *ad hoc*-ery: a sentence is true iff it's true on some admissible precisification; and so it's true iff it's true on some admissible precisification, for some admissible precisification of 'admissible precisification'. That is, any sentence whose value on a model has a 1 in it anywhere—any sentence whose value isn't $\{\{0\}\}$—is true on that model.

[19]For simplicity, we look at only one predicate: N for 'noonish'. This set is then a set of sets of precisifications for 'noonish'. Let $\{x–y\}$ be the set of times between x and y inclusive.

[20]Actually I don't see that anything does.

18.4.1.1 Connectives

So that's how our atomics get their values. Of course, we need some way to assign values to compound sentences as well, and the familiar LP operations (call them $\wedge_1$, $\vee_1$, and $\neg_1$) won't work—they're defined only over V_1, but our atomics take values from V_2. Fortunately, a simple tweak will work, getting us sensible level-2 operations $\wedge_2$, $\vee_2$, and $\neg_2$ defined over V_2.

Recall one of our earlier observations about the LP connectives: in a conjunction, every precisification of one conjunct sees every precisification of the other conjunct (*mutatis mutandis* for disjunction). We can use this to define our level-2 connectives.

Consider the conjunction of two V_2 values u and v. Remember, values from V_2 are sets of values from V_1, and we already have well-behaved connectives over V_1. To get one potential V_1 value for the conjunction, we can pull a V_1 value from u and one from v, and conjoin them. If we do that in every possible way, and collect all the results into a set, we get a V_2 value appropriate to be that value of the conjunction. More formally: $u \wedge_2 v = \{u' \wedge_1 v' : u' \in u, v' \in v\}$. The same idea will work for disjunction—$u \vee_2 v = \{u' \vee_1 v' : u' \in u, v' \in v\}$—and negation—$\neg_2 u = \{\neg_1 u' : u' \in u\}$. So let's simply adopt these as our level-2 connectives.

18.4.1.2 Consequence

Level-2 models now assign values to every sentence; first the atomics, via the sets of sets of precisifications, and then to all sentences, via the level-2 connectives. What's more, we have a set $D_2 \subseteq V_2$ of *designated* values—values that count as true. (Some of them also count as false, of course.) This means that we're in a position to define level-2 consequence. We do it in the expected way: $\Gamma \vDash_2 \Delta$ iff, for every level-2 model M, either $M(\delta) \in D_2$ for some $\delta \in \Delta$, or $M(\gamma) \notin D_2$ for some $\gamma \in \Gamma$.

So we have a full logic erected 'up a level' from LP, as it were. At first blush, this might seem like not much of a response to the challenge of first-order vagueness. After all, it seems that we simply abandoned the initial theory and adopted another. That would hardly be a persuasive defence. But in fact that's not quite what's happened; as it turns out, $\vDash_2 = \vDash_{LP}$.[21] We haven't actually abandoned the initial theory—we've just offered an alternative semantics for it, one that fits the structure of second-order vagueness quite naturally. What's more, we haven't had to invoke any special machinery to do it. Simply re-applying the first-order theory to itself yields this result.

18.4.2 Generalizing the Construction

Of course, the above construction only works for second-order vagueness, and there is much more to higher-order vagueness than that. In particular, just as it was vague

[21] For proof, see Priest (1984).

which precisifications are admissible precisifications of ‘noonish’, it’s vague which precisifications are admissible precisifications of ‘admissible precisification’. From the above construction, of course, one can predict the reply: we’ll look at admissible precisifications of ‘admissible precisification’ of ‘admissible precisification’, which is of course itself vague, and so on and so on. Let’s lay out a general picture here.

Let an n-set of precisifications be defined as follows: a 0-set of precisifications is just a precisification, and a $(k + 1)$-set of precisifications is a set of k-sets of precisifications. Let sets V_n of values be defined as follows: $V_0 = \{1, 0\}$, and $V_{k+1} = \wp(V_k) - \emptyset$. A level-$n$ model M_n is then a tuple $\langle D, I, P_n \rangle$ such that D is a domain of objects, I is a function from terms to members of D, and P_n is an n-set of precisifications. Consider an atomic sentence A. We build up its value $M_n(A)$ as follows: in concert with I, every precisification p assigns A a value $p(A)$ from V_0, and every $(k + 1)$-set P_{k+1} of precisifications assigns A a value $P_{k+1}(A) = \bigcup_{P_k \in P_{k+1}} \{P_k(A)\}$ from V_{k+1}. $M_n(A)$ is then just $P_n(A)$. For the level-1 and level-2 cases, this is just the same as the above setup, but of course it extends much farther.

Which values count as true? By parallel reasoning to our earlier cases, any value that contains a 1 at any depth. More precisely, we can define a hierarchy D_n of sets of designated values as follows: $D_0 = \{1\}$, and $D_{k+1} = \{v \in V_{k+1} : \exists u \in v(u \in D_k)\}$.

For the connectives: we define a hierarchy $\wedge_n, \vee_n, \neg_n$ of operations as follows: $\wedge_0, \vee_0$, and $\neg_0$ are simply classical conjunction, disjunction, and negation. For values $u_{k+1}, v_{k+1} \in V_{k+1}$, $u_{k+1} \wedge_{k+1} v_{k+1} = \{u_k \wedge_k v_k : u_k \in u_{k+1}, v_k \in v_{k+1}\}$. That is, the a conjunction of sets of values is the set of conjunctions of values from those sets. Similarly for disjunction: $u_{k+1} \vee_{k+1} v_{k+1} = \{u_k \vee_k v_k : u_k \in u_{k+1}, v_k \in v_{k+1}\}$, and for negation: $\neg_{k+1} u_{k+1} = \{\neg_k u_k : u_k \in u_{k+1}\}$. Again, this gives us just what we had before in the case where $n = 1$ or 2, but extends much farther.

We are now in a position to define a hierarchy of consequence relations $\vDash_n$ as follows: $\Gamma \vDash_n \Delta$ iff, for every n-level model M_n, either $M_n(\delta) \in D_n$ for some $\delta \in \Delta$, or $M_n(\gamma) \notin D_n$ for some $\gamma \in \Gamma$. Of course, this simply extends our earlier definition to the new level-n framework.

18.4.3 Good News

Just as in the case of second-order vagueness, this construction allows us to fully map the structure of nth-order vagueness for any n. By collecting up (with a bit of jiggering), we can come to an ω-valued model that fully maps the structure of all higher-order vagueness. What’s more, just as in the second-order case, we haven’t affected our consequence relation at all; for every $n \geq 1$ (including ω), $\vDash_n = \vDash_1$.[22] This shows us that, although we can fully map this structure, there is in fact no need

[22]Proof and details can be found in Priest (1984). Note as well that the result can be iterated past ω into the transfinite; I don’t think that’ll be necessary here, since every new level is created to address the vagueness of some finite predicate.

to for logical purposes; the logic we define remains unchanged. We may as well stick with the simple three-valued version. (Or any other version we like. I like the three-valued version for its simplicity, but if there's some reason to prefer another version, then by all means.) It's worth noting that the defender of K_3 can make precisely the same defence here.

18.5 Conclusion: The Sorites

We should close by examining the familiar paradox that arises from vague language: the sorites. Here's a sample, built on 'noonish' (still written N):

1. $N12{:}00$
2. $\forall x[(Nx) \rightarrow (N(x + 0{:}00{:}01))]$
3. $N20{:}00$

Now, before we do any logic at all, we know a few things about this argument. We know that the first premise is true, and that the conclusion is not. That leaves us just a few options: we can deny the second premise, or we can deny that $\rightarrow$ supports modus ponens.[23]

Well, what is $\rightarrow$, anyway? (This question is raised forcefully in Beall and Colyvan (2001).) There are many possibilities. Each possibility creates a distinct sorites argument. And, while we may think that the sorites argument must be solved in a uniform way no matter which vague predicate it's built on, we certainly should not think that it must be solved in a uniform way no matter which binary connective $\rightarrow$ it's built on.

Some examples. (1) Suppose $A \rightarrow B$ is just $A \wedge B$. Then surely modus ponens is valid for $\rightarrow$; after all, B would follow from $A \rightarrow B$ alone on this supposition. But the second premise in the sorites is obviously not true, given this interpretation: it simply claims that every moment, along with the moment one second later, is noonish. That's not at all plausible. So on this interpretation, the sorites is to be answered by rejecting the second premise. (2) Suppose on the other hand that $A \rightarrow B$ is $\neg(A \wedge \neg A)$. Then the second premise is clearly true, at least given most theorists' commitments about the law of non-contradiction (including the commitments of the LP-based approach), but it just as clearly does not support modus ponens. From A and $\neg(A \wedge \neg A)$, we cannot conclude B.

[23]Some have accepted the conclusion or rejected the first premise (e.g., Unger 1979), but to take such a position seriously is to remove much of the sense of 'noonish'. And to take it seriously for every vague predicate would make it very hard indeed to talk truly at all. There are other more radical approaches, too: we might reject transitivity of entailment (as in Zardini 2008 or Cobreros et al. 2012), or universal instantiation (as in Kamp 1981). The LP-based solution offered here keeps to the more conservative side of the street.

Of course, $A \wedge B$ and $\neg(A \wedge \neg A)$ are silly conditionals. But they make a point: whatever our commitment to uniform solution, it does not hold when we vary the key connective in the sorites argument. We are free to reject the second premise for some readings of $\rightarrow$ and deny modus ponens for others, and this does not make our solution non-uniform in any way worth avoiding. We might both reject the premise and deny modus ponens for some readings of $\rightarrow$, for example if we read $A \rightarrow B$ simply as A. The one thing we cannot do is accept both the premise and the validity of modus ponens, on any single reading of $\rightarrow$.

LP as presented here includes at best a very weak conditional. Its material conditional, defined as $A \supset B := \neg(A \wedge \neg B)$, does not support modus ponens.[24] Given the theory of vague predicates advanced here, the second premise of the sorites is true if we read $\rightarrow$ as $\supset$. So the present account doesn't run into any trouble on that version of the sorites. What's more, as mentioned in Hyde (2001), the Stoics sometimes used this form of the argument (the form using 'for any moment, it's not the case both that moment is noonish and that one second later isn't noonish'), precisely to avoid debates about the proper analysis of conditionals. If we do the same, no trouble ensues.

On the other hand, the most compelling versions of the sorites use the 'if...then' of natural language. $\supset$ isn't a very promising candidate for an analysis of a natural-language conditional, in LP or out of it, because of the well-known paradoxes of material implication (see e.g., Routley et al. 1982 for details). What is the right analysis of natural-language conditionals is a vexed issue (to say the least!) and not one I'll tackle here, so this is not yet a response to the sorites built on 'if...then'. For now, we can see that the LP-based approach answers the material-conditional version of the sorites handily.

What's more, it embodies the picture of vague language underlying subvaluationist and supervaluationist motivations in a more natural way than SB and SP themselves do. It also verifies much of our ordinary talk about borderline cases, contradictory and otherwise, and provides a satisfying non-*ad hoc* account of higher-order vagueness. In short, LP should be considered a serious contender in the field of non-classical approaches to the phenomenon of vagueness.

Acknowledgements Research partially supported by the French government, Agence Nationale de la Recherche, grant ANR-07-JCJC-0070, program "Cognitive Origins of Vagueness", and by the Spanish government, grant "Borderlineness and Tolerance" ref. FFI2010-16984, MICINN. Many thanks as well to Jc Beall, Rachael Briggs, Mark Colyvan, Dominic Hyde, Joshua Knobe, William Lycan, Ram Neta, Graham Priest, Greg Restall, Keith Simmons, Mandy Simons, and Zach Weber, as well as audiences at the Fourth World Congress of Paraconsistency, the University of Queensland, and Carnegie Mellon University for valuable discussion, insight, and support.

[24]This is because modus ponens on $\supset$ is equivalent to disjunctive syllogism, which anyone who takes contradictions seriously ought to reject. See Priest (1979) for discussion.

References

Beall, J.C. 2008. Prolegomenon to future revenge. In *Revenge of the liar: New essays on the paradox*, ed. J.C. Beall, 1–30. Oxford: Oxford University Press.
Beall, J.C., and M. Colyvan. 2001. Heaps of gluts and hyding the sorites. *Mind* 110: 401–408.
Beall, J.C., and D. Ripley. 2004. Analetheism and dialetheism. *Analysis* 64(1): 30–35.
Cobreros, P., P. Égré, D. Ripley, and R. van Rooij. 2012. Tolerant, classical, strict. *Journal of Philosophical Logic* 41(2): 347–385.
Colyvan, M. 2008. Vagueness and truth. In *From truth to reality: New essays in logic and metaphysics*, ed. H. Duke, 29–40. New York: Routledge.
Fine, K. 1975. Vagueness, truth, and logic. *Synthèse* 30: 265–300.
Hyde, D. 1997. From heaps and gaps to heaps of gluts. *Mind* 106: 641–660.
Hyde, D. 2001. A reply to beall and colyvan. *Mind* 110: 409–411.
Hyde, D. 2010. The prospects of a paraconsistent approach to vagueness. In *Cuts and clouds: Vagueness, its nature, and its logic*, ed. R. Dietz and S. Moruzzi, 385–406. Oxford: Oxford University Press.
Hyde, D., and M. Colyvan. 2008. Paraconsistent vagueness: Why not? *Australasian Journal of Logic* 6: 107–121.
Kamp, H. 1981. Two theories about adjectives. In *Aspects of philosophical logic*, ed. U. Mönnich. Dordrecht: Reidel.
Keefe, R. 2000. *Theories of vagueness*. Cambridge: Cambridge University Press.
Kripke, S. 1975. Outline of a theory of truth. *Journal of Philosophy* 72(19): 690–716.
Lewis, D. 1970. General semantics. *Synthèse* 22: 18–67.
Lewis, D. 1982. Logic for equivocators. *Noûs* 16(3): 431–441.
Machina, K.F. 1972. Vague predicates. *American Philosophical Quarterly* 9: 225–233.
Parsons, T. 1984. Assertion, denial, and the liar paradox. *Journal of Philosophical Logic* 13: 137–152.
Priest, G. 1979. Logic of paradox. *Journal of Philosophical Logic* 8: 219–241.
Priest, G. 1984. Hyper-contradictions. *Logique et Analyse* 107: 237–243.
Priest, G. 2002. *Beyond the limits of thought*. Oxford: Oxford University Press.
Raffman, D. 1994. Vagueness without paradox. *The Philosophical Review* 103(1): 41–74.
Recanati, F. 2004. *Literal meaning*. Cambridge: Cambridge University Press.
Restall, G. 2005. Multiple conclusions. In *Logic, methodology, and philosophy of science: Proceedings of the twelfth international congress*, ed. P. Hajek, L. Valdes-Villanueva, and D. Westerståhl, 189–205. Kings' College, London.
Ripley, D. 2011. Contradictions at the borders. In *Vagueness and communication*, ed. R. Nouwen, R. van Rooij, H.-C. Schmitz, and Uli Sauerland, 169–188. Heidelberg: Springer.
Routley, R., R.K. Meyer, V. Plumwood, and R.T. Brady. 1982. *Relevant logics and their rivals*, vol. 1. Atascadero: Ridgeview.
Smith, N.J.J. 2008. *Vagueness and degrees of truth*. Oxford: Oxford University Press.
Sorensen, R. 2001. *Vagueness and contradiction*. Oxford: Oxford University Press.
Tappenden, J. 1993. The liar and sorites paradoxes: Toward a unified treatment. *The Journal of Philosophy* 90(11): 551–577.
Unger, P. 1979. There are no ordinary things. *Synthèse* 41: 117–154.
Varzi, A. 2000. Supervaluationism and paraconsistency. In *Frontiers in paraconsistent logic*, ed. D. Batens, C. Mortensen, G. Priest, and J.-P. Van Bendegem, 279–297. Baldock: Research Studies Press.
Varzi, A. 2007. Supervaluationism and its logics. *Mind* 116: 633–676.
Williamson, T. 1994. *Vagueness*. London/New York: Routledge.
Wilson, D., and D. Sperber. 2002. Truthfulness and relevance. *Mind* 111: 583–632.
Zardini, E. 2008. A model of tolerance. *Studia Logica* 90: 337–368.

Chapter 19
Are the Sorites and Liar Paradox of a Kind?

Dominic Hyde

19.1 On Unification

Might the sorites paradox and liar paradox admit of a uniform solution? On the face of it there is little to recommend their unification. Of course, they may be of a kind simply in so far as, when all is said and done, a defensible response to each paradox can be found by appeal to a logical framework which in each case is the same. In fact, simplicity of theory choice might push us to attempt a solution of the one by means already invoked for the other, so that having paid the price thought necessary to accommodate the one paradox we achieve the virtue of having to pay no additional price to accommodate the other.

Beall and Colyvan (2001), for example, suggest looking to a paraconsistent response to the sorites paradox on the assumption that the liar paradox is best solved paraconsistently. In this spirit, if one finds it appealing to respond to the liar paradox, and the semantic paradoxes, more generally, by appeal to Priest's *LP* (say) then one should look to the sorites paradox with this logical framework already in mind as an economical response. After all, all other things being equal, one pays no price higher for applying such a logical framework to vagueness and the associated sorites paradox than that already paid for its advocacy in the face of the semantic paradoxes.

We need to be careful though about just what can be established in this way. Even assuming that one endorses a paraconsistent response to the liar paradox, such reasoning still only gets us to conclusion that sorites should, if at all possible, be addressed by use of *some* logic already accepted within one's overall philosophy to deal with problems elsewhere. So that, for example, if one takes the view that, while the liar paradox warrants a paraconsistent approach, nonetheless theproblem

D. Hyde (✉)
School of History, Philosophy, Religion and Classics, The University of Queensland, Brisbane, Australia
e-mail: d.hyde@uq.edu.au

K. Tanaka et al. (eds.), *Paraconsistency: Logic and Applications*, Logic, Epistemology, and the Unity of Science 26, DOI 10.1007/978-94-007-4438-7_19,

of reference-failure warrants a paracomplete (e.g. truth-value gap) approach then the argument from theoretical economy does not support a paraconsistent approach any more than a paracomplete one (and *vice versa*).

This being said, it may well be that one takes paraconsistency, and only paraconsistency, to provide an adequate logical framework for handling all and any logical puzzles thrown up by philosophy. And it is this that Beall and Colyvan should be taken to have proposed—to be sure, classical logic must be weakened to a paraconsistent one (from which classical logic can be recovered as a special case given certain additional assumptions where permissible) but no weakening of completeness (in the sense of bivalence) is required. Not everything is of a paraconsistent nature, much ordinary reasoning is perfectly well handled by classical logic, but where difficulties arise, as with the paradoxes, only a weakening to a paraconsistent framework is necessary.

Unification in this sense, however, is something that is achieved after the fact, as it were. We analyse each of the problems separately and come to see that, when a satisfactory account of each is arrived at, they turn out to be similar. A more interesting sense in which the two paradoxes might be unified is by virtue of the paradoxes themselves being of a kind. That is to say, the paradoxes might admit of a uniform solution since they are in some sense or other similar in their nature, generating puzzlement by virtue of some likeness in their underlying cause.

In this vein, Field (2003), for example, argues that both the semantic and sorites paradoxes are to be dealt with by way of a paracomplete logic that rejects the law of excluded middle—the strongly paracomplete logic K_3 supplemented with a non-truth-functional conditional—since there are "some rather general considerations that suggest the naturalness of "unification" (Field 2003, p. 308) and among them are considerations pertaining to the *source* of paradox in each case.

> The semantic paradoxes seem to arise from the fact that the standard means for explaining 'True' (namely, the truth schema) fails to uniquely determine the application of the term to certain sentences (the "ungrounded" ones); and this seems to be just the sort of thing that gives rise to other sorts of vagueness and indeterminacy.

Thus, it is suggested, the extension of 'True' seems to be indeterminate (i.e. not uniquely determined) in just the same way that vague predicates, for example, are.

Of course, the indeterminacy exhibited by 'True', that is the inability "to uniquely determine the application of 'True' to certain sentences", is consistent with either of two options: (1) all determinations of the application of 'True' must be unique, so it is not determined at all for the sentences in question—i.e. such sentences are not determined to be true nor to be false, and are thus underdetermined in respect of truth; or (2) though not uniquely determined it is determined nonetheless—i.e. such sentences are determined to be true and determined to be false, and are thus overdetermined in respect of truth. Only the former leads to a unified *paracomplete* solution.

Attempts at paracomplete solutions to the liar are, of course, well known but attempts at unified paracomplete solutions are less remarked upon. In addition to Field (2003), such an approach to the liar was earlier pursued in McGee (1990) and Tappenden (1993). McGee proposes that though truth seems overdetermined in the case of liar sentences we should nonetheless treat it as if it is underdetermined, thus assimilating such cases to cases of vagueness, thereby avoiding paradox and inconsistency.

> If, contrariwise, we attempted to eliminate vagueness as well as contradiction, replacing our traditional way of using 'true' by a reformed usage that was perfectly precise as well as perfectly consistent, the logical structure of our everyday usage of 'true' would, I claim, be damaged beyond repair. (McGee 1990, p. 8)

The proposal is, therefore, that we adopt a methodological approach to the semantic paradoxes that sees them as having their source in the vagueness of the predicate 'true'. Just as a borderline case of 'bald' is such that it is not settled whether the predicate applies, liar sentences present situations in which it is left unsettled whether the sentence is true or not.[1]

In a similar vein, Tappenden pursues a unified treatment of the liar and sorites paradoxes by developing the idea that "vague predicates may usefully be seen as analogous to the truth predicate" (Tappenden 1993, p. 551).

In each case there is a species of indeterminacy involved, or so it is proposed. In neither case is the predicate's application uniquely determined in problematic cases. It may be indeterminate whether Harry is bald, and so too "ungrounded" sentences like the liar sentence are such that it is indeterminate whether they are true.

Field makes the point in respect of 'true' by way of the T-schema. If one pursues the idea that the extension of true be settled by appealing to Tarski's material adequacy condition—so that, for a sentence p, True(p) if and only if p—then one finds that only "grounded" sentences are determined as uniquely being either in the predicate's extension or in its anti-extension. Liar sentences are not determined either way, and the extension of 'true' is therefore indeterminate. Of course, the reliance on the T-schema here is inessential. It is enough for paradox that p entail and be entailed by True(p), and thus these entailments alone are sufficient for establishing that only "grounded" sentences are determined as uniquely being either in the extension of the truth predicate or its anti-extension.[2]

So the paradoxes might be said to be analogous by virtue of their both giving rise to indeterminacy—in the case of the liar paradox truth is shown (it is said) to be indeterminate, and in the case of the sorites paradox we are similarly confronted with the indeterminacy of the predicate involved.

In fact, McGee argues more strongly that we must recognise 'true' as not merely *analogous* to a vague predicate by virtue of "ungrounded" sentences giving rise to something akin to indeterminacy characteristic of vagueness, but that 'true' itself is *in fact* already vague since it takes vague sentences as arguments. The T-schema

---

[1]Caution is required here. The proposal is paracomplete in the sense that not every sentence is determined to be true or determined to be false; some sentences (e.g. vague ones, liar sentences) remain undetermined. The paracomplete logic advocated is the popular supervaluationism adapted from Van Fraassen (1966).

[2]See Field (2008, Part I, §3).

shows that the vagueness of a sentence p generates a borderline case for 'true'. (It is for this reason, presumably, that he claimed that in attempting to reform our usage of 'true' so that it was perfectly precise its logical structure would be "damaged beyond repair".) So, it is argued, 'true' is *already* vague. It might seem then to strengthen the argument that liar sentences be accommodated by assimilating the species of indeterminacy that they generate to that species already recognised as affecting the predicate. Of course, strictly speaking it does *not* follow from p's being a borderline case for 'true' that 'true' is vague; the vagueness is due to the vagueness of the sentence p, not the vagueness of 'true'. (cf. Rolf 1980.) However, the T-schema does nonetheless generate borderline cases for truth, i.e. cases for which it is indeterminate whether they are true.[3] With truth shown to be indeterminate in this way, the case for treating "ungrounded" sentences as being indeterminate in a relevantly similar sense is now strengthened since it no longer requires the recommendation that truth be treated as sometimes indeterminate—this is already required—but, rather, that the species of indeterminacy exhibited by truth in respect of "ungrounded" sentences be treated as the same as that species of indeterminacy exhibited by truth in respect of vague sentences.

Of course, again, nothing so strong as the T-schema is required to establish that there are sentences for which it is indeterminate whether they are true. While the schema will establish that indeterminate sentences will generate immediate indeterminacy in respect of their truth, non-disquotationalists—e.g. gap theorists who abandon the T-schema to make room for truth-value gaps—will nonetheless recognise the existence of sentences that are borderline cases for 'true'. Higher-order vagueness shows (at least in the view of many gap-theorists, for example supervaluationists) that a tripartite division of sentences into the True, False and Indeterminate is unacceptable. Just as there is no sharp boundary between the True and the False, there is no sharp boundary between the True and the non-True, on pain of admitting sharp boundaries to the class of sentences that take the value Indeterminate (with the untoward consequence of admitting sharp boundaries to the class of borderline cases for the vague predicate in question).

Irrespective of acceptance or otherwise of the T-schema then one ought admit that truth is "vague" (at least in the sense that it admits of borderline cases, be this "vagueness" proper or not—see comments leading to footnote 3 above). The suggestion then is that one seek to accommodate the problem of "ungrounded' sentences and the associated liar paradox within the semantic framework developed to respond to the problem of vague (including higher-order vague) sentences and the associated sorites paradox by virtue of the supposition that both problems arise from the same source—indeterminacy.

This push towards unification of the two problems from a paracomplete perspective might be thought to be further strengthened by considerations bearing on the structure of the respective paradoxes that they engender—in particular, their both

[3] We are, of course assuming a semantic conception of vagueness—what Burgess (1998) calls an "indeterminist conception".

giving rise to revenge problems. In short, the idea is that each apparent paradox—the sorites and the liar—can be thought of as admitting *prima facie* solutions against which the phenomena in question—vagueness and self-reference—exact revenge by way of further paradox. Paradoxical arguments for which solutions have been offered are, as we shall see, able to be recast with paradox reappearing at higher orders. In this light, the recommendation that the indeterminacy arising from each paradox be seen as the same species is further strengthened by our noticing that in each case the indeterminacy in question exhibits similar higher-order structure.

19.2 The Sorites Paradox

Consider, firstly, the sorites paradox, in particular the line drawing form of the sorites paradox, and the associated problem of higher-order vagueness. Let us take as our example the paradox of the bald man. A man with one hair on his head is surely bald, yet a man with a million hairs on is head is not bald. Since, for any number n, a man with n hairs on his head is either bald or not bald, it follows that there must be some n for which a man with n hairs on his head is bald while a man with $n + 1$ hairs on his head is not bald (i.e. hirsute).[4] Where $Bald(n)$ represents the predicate 'a man with n hairs on his head is bald':

$Bald(1)$
$\sim Bald(10^6)$
$\forall n[Bald(n) \vee \sim Bald(n)]$
So, $\exists n[Bald(n) \,\&\, \sim Bald(n+1)]$

But the vagueness of predicates like 'bald' speak against supposing there to be such a sharp line between their application and the application of their negation. No single hair can make the difference between baldness and non-baldness; no single grain of sand can make the difference between a heap and a non-heap; and no single millisecond can make the difference between being alive and not alive. The foregoing argument results in the unacceptable.

Avoidance of the unacceptable follows from the recognition that vague predicates admit of borderline cases—cases to which neither the predicate nor its negation are uniquely determined to apply. There are people who are borderline cases of baldness, so that they are, paraphrasing Field, neither uniquely determined to be bald nor uniquely determined to be not bald; and so too with heaps and life. Recall Harry, mentioned earlier; suppose that the number of hairs Harry has on his head is h. It is simply indeterminate whether he is bald or not; i.e. Harry is not uniquely determined to be bald, $\sim DBald(h)$, and not uniquely determined to benot bald, $\sim D\sim Bald(h)$.

---

[4]If we further suppose that for any m, n such that $m < n$, if a man with m hairs is not bald then a man with n hairs is not bald, then it follows that there must be some *unique* n for which a man with n hairs on his head is bald while a man with $n + 1$ hairs on his head is not.

As a borderline case of baldness then, Harry is evidence of the fact that it is not uniquely determined to be true that a man with h hairs on his head is bald nor uniquely determined to be false that such a man is bald:

(1) $\sim DTrue[Bald(h)]$ and $\sim DTrue\sim[Bald(h)]$.

On paracomplete approaches to vagueness this lack of uniquely determined truth or falsity in respect of a sentence p results in $q \nvDash p, \sim p$ for some sentence q.[5] This follows immediately given, firstly, that from a paracomplete perspective any such determination is unique thus any mention of uniqueness in the foregoing is redundant—lack of uniquely determined truth or falsity amounts then simply to lack of determinate truth or falsity.[6] Given then, secondly, either (1) a definition of validity in terms of truth preservation and the subsequent identification of "determinate truth" with "truth" (as developed, for example, by: supervaluationism in Fine (1975) and Keefe (2000); many-valued theories like fuzzy logic and standard three-valued theories built on K_3) or (2) a definition of validity in terms of determinate truth preservation (as developed, for example, by Field 2003), the result follows.

The paradox is taken to be defused once we recognise this indeterminacy associated with the vagueness of the predicate involved, 'bald'. This is variously achieved depending on whether the failure of completeness ($\nvDash p, \sim p$) is taken to undermine the law of excluded middle—a premise of the foregoing sorites paradox—or not. Many-valued theories, including standard three-valued theories built on K_3, and Field's variant on K_3, K_3^+, in endorsing subjunction ($p \vee \sim$p $\vDash p, \sim p$), *do* reject the law of excluded middle with the consequence that the sorites paradox is diagnosed as having a non-true premise.

Standard supervaluationism, on the other hand, rejects subjunction and endorses excluded middle thus the paradox is said to have a true (i.e. determinately true) conclusion but this is explained as acceptable. The theory recognises the unacceptability of postulating sharp boundaries to the application of vague predicates but denies that the determinate truth of the conclusion:

(2) $DTrue\exists n[Bald(n)\, \&\, \sim Bald(n+1)]$

expresses the existence of a sharp boundary for 'bald'. Such a sharp boundary is said to be expressed, instead, by the claim:

(3) $\exists n[DTrueBald(n)\, \&\, DTrue\sim B(n+1)]$

and the existence of borderline cases for 'bald' somewhere along the sorites series is taken to show that there can be no such n. The transition along the series from members for which it is determinately true that they are bald to members for which it is determinately false that they are bald proceeds by way of *intermediate* cases

[5] Where '$\vDash$' represents the generalised multiple-conclusion consequence relation.

[6] This presumption of uniqueness thus amounts to a presumption of consistency and cannot be assumed in a paraconsistent setting, on which more later.

for which it is neither determinately true nor determinately false that they are bald. The unacceptable claim (3) is rejected, even though the theory notoriously validates the apparently equivalent claim (2). ("Determinate truth", i.e. "truth" for the supervaluationist, does not distribute over '∃'.)

All well and good, one might think, but the phenomenon that underwrites the sorites paradox, vagueness, exacts revenge by way of higher-order vagueness. The paradox can simply be rerun in the language extended to include D, using the predicate $DBald(x)$ as follows. Since a man with one hair on his head is determinately bald and a man with h hairs on his head is not, and for any n a man with n hairs on his head is either determinately bald or not determinately bald, it follows that there must be some n such that a man with n hairs on his head is determinately bald while a man with $n + 1$ is not.
That is:

$DBald(1)$
$\sim DBald(h)$
$\forall n[DBald(n) \vee \sim DBald(n)]$
So, $\exists n[DBald(n) \,\&\, \sim DBald(n + 1)]$

Thus, a sharp line is now seemingly postulated between the application of the predicate $DBald(x)$ and $\sim DBald(x)$. But higher-order vagueness points to the fact that there is no more a sharp boundary between the application of 'determinately bald' and the application of 'not determinately bald' than there is between the cases of baldness and the cases of non-baldness. An unacceptably paradoxical conclusion is again arrived at.

Avoidance of the unacceptable again follows from the recognition that vague predicates like 'determinately bald' admit of borderline cases just as 'bald' does. There are people who are borderline cases of determinate baldness, so that they are neither determinately determinately bald nor determinately not determinately bald. Suppose Fred is one such and that the number of hairs Fred has on his head is f. It is simply indeterminate whether he is determinately bald or not determinately bald; i.e. Fred is not determinately determinately bald and not determinately not determinately bald. So, $\sim DDBald(f)$ and $\sim D\sim DBald(f)$. As a borderline case of determinate baldness then (or, if you prefer, a borderline case of borderline baldness), Fred is evidence of the fact that it is not determinately true that a man with f hairs on his head is determinately bald nor determinately false that such a man is determinately bald:

(4) $\sim DTrue[DBald(f)]$ and $\sim DTrue\sim[DBald(f)]$.

The paracomplete response results in $q \nvDash Dp, \sim Dp$. As before, this follows immediately given a definition of validity in terms of truth preservation and the subsequent identification of "determinate truth" with "truth" or by defining validity in terms of determinate-truth preservation.

The higher-order paradox is defused once we recognise this indeterminacy associated with the vagueness of the predicate involved, 'determinately bald'. As before, subjunctive theories like many-valued theories will reject the law of excluded middle, but now the thought is that excluded middle claims even for sentences asserting determinateness fail:

$\not\models Dp \vee {\sim}Dp$

with the consequence that the higher-order sorites paradox is diagnosed as having a false premise. A non-subjunctive theory like standard supervaluationism, on the other hand, endorses excluded middle focussing instead on the acceptability of the argument's conclusion. The theory again recognises the unacceptability of postulating sharp boundaries to the application of vague predicates, including those involving determinacy, but denies that the determinate truth of the conclusion:

(5) $DTrue\exists n[DBald(n) \,\&\, {\sim}DBald(n+1)]$

expresses the existence of a sharp boundary for 'bald'. Such a sharp boundary is said to be expressed, instead, by the claim:

(6) $\exists n[DTrueDBald(n) \,\&\, DTrue{\sim}DB(n+1)]$

and the existence of borderline cases for 'determinately bald'—i.e. borderline borderline cases of baldness—somewhere along the sorites series is taken to show that there can be no such n. The transition along the series from members for which it is determinately true that they are determinately bald (clear cases of determinate baldness) to members for which it is determinately false that they are determinately bald (clear cases of non-determinate baldness, e.g. clear borderline cases of baldness) proceeds by way of further intermediate cases for which it is neither determinately true nor determinately false that they are determinately bald. The unacceptable claim (6) is rejected, even though the theory validates the apparently equivalent claim (5).

Still higher orders of vagueness will generate further sorites paradoxes using the predicate 'determinately determinately bald', $DDBald(x)$, for example. And the responses offered above can be marshalled here too. In each case where a vague predicate is employed to generate a sorites paradox, one can respond by appealing to the existence of borderline cases at the appropriate level. First-order indeterminacy in respect of baldness is evidence of borderline cases as characterised in (1), cases whose existence is thought sufficient to dispel puzzlement surrounding the first-order sorites using the predicate 'bald'. Second-order indeterminacy in respect of baldness is evidence of borderline cases as characterised in (4), cases whose existence is thought sufficient to dispel puzzlement surrounding the second-order sorites using the predicate 'determinately bald'. And so on.[7]

---

[7]This is somewhat simplistic as regards both higher-order vagueness and higher-order sorites arguments. When discussing second-order vagueness one should, I think, recognise not only indeterminacy in respect of determinate baldness but also indeterminacy in respect of determinate non-baldness and indeterminacy in respect in borderline baldness, i.e. indeterminate baldness. Correlatively, one should recognise the existence of second-order sorites arguments using 'determinately not bald' and 'borderline bald', i.e. 'indeterminately bald'. And so on for higher orders. The simple discussion offered above, however, is sufficient. It is easy to see how the

There is then a clear sense in which the response to a sorites paradox using a predicate F, in appealing to talk of being *determinately* F to express the idea of borderline cases (by means of which the paradox is to be defused), makes available the expression of further paradox in respect of *determinately* F, in turn requiring talk of being *determinately determinately* F, which in turn makes available the expression of further paradox, and so on... The phenomenon of vagueness exacts revenge, drawing on the very resources invoked to defuse paradox to generate yet more paradox.

19.3 The Liar Paradox

Such revenge problems are familiar from discussions of the liar paradox. Consider the liar sentence λ_1, 'This sentence is not true'. Paradox is taken to ensue given the assumption that λ_1 is either true or not true, as follows. Assume $True(\lambda_1) \vee {\sim}True(\lambda_1)$. If $True(\lambda_1)$ then λ_1. But λ_1 says of itself that it is *not* true, hence it follows that ${\sim}True(\lambda_1)$, and so $True(\lambda_1)$ & ${\sim}True(\lambda_1)$. Contradiction. Suppose instead then that ${\sim}True(\lambda_1)$. Since λ_1 says of itself that it is not true, it follows that $True(\lambda_1)$, and so ${\sim}True(\lambda_1)$ & $True(\lambda_1)$. Contradiction again. Thus we have shown by dilemma that contradiction follows.

Since the assumption of either $True(\lambda_1)$ or ${\sim}True(\lambda_1)$ leads to contradiction, we ought conclude that, as with vague predications of borderline cases, it is indeterminate whether or not truth can be predicated of the liar sentence λ_1. That is to say, it is not uniquely determined whether or not $True(\lambda_1)$, i.e.:

(7) ${\sim}DTrue[True(\lambda_1)]$ and ${\sim}DTrue{\sim}[True(\lambda_1)]$.[8]

With the indeterminacy understood in the manner of a paracomplete response analogous to the liar paradox as outlined in Sect. 19.1, i.e. where truth is uniquely determined to apply if at all, it then follows that is neither determinately true nor determinately false that $True(\lambda_1)$.

As was the case when offering a paracomplete model of vagueness, we might as Field (2003) does, define validity in terms of determinate truth preservation with the result that the liar sentence proves the existence of a sentence p—namely, the sentence $True(\lambda_1)$—such that $\nvDash p, {\sim}p$. Assuming that some sentence q is nonetheless evaluated as determinately true, it follows that $q \nvDash p, {\sim}p$. Alternately, validity might be defined in the usual way as truth-preservation and our ordinary concept of truth be identified with "determinate truth". According to this response the liar sentence proves the existence of a sentence p that is neither true nor

newly recognised categories of higher-order vagueness might be employed in an attempt to dispel puzzlement engendered by each of the newly recognised higher-order arguments.

[8]Given that λ_1 says of itself that it is not true, $\lambda_1 \leftrightarrow {\sim}True(\lambda_1)$. So (7) then simplifies to the claim that: ${\sim}DTrue({\sim}\lambda_1)$ and ${\sim}DTrue(\lambda_1)$.

false—i.e. a truth-value gap theory is advocated—and for this reason $\nvDash p, \sim p$. Given the fact the some sentence q is nonetheless evaluated as true, we again have that $q \nvDash p, \sim p$. On either account the response is a paracomplete one.

The paradox itself is then variously defused depending again, as with the sorites paradox, on whether or not the paracomplete logic proposed accepts subjunction as valid. Subjunctive theories like those based on K_3, L_3 or L_∞ as before, reject the law of excluded middle as invalid and, more particularly, the particular instance on which the paradox depends, $True(\lambda_1) \vee \sim True(\lambda_1)$. Non-subjunctive theories like supervaluationism on the other hand, validate excluded middle but reject proof by cases thus declaring the paradox invalid.

Both McGee and Field reject the aforementioned approach according to which liar sentences are neither true nor false, rejecting the identification of ordinary truth with "determinate truth" (McGee 1990, p. 6; Field 2003, p. 270). Each rejects a paracomplete truth-value gap approach to the semantic paradoxes, with Field pursuing a paracomplete solution compatible with the retention of the T-schema (T). The retention of (T) is only possible, according to Field, if we reject such an approach; to accept the existence of truth-value gaps is to commit to $\sim[True(p) \vee True(\sim p)]$ which in conjunction with (T)—True(p) if and only if p—is equivalent to $\sim(p \vee \sim p)$ which entails $\sim p$ & $\sim\sim p$.[9] In the currently assumed paracomplete, consistent setting this leads to triviality by *ex contradictione quodlibet*.

Of course, the desire to retain (T) is reconcilable with the existence of gaps if the argument for incompatibility just given represents the rejection of bivalence—$not[True(p) \vee True(\sim p)]$—by means of an alternate negation operator, exclusion negation '$\neg$' which, like choice negation '$\sim$', takes truths to falsehoods and falsehoods to truths but, unlike choice negation, takes gaps to truths instead of to gaps. Thus (T) is compatible with gaps expressed as $\neg[True(p) \vee True(\sim p)]$. As Beall (2002) shows, all that follows is $\neg p$ & $\neg\sim p$ which is not a contradiction. So we can either identify truth with determinate truth and thus commit to gaps while retaining (T), as Beall shows, by distinguishing two kinds of negation or we can employ a univocal concept of negation while retaining (T) by distinguishing truth and determinate truth. For all that has been said so far then there is little to differentiate Field's position from the gap-view as described by Beall (2002, fn 6).

However, where gaps evidenced by the liar sentence are expressed by way of the stronger claim that Field takes exception to—$\sim[True(p) \vee True(\sim p)]$—we can note that (T) is indeed untenable. This being said, nothing as strong as (T) was used in the derivation of paradox above. 'True' as used in the derivation might be rejected as an adequate truth predicate by virtue of its failing what Field takes as a minimal constraint on truth, namely (T), but it nonetheless is sufficiently truth-like to validate the inference from '$True(p)$' to 'p' and vice versa and it is this that the paradox relies on. Irrespective of issues surrounding truth and theT-schema, the

[9] We are assuming that 'True($\sim p$)' is equivalent to 'False(p)', that the biconditional in (T) is contraposible and De Morgan's laws.

foregoing liar paradox is avoided by noting that the problematic sentence is neither determinately true nor determinately not true—i.e. it is indeterminate whether the sentence is true.

Now, as is well known, such a response will, even if considered an adequate response to the liar paradox generated by λ_1, immediately encounter the weakened liar sentence λ_2, 'This sentence is not determinately true', that generates the strengthened liar paradox. As with the sorites paradox, the very means employed for escaping the initial paradox makes available the possibility of revenge by way of the strengthened liar. However, *unlike the sorites paradox and higher-order revenge variants* any attempt to avoid this paradox by reapplying the response to the initial paradox will fail.

To see this, consider again the weakened liar sentence λ_2. Just as paradox follows from the assumption that λ_1 either is or is not in the extension of 'true', paradox now follows from the assumption that λ_2 either is or is not in the extension of 'determinately true'. Assuming the sentence is determinately true, it follows that things are as the sentence describes them as being so, since the sentence says of λ_2 that it is not determinately true, it follows that λ_2 is not determinately true after all. So if λ_2 is determinately true then it is both determinately true and not determinately true. Contradiction. Suppose then, contrariwise, that the sentence is *not* determinately true. Since this is what λ_2 says, it follows that λ_2 is therefore true. Moreover, this has just been shown to be the case, the foregoing reasoning settles the matter, hence λ_2 is determinately true.[10] If λ_2 is not determinately true then it is both determinately true and not determinately true. Contradiction. On the assumption then that λ_2 is either determinately true or not, it is both determinately true and not determinately true. Thus paradox reasserts itself in respect of λ_2.

In respect of the strengthened liar we now face renewed contradiction on the assumption that λ_2 is either determinately true or not determinately true and so we might seek to avoid contradiction, as before, by claiming that it is indeterminate whether or not determinate truth can be predicated of λ_2. That is to say, it is indeterminate whether or not $DTrue(\lambda_2)$, i.e. it is neither determinately true nor determinately false that $DTrue(\lambda_2)$.

Just as the liar sentence λ_1 was seen to exhibit first order indeterminacy, satisfying:

$$\sim DTrue[True(x)] \text{ and } \sim DTrue \sim[True(x)]$$

the weakened liar sentence λ_2 might now appear to exhibit second order indeterminacy, satisfying:

$$\sim DTrue[DTrue(x)] \text{ and } \sim DTrue \sim[DTrue(x)].$$

[10]McGee, in effect, contests the inference from 'λ_2 is true' to 'λ_2 is determinately true', thus avoiding contradiction (McGee 1990, p. 7), but many including Field endorse the inference (Field 2003, p. 298). Suffice to say there is a debate to be had here, but space precludes its presentation. That is a story for another day.

In fact, we might seek to generalise. Where we have a truth or truth-like predicate τ (e.g. 'true', 'determinately true', 'determinately determinately true', etc.) satisfying:

(τ – **Elim**) p follows from $\tau(p)$

and

(τ – **Intro**) $\tau(p)$ follows from p,

then sentences of the form 'This sentence is not τ' can be shown to lead to paradox on the assumption that the sentence itself is either τ or not τ. On the approach being advocated, paradox is said to be avoided by noting that it is neither determinately true nor determinately false whether the problematic sentence is τ—i.e. it is indeterminate whether the sentence is τ. Each truth-like predicate is shown, on pain of paradox, to have indeterminate extension.

Indeed, if one thinks that the liar paradox and subsequent strengthenings are to be handled in the same way that the sorites paradox and subsequent higher-order paradoxes were handled then such a view makes sense. Just as each appropriately chosen member of a sorites series was found, on pain of paradox, to be such that it was indeterminate whether it satisfied the relevant higher-order vague predicate, each of the respective weakened liar sentences—'This sentence is not true', 'This sentence is not determinately true', 'This sentence is not determinately determinately true', etc.–will be said, on pain of paradox, to be such that it is indeterminate whether it satisfies the relevant higher-order truth-like predicate—'true', 'determinately true', 'determinately determinately true', etc. respectively. On this view, the initial and subsequent revenge paradoxes of both vagueness and self-reference alike are similarly answered and the logic of each is the same. The paradoxes are indeed unified and the lesson to drawn from each class of paradoxes is that the relevant predicates are indeterminate as regards their extensions.

However, the close similarity is an illusion. If the strengthened liar is to be solved at all within a paracomplete approach being considered then it is *not* solved in the same way. As the reader may have already surmised, the attempt to respond to the strengthened paradox involving λ_2 by claiming that it exhibits a kind of second-order indeterminacy satisfying '$\sim DTrue[DTrue(x)]$ and $\sim DTrue\sim[DTrue(x)]$' will not, in fact, avoid paradox. To see this, consider the claim that:

(8) $\sim DTrue[DTrue(\lambda_2)]$ and $\sim DTrue\sim[DTrue(\lambda_2)]$.

From this it follows that $\sim DTrue\sim[DTrue(\lambda_2)]$ and so, given that $\lambda_2 \leftrightarrow \sim DTrue(\lambda_2)$, by substitution $\sim DTrue(\lambda_2)$. Since λ_2 then follows by further substitution, we can further infer by τ-Intro that $DTrue(\lambda_2)$. Contradiction thus ensues.

The initially tempting thought that we might unify the two paradoxes by way of a paracomplete approach must be rejected.

19.4 The Paraconsistent Turn

What then of a paraconsistent approach to the paradoxes? Recognising each as involving a species of indeterminacy and recognising that truth must already be admitted as vague, might the paradoxes be united if considered from a paraconsistent perspective?

Consider, again, the sorites paradox, in particular the line drawing form of the sorites paradox. A man with one hair on his head is surely bald, yet a man with a million hairs on is head is not bald. Since, for any number n, a man with n hairs on his head is either bald or not bald, it follows that there must be some n for which a man with n hairs on his head is bald while a man with $n + 1$ hairs on his head is not bald (i.e. hirsute). As before:

$Bald(1)$
$\sim Bald(10^6)$
$\forall n[Bald(n) \vee \sim Bald(n)]$
So, $\exists n[Bald(n)\ \&\ \sim Bald(n+1)]$

Again, the vagueness of predicates like *Bald(x)* speak against supposing there to be such a sharp line between their application and the application of their negation. No single hair can make the difference between baldness and non-baldness; etc. The foregoing argument results in the unacceptable.

From a paraconsistent perspective, avoidance of the unacceptable again follows from the recognition that vague predicates admit of indeterminacy, i.e. borderline cases. On this approach, the indeterminacy of application of the predicate *Bald(x)*—its application not being uniquely determined—now results in cases to which neither the predicate nor its negation is uniquely determined to apply by virtue of both the predicate and its negation truly applying.

Recall Harry, mentioned earlier; suppose that the number of hairs Harry has on his head is h. It is simply indeterminate whether he is bald or not; i.e. Harry is not uniquely determined to be bald, $\sim DBald(h)$, and not uniquely determined to be not bald, $\sim D\sim Bald(h)$. As a borderline case of baldness then, Harry is evidence of the fact that it is not uniquely determined to be true that a man with h hairs on his head is bald nor uniquely determined to be false that such a man is bald:

(1)$\sim DTrue[Bald(h)]$ and $\sim DTrueDTrue\sim[Bald(h)]$.

This much is common data for both paracomplete and paraconsistent approaches. On a paraconsistent approach to vagueness, however, this lack of uniquely determined truth or falsity in respect of a sentence p results in $p, \sim p \nvDash q$ for some sentence q. To see this notice firstly that, from a paraconsistent perspective, every sentence is determined to have a truth-value and any talk of determinateness is redundant—lack of uniquely determined truth or falsity amounts then simply to lack of unique truth or falsity.[11] Given then that lack of unique truth entails falsehood and lack of unique falsehood entails truth, it follows that:

(9)*True*[$\sim$*Bald*(h)] and *True*[*Bald*(h)].

Subsequently defining validity as truth preservation and recognising that not every sentence is true, the result follows.

The paradox is taken to be defused once we recognise this indeterminacy associated with the vagueness of the predicate involved, *Bald(x)*. Notice that on the approach being pursued any borderline case n of the predicate *Bald(x)* is such that *Bald*(n) is true and so too $\sim$*Bald*(n). Thus, where adjunction holds (p, $\sim p \vDash p \,\&\, \sim p$), e.g. in *LP*, we see that it is true for some n that *Bald*(n) & $\sim$*Bald*(n), i.e. it is true that $\exists n$[*Bald*(n) & $\sim$*Bald*(n)]. In fact, this is true for any borderline case of baldness. Given monotonicity (i.e. anyone with more hair than a non-bald person is non-bald) it then follows from $\sim$*Bald*(n) that $\sim$*Bald*($n + 1$) and so it is obviously true that $\exists n$[*Bald*(n) & $\sim$*Bald*($n + 1$)]. Adjunctive theories will endorse the seemingly paradoxical conclusion of the sorites. But, of course, its negation is also true, i.e. it is true that $\forall n$[$\sim$*Bald*(n) $\vee$ *Bald*($n + 1$)]; since every pair of adjacent items $\langle n, n + 1\rangle$ in the sorites series are such that they are both true or both false (and sometimes both), every pair instantiates the tolerance principle that underwrites the sorites paradox in the form of the universally quantified claim just cited. The indeterminacy that attends the application of the vague predicate *Bald(x)* is evidenced by its being both true and false that there is a cut-off to its application.[12]

On such an approach the truly paradoxical claim expressing the existence of a sharp cut-off between the bald and the non-bald is the claim that it is true and true *only* that $\exists n$[*Bald*(n) & $\sim$*Bald*($n + 1$)] and this of course avoided.

Reminiscent of paracomplete responses to the sorites paradox though, the phenomenon that underwrites the sorites paradox, vagueness, might again be thought to exact revenge by way of higher-order vagueness. Having already admitted into our language an operator D, 'it is uniquely determined that', for the expression of borderline cases, a new sorites can be expressed in the extended language. The paraconsistent modelling of the extension of the vague predicate *Bald(x)* partitions it into those items to which the predicate is uniquely determined to apply (i.e. those for which the predication is uniquely true and thus true only), those to which neither the predicate nor its negation is uniquely determined to apply (i.e. the borderline cases for which the predication is neither uniquely true nor uniquely false and thus is both true and false) and those to which the predicate's negation is uniquely determined to apply (i.e. those for which the predication is uniquely false and thus false only). And so, as before, we might wonder as to the existence of a sharp cut-off between those uniquely determined to be bald and those not uniquely determined to be bald. Higher-order vagueness speaks against the existence of such a boundary yet it appears that its existence can be proven.

[11]This presumption of determinateness thus amounts to a presumption of completeness.

[12]We shall leave discussion of non-adjunctive paraconsistent approaches for another day.

Thus it appears that further paradox threatens using the predicate of the extended language, $DBald(x)$, as follows. Since a man with one hair on his head is uniquely determined to be bald and a man with h hairs on his head is not (*ex hypothesi* he is both bald *and* not bald), and for any n a man with n hairs on his head is either uniquely determined to be bald or not uniquely determined to be bald, it follows that there must be some n such that a man with n hairs on his head is uniquely determined to be bald while a man with $n + 1$ is not.

$DBald(1)$
$\sim DBald(h)$
$\forall n[DBald(n) \vee \sim DBald(n)]$
So, $\exists n[DBald(n) \& \sim DBald(n + 1)]$

Thus, a sharp line is seemingly postulated between the application of the predicate $DBald(x)$ and $\sim DBald(x)$ corresponding to a sharp boundary between the uniquely true and the rest. But higher-order vagueness establishes that there is no more a sharp boundary between those people who are uniquely determined to be bald and those who are not uniquely determined to be bald than there was between those people who are bald and those who are not bald. Paradox appears to reassert itself.

The response is exactly the same as it was before; there is no paradox here. It is indeed true that $\exists n[DBald(n) \& \sim DBald(n + 1)]$ but is also false, reflecting the indeterminacy that attends the application of the vague predicate $DBald(x)$ as evidenced by the existence of borderline cases for the predicate (borderline cases which are, in turn, borderline cases of borderline cases of *Bald(x)*). Such a borderline case, f say, is such that $\sim DDBald(f)$ and $\sim D\sim DBald(f)$. It is neither uniquely determined to be true that $DBald(f)$ nor is it uniquely determined to be false; thus it is both true and false that $DBald(f)$. Hence, it is true that $DBald(f)$ and true that $\sim DBald(f)$, therefore true that $DBald(f)$ & $\sim DBald(f)$, from which it follows that it is true that $\exists n[DBald(n) \& \sim DBald(n)]$. Since adding more hairs to someone not uniquely determined to be bald results in their remaining not uniquely determined to be bald, some simple logic establishes the truth of $\exists n[DBald(n) \& \sim DBald(n + 1)]$. And for reasons exactly analogous to those given earlier for the falsehood of $\exists n[Bald(n) \& \sim Bald(n + 1)]$ we can establish the falsehood of $\exists n[DBald(n) \& \sim DBald(n + 1)]$, in addition to its truth. The moral is that *any* such paradox generated by a vague predicate is answerable in this way.

It might be objected, however, that the revenge paradox has not been properly expressed by the foregoing argument and revenge problems have been simply sidestepped. The foregoing establishes that it is both true and false that $DBald(f)$, i.e. that it is both true and false that a man with f hairs on his head is uniquely determined to be bald. But, the objection goes, if that claim is true then f satisfies $Bald(x)$ and does so *uniquely*, hence it is precluded from satisfying its negation yet if the claim is false (as established) then it does satisfy its negation. So $DBald(x)$ does not properly capture the requisite uniqueness required for the revenge paradox. Revenge is properly exacted by a (higher-order) sorites argument that invokes a vague predicate capable of expressing the concept of being a true and undeniable satisfier of $Bald(x)$ in the sense that precludes *anything* to the contrary... including anything to the contrary of precluding anything to the contrary, and so on.

To put the point another way, the objection is founded on the idea that where there is a distinction drawn between the uniquely true (predications involving clear exemplars) and the true-and-false (predications involving borderline cases), this distinction must, paradoxically, be sharply drawn since the only means available for describing borderline cases for the distinction is to claim that there are cases for which it is both true and false that they are uniquely true yet this is impossible. If it is false that it is uniquely true then it is not uniquely true, yet it is also uniquely true since, *ex hypothesi*, it is true that it is uniquely true. Thus there can not be borderline cases between the uniquely true and the true-and-false on pain of contradiction. Sweet revenge? Not yet since paraconsistency treats the "pain of contradiction" as a phantom pain and the contradiction involved in claiming that some predications are uniquely true and not uniquely true is one such. What *will* be distinguished by a paradoxically sharp boundary then, it might be thought, are those truths that are nowise false, at all, in any way and those that are true and false. If we can express this very stringent, exclusive notion of truth then the objector will have their revenge.

But a paraconsistent approach can admit that something might be both "a truth that is nowise false at all in any way" and not such a truth by virtue of also being false. In fact, regardless of how stringent and exclusive a notion of truth we invoke, it will always remain open to the paraconsistentist to admit that something both satisfies this notion and also does not satisfy it. Thus I think we can see why revenge cannot, *contra* the objector, be exacted on the paraconsistent approach being outlined. Attempts at revenge simply produce ever more sorites paradoxes disarmed in exactly the same way as those which have come before. The means for their solution cannot be used to construct another sorites paradox which by its very nature is immune to solution.

Of course, just such a point has been made in respect of attempts to exact revenge on paraconsistent approaches to the liar paradox. Consider the liar sentence λ_1 again, 'This sentence is not true'. Assuming $True(\lambda_1) \vee {\sim}True(\lambda_1)$ it follows that $True(\lambda_1)$ and ${\sim}True(\lambda_1)$. Mirroring the indeterminacy of application of the predicate $Bald(x)$, the application of $True(x)$ is not uniquely determined—λ_1 now presents us with a sentence to which neither $True(x)$ nor its negation is uniquely determined to apply by virtue of both the predicate and its negation truly applying.

As in the paracomplete case, λ_1 is not uniquely determined to be true, ${\sim}DTrue(\lambda_1)$, and not uniquely determined to be not true, ${\sim}D{\sim}True(\lambda_1)$; i.e. it is indeterminate whether λ_1 is true. It is easy to establish again therefore that:

(7)${\sim}DTrue[True(\lambda_1)]$ and ${\sim}DTrue{\sim}[True(\lambda_1)]$.

Given then that lack of unique truth entails falsehood and lack of unique falsehood entails truth, it follows, mimicking (9), that:

(10)$True[{\sim}True\lambda_1)]$ and $True[True(\lambda_1)]$.

In a paraconsistent setting then λ_1 stands in the same relation to $True(x)$ that h stands in to $Bald(x)$, and the liar and sorites paradoxes are, to this extent, of a kind. Moreover, just as attempts at revenge in respect of the sorites were seen to founder, so too (as is well known) with the liar. On the paracomplete approach to the initial liar paradox a new semantic category (the category of being neither determinately true nor determinately false) was invoked into which the liar sentence λ_1 was said to fall. Thus, it was said, if we admit that sentences might fall into this new category paradox can be avoided in respect of λ_1. Revenge was sought by way of a weakened liar sentence, λ_2, by means of which paradox was said to follow *even admitting this new category*. Analogously then, on the paraconsistent approach, having admitted a new semantic category (neither uniquely true nor uniquely false, i.e. both true and false) to avoid paradox in respect of λ_1, revenge may be sought by way of a weakened liar sentence λ_3 from which paradox might be taken to follow even admitting this new semantic category. But the attempt fails for the same reason that it did in respect of the attempted revenge sorites.

An obvious candidate for exacting revenge is the liar sentence λ_3, 'This sentence is not-true only', i.e. 'This sentence is uniquely determined to be not true', expressible in the extended language as '$D{\sim}True(\lambda_3)$'. Bearing in mind the semantic innovation of the paraconsistent approach, paradox is taken to follow given the assumption that $D{\sim}True(\lambda_3) \vee {\sim}D{\sim}True(\lambda_3)$. If $D{\sim}True(\lambda_3)$ then since λ_3 says of itself that it is uniquely determined to be not true it follows that what it says is true, i.e. *True*(λ_3). But then, if true, it is not uniquely determined to be not true. Hence, on the assumption that $D{\sim}True(\lambda_3)$, it follows that $D{\sim}True(\lambda_3)$ and ${\sim}D{\sim}True(\lambda_3)$. Suppose, on the other hand, then that ${\sim}D{\sim}True(\lambda_3)$. If not uniquely determined to be not true, it must be true (though not necessarily uniquely so). Since λ_3 says of itself that it is uniquely determined to be not true and this is supposed to be true it follows that λ_3 *is* uniquely determined to be not true and so, on the assumption that ${\sim}D{\sim}True(\lambda_3)$, it follows that ${\sim}D{\sim}True(\lambda_3)$ and $D{\sim}True(\lambda_3)$. By dilemma then $D{\sim}True(\lambda_3)$ and ${\sim}D{\sim}True(\lambda_3)$. Contradiction.

Just as the higher-order sorites established that it was both true and false that *DBald*(x) applied to borderline case f, the extended liar just considered establishes that it is both true and false that $D{\sim}True(x)$ applies to λ_3. On the paraconsistent approach being advocated, the resulting contradiction can again simply be embraced. Moreover, since $\lambda_3 \leftrightarrow D{\sim}True(\lambda_3)$, by substitution we see that the extended liar sentence λ_3 itself provides yet another example of a sentence that is both true and false.

Of course, the proponent of the revenge paradox might, as with the attempt at revenge by way of higher-order vagueness, complain that the revenge problem has not been properly posed. When constructing a sentence that says of itself that it is *uniquely* determined to be not true, i.e. not-true *only*, this ought preclude its being true and so the possibility that the sentence is both uniquely determined to be not true yet also true by virtue of being not uniquely determined to be not true should be ruled out. Yet the response to the revenge problem described above embraces this seemingly impossible position.

As before, however, the attempt at revenge tries to appeal to a sentence whose truth conditions rule out inconsistent semantic evaluation by invoking the idea of being uniquely not true (and thus not also true). But the paraconsistentist can admit a sentence as being uniquely evaluated in some way while also admitting that the sentence is not uniquely evaluated in this way. To be sure this is yet another contradiction but that has yet to be shown to be any more problematic than admitting contradiction at the outset. Revenge no more dogs the liar by way of strengthened paradox than it dogs the sorites by way of higher-order paradox.

19.5 Conclusion

We have considered attempts to unify the liar and sorites paradoxes and found that while they both may be said to exhibit indeterminacy and be alike in this respect, attempts to model the indeterminacy by way of a paracomplete logic result in the two paradoxes diverging in their logical structure in the face of extended paradoxes. If, on the other hand, a paraconsistent logic is invoked then the paradoxes are both of a kind in having their source in the indeterminacy of the relevant predicates involved and, moreover, the treatment of the extended paradoxes generated in each case are also of a kind. Only paraconsistency then offers the prospect of a unified treatment of these vexing puzzles.

References

Beall, J.C., and M. Colyvan. 2001. Heaps of gluts and hyde-ing the sorites. *Mind* 110: 401–408.

Beall, J.C. 2002. Deflationism and gaps: Untying 'not's in the debate. *Analysis* 62: 299–305.

Burgess, J.A. 1998. In defence of an indeterminist theory of vagueness. *Monist* 81: 233–252.

Field, H. 2003. Semantic paradoxes and the paradoxes of vagueness. In *Liars and heaps*, ed. J.C. Beall and M. Glanzberg, 262–311. Oxford: Oxford University Press.

Field, H. 2008. Solving the paradoxes, escaping revenge. In *Revenge of the liar*, ed. J.C. Beall, 78–144. Oxford: Oxford University Press.

Fine, K. 1975. Vagueness, truth and logic. *Synthese* 30: 265–300.

Keefe, R. 2000. *Theories of vagueness*. Cambridge: Cambridge University Press.

McGee, V. 1990. *Truth, vagueness and paradox*. Indianapolis: Hackett.

Rolf, B. 1980. A theory of vagueness. *Journal of Philosophical Logic* 9: 315–325.

Tappenden, J. 1993. The liar and sorites paradoxes: Toward a nified treatment. *Journal of Philosophy* 90: 551–77.

Van Fraassen, B.C. 1966. Singular terms, truth-value gaps, and free logic. *Journal of Philosophy* 63: 481–85.

Chapter 20
Vague Inclosures

Graham Priest

20.1 Introduction: Vagueness and Self-reference

Sorites paradoxes and the paradoxes of self-reference are quite different kinds of creature. The first are generated by the fact that some predicates have a certain kind of tolerance to small changes in their range of application. The second are generated by the fact that some things can refer, directly or indirectly, to themselves. Or so it seemed to me until recently. I am now inclined to think differently. The paradoxes of self-reference can naturally be seen as having a form given by the Inclosure Schema. In the Schema, a construction is applied to collections of a certain kind to produce a different object of the same kind. Contradiction arises at the limit of all things of that kind. Sorites paradoxes can be seen as having exactly the same form. In this paper, I will start by explaining how. Given that paradoxes of sorites and self-reference are of the same kind, they should have the same kind of solution. I hold that a dialetheic solution is the correct one for paradoxes of self-reference. It follows that a dialetheic solution is therefore appropriate for sorites paradoxes. The rest of the paper investigates what such a solution is like, and especially so called "higher order" vagueness.

G. Priest (✉)
Departments of Philosophy, University of Melbourne, Australia, University of St Andrews, Scotland

The Graduate Center, City University of New York, USA
e-mail: g.priest@unimelb.edu.au

K. Tanaka et al. (eds.), *Paraconsistency: Logic and Applications*, Logic, Epistemology, and the Unity of Science 26, DOI 10.1007/978-94-007-4438-7_20,

20.2 The Inclosure Schema

Let us start with the Inclosure Schema and its application to the paradoxes of self-reference.[1] An inclosure paradox arises when for some monadic predicates φ and θ, and a one place function, δ, there are principles which appear to be true, or a priori true (see Priest 1995, p. 277), and which entail the following conditions. (It is not required, note, that the arguments entailing the conditions be sound, though dialetheism prominently allows for this possibility.)

1. There is a set, Ω, such that $\Omega = \{x : \varphi(x)\}$, and $\theta(\Omega)$ (**Existence**)
2. If $X \subseteq \Omega$ and $\theta(X)$:

   (a) $\delta(X) \notin X$ (**Transcendence**)
   (b) $\delta(X) \in \Omega$ (**Closure**)

(A special case of an inclosure is when $\theta(X)$ is the vacuous condition, $X = X$, and so mention of it may be dropped.) Given these conditions, a contradiction occurs at the limit when $X = \Omega$. For then we have $\delta(\Omega) \notin \Omega \wedge \delta(\Omega) \in \Omega$.

To illustrate: In the Burali-Forti paradox, $\varphi(x)$ is 'x is an ordinal', so that Ω is the set of all ordinals, On—defined, let us assume, as von Neumann ordinals. $\theta(X)$ is the vacuous condition; and $\delta(X)$ is the least ordinal greater than every member of X. By definition $\delta(X)$ satisfies Transcendence and Closure. The brunt of the Burali-Forti paradox is exactly in showing that $\delta(X)$ is well defined, even when $X = \Omega$. The reasoning shows that On is itself an ordinal—an ordinal greater than all ordinals.

In the liar paradox, $\varphi(x)$ is the predicate Tx, 'x is true', so that Ω is the set of true sentences; $\theta(X)$ is the predicate 'X is definable', i.e., is a set that is referred to by some name; if X is definable, let N be an appropriate name; then $\delta(X)$ is a sentence, σ, constructed by an appropriate self-referential construction, of the form $\langle \sigma \notin N \rangle$. (I use angle brackets as a name-forming device.) Liar-type reasoning establishes Transcendence and Closure. The liar paradox arises in the limit. $\Omega = \{x : Tx\}$, and $\delta(\Omega)$ is a sentence, σ, of the form $\langle \sigma \notin \{x : Tx\} \rangle$, i.e., '$\sigma$ is not true'.

20.3 Sorites and Inclosures

Let us now see how sorites paradoxes fit the Schema.

In a sorites paradox there is a sequence of objects, $a_0, \ldots, a_n$, and a vague predicate, P, such that Pa_0 and $\neg Pa_n$; but for successive members of the sequence there is very little difference between them with respect to their P-ness, so that if one satisfies P, so does the other—the principle of tolerance.

[1]For details of the following, see especially Priest (1995, Part 3).

For the Inclosure Schema, let $\varphi(x)$ be Px, so $\Omega = \{x : Px\}$; $\theta(X)$ is the vacuous condition. Ω is a subset of $A = \{a_0, \ldots, a_n\}$—indeed, a proper subset, since a_n is not in it—and so we have Existence. If $X \subseteq \Omega$ then, since X is a proper subset of A, there must be a first member of A not in it. Let this be $\delta(X)$. By definition, $\delta(X) \notin X$. So we have Transcendence. Now, either $\delta(X) = a_0$ (if $X = \phi$), and so $P\delta(X)$; or (if $X \neq \phi$) $\delta(X)$ comes immediately after something in $X \subseteq \Omega$, so $P\delta(X)$, by tolerance. In either case, $\delta(X) \in \Omega$, so we have Closure.

The inclosure contradiction is of the form $\delta(\Omega) \notin \Omega \wedge \delta(\Omega) \in \Omega$. In the case of the sorites paradox, the contradiction is that the first thing in the sequence that is not P is P. Diagonalisation takes us out of X; and tolerance keeps us within Ω. We see why a contradiction occurs at the limit of P-things.

If the self-referential paradoxes and sorites paradoxes are of the same kind, the Principle of Uniform Solution—'same kind of paradox, same kind of solution'—tells us that we should expect the same kind of solution (see Priest 1995, §§ 11.5, 11.6, 17.6). I take the correct solution to the paradoxes of self-reference to be a dialetheic one (see Priest 1987, 1995). It follows that the solution to the sorites paradoxes should be so too. A simple-minded thought is this: In the case of the paradoxes of self-reference we endorse the soundness of the arguments. These establish certain contradictions, the trivialising consequences of which are avoided by not endorsing Explosion. We should just do the same in sorites paradoxes: endorse the soundness of the arguments. But this cannot be right. Sorites paradoxes are, in their own right, as near triviality-making as makes no difference. One can prove that an old thing is young, that a red thing is blue, and anything else for which one can postulate an appropriate sorites progression. We must be less simple-minded. What follows is, hopefully, so.

20.4 The Structure of Sorites Transitions

Come back to the sorites progression of Sect. 20.3. Pa_0 is true (and true only); Pa_n is false (and false only). If we write the least-number operator as μ then $\delta(\{x : Px\})$ is $\mu h(a_h \notin \{x : Px\}$, that is, $\mu h \neg Pa_h$.[2] Let this be a_i. The Inclosure Schema tells us that $a_i \in \{x : Px\}$ and $a_i \notin \{x : Px\}$, that is, $Pa_i \wedge \neg Pa_i$. So we know that there is at least one h for which Pa_h is both true and false. For all we have seen so far, there may be more than one. If there are, there is no reason, in principle, why these should be consecutive,[3] but the uniform nature of a sorites progression at least

[2]In naive set theory, the comprehension schema gives: $y \in \{x : Px\} \leftrightarrow Py$, and contraposition gives $y \notin \{x : Px)\} \leftrightarrow \neg Py$.

[3]The Technical Appendix to Part 3 of Priest (1995) constructs models of the Inclosure Schema where some ordinals are consistent and some are not. Sect. 4 of the Appendix gives a model in which inconsistent ordinals need not be consecutive.

suggests this. Assuming it to be so, the structure of a sorites progression will look like this, where a_k is the last thing that is P, and $i \leq k$.[4]

$$\begin{array}{llllllll} a_0 & \ldots & a_i & \ldots & a_k & \ldots & a_n & \\ [- & - & P & - & -] & & & \\ & & [- & - & \neg P & - & -] & \end{array}$$

We can think of the sequence of dialetheic objects as providing a transition from the things that are definitely P to the things that are definitely not P. Many have argued that in sorties progressions there is a borderline area where the relevant statements have truth value gaps. What intuition actually tells us is that in the middle of the progression, things are symmetric with respect to the ends. The statements about the transition objects should therefore be symmetric with respect to the statements about the ends. And from this point of view, being *both* true and false is as good as being *neither* (see Hyde 1997).

Most importantly, however, note the position of a_i. It might be thought that the first thing that is not P should be a_{k+1}, but it is not. The first thing that is *not* P is actually identical with, or to the left of, the last thing which is P! a_{k+1} is not the first thing that is not P, but the first thing of which P is not true. We are in the territory of higher order vagueness here. We will turn to that matter later.

20.5 Sorites Arguments

What does this tell us about sorites arguments? What tolerance tells us is that for some appropriate biconditional, $\Leftrightarrow$, $Pa_h \Leftrightarrow Pa_{h+1}$ (for $0 \leq h < n$). The sorites argument is then of the form:

$$\cfrac{\cfrac{Pa_0 \quad Pa_0 \Leftrightarrow Pa_1}{Pa_1} \quad Pa_1 \Leftrightarrow Pa_2}{Pa_2}$$

$$\ddots$$

$$\cfrac{Pa_{n-1} \quad Pa_{n-1} \Leftrightarrow Pa_n}{}$$

The next question is what this biconditional is. The correct understanding is, I take it, that it is a material biconditional, $\equiv$: consecutive sorites statements have the same truth value. This is what, it seems to me, tolerance is all about. Thus, where $\alpha \supset \beta$ is $\neg\alpha \vee \beta$, we have $(Pa_h \supset Pa_{h+1}) \wedge (Pa_h \supset Pa_{h+1})$. This is true if Pa_h and Pa_{h+1} are both true or both false. (If one is true and the other is false, it is false as well.)

[4]It is clear from the diagram that $\{x : Px\} \cap \{x : \neg Px\}$ is not empty. But since this set is $\{x : Px\} \cap \overline{\{x : Px\}}$, it is empty as well. It is difficult to represent this fact in a consistent diagram!

Given this understanding of the conditional, every major premise of the argument is true. For every h, Pa_h and Pa_{h+1} are both true or both false. But assuming an appropriate paraconsistent logic,[5] the disjunctive syllogism (DS)—*modus ponens* (MP) for the material conditional—is invalid: $\alpha, \alpha \supset \beta \nvdash \beta$; and of course, exactly the same is true of the material biconditional.

It should be noted that, though the sorites argument itself is invalid, the situation is still inconsistent. The sorites is generated by the sentences:

$$Pa_0$$

$$\neg Pa_n$$

$$(Pa_h \equiv Pa_{h+1}) \qquad (0 \leq h < n)$$

From these, we cannot prove $Pa_h \wedge \neg Pa_h$, for any particular h; but can prove:

$$\bigvee_{0 \leq h \leq n} (Pa_h \wedge \neg Pa_h)$$

To see this, write α_h for Pa_h. Then α_0 and $\alpha_0 \equiv \alpha_1$ give $(\alpha_0 \wedge \neg\alpha_0) \vee \alpha_1$. This, plus $\alpha_1 \equiv \alpha_2$, give $(\alpha_0 \wedge \neg\alpha_0) \vee (\alpha_1 \wedge \neg\alpha_1) \vee \alpha_2$; and so on till we have $(\alpha_0 \wedge \neg\alpha_0) \vee \ldots \vee (\alpha_{n-1} \wedge \neg\alpha_{n-1}) \vee \alpha_n$. Whence $\neg\alpha_n$ gives the result. In other words, this information tells us that the inclosure is located somewhere along the track; but it, itself, does not tell us exactly where.

20.6 "Extended" Paradoxes of Self-reference

We now come to the vexed question of so called higher order vagueness. Let me start, for reasons that will become clear later, by talking about an apparently different issue: "extended paradoxes" in the context of the semantic paradoxes. When people offer solutions to the semantic paradoxes of self-reference, it always seem to turn out that the machinery that they deploy to solve them allows the formulation of paradoxes equally virulent—or maybe better, simply moves the old paradox to a new place. Let me illustrate with respect to the liar and truth value gaps.

The semantic paradoxes deploy the T-schema. If we write T for the truth predicate, and angle brackets for naming, then the T-schema is the principle that:

$$T\langle\alpha\rangle \leftrightarrow \alpha$$

for every closed sentence, α (where $\leftrightarrow$ is an appropriate, detachable, biconditional.[6]). Writing F for the falsity predicate, so that $F\langle\alpha\rangle$ is $T\langle\neg\alpha\rangle$, the simple

[5]In what follows, we will take this to be the logic LP of Priest (1987, Chap. 5); but matters are much the same in virtually every paraconsistent logic.

[6]For the sake of definiteness, let this be the conditional of Priest (1987, §19.8).

liar paradox is a sentence, λ_0, obtained by some technique of self-reference, of the form $F\langle\lambda_0\rangle$. Substituting in the T-schema, we get:

$$T\langle\lambda_0\rangle \leftrightarrow F\langle\lambda_0\rangle$$

The Principle of Bivalence tells us that for all α:

$$T\langle\alpha\rangle \vee F\langle\alpha\rangle$$

and applying this to λ_0, we infer $T\langle\lambda_0\rangle \wedge F\langle\lambda_0\rangle$: λ_0 is both true and false.

A standard suggestion is to avoid this conclusion is to deny the Principle of Bivalence. Sentences are not necessarily true or false; some are neither (N). So the Principle is replaced by:

$$T\langle\alpha\rangle \vee F\langle\alpha\rangle \vee N\langle\alpha\rangle$$

True, we can no longer infer that λ_0 is both true and false, but now we can construct the "extended liar paradox", a sentence λ_1 of the form $F\langle\lambda_1\rangle \vee N\langle\lambda_1\rangle$. Substituting this in the T-schema, we get:

$$T\langle\lambda_1\rangle \leftrightarrow (F\langle\lambda_1\rangle \vee N\langle\lambda_1\rangle)$$

And all three of the possibilities lead to trouble.

Such a conclusion is obviously fatal to gap-theories of this kind. Some have thought that extended paradoxes of the same kind sink dialetheic ("glut") theories. That $T\langle\lambda_0\rangle \wedge F\langle\lambda_0\rangle$ is obviously no problem for a glut theory. The extended liar is now a sentence, λ_2, of the form: $F\langle\lambda_2\rangle \wedge \neg T\langle\lambda_2\rangle$; or given that there are no gaps, so that anything not true is false, just $\neg T\langle\lambda_2\rangle$. Substituting in the T-schema gives:

$$T\langle\lambda_2\rangle \leftrightarrow \neg T\langle\lambda_2\rangle$$

and so, given the Law of Excluded Middle, $T\langle\lambda_2\rangle \wedge \neg T\langle\lambda_2\rangle$. But only a little thought suffices to show that this is no problem for a dialetheist. Dialetheism was never meant to give a consistent solution to the paradoxes. (Even in the case of the simple liar, things are inconsistent, since we have $\lambda_0 \wedge \neg\lambda_0$.) The point was to allow contradictions, but in a controlled way. The "extended" argument does show, however, that the very categories deployed in a dialetheic account of the paradoxes are themselves subject to the very sort of inconsistency they characterise. This is, indeed, to be expected. We may show, moreover, that all the inconsistencies generated are under control, by constructing a single "semantically closed" theory, which is inconsistent, but in which the inconsistencies are quarantined. Specifically, we can take a first-order language with a truth predicate, T, and some form of naming device, $\langle . \rangle$. We can then formulate a theory in this language, which contains all instances of the T-schema, and an appropriate form of self-reference. The theory can be shown to be inconsistent, but non-trivial.[7]

[7] Specifically, no inconsistencies involving only the grounded sentences of the language (in the sense of Kripke) are provable. See Priest (2002, § 8.2).

20.7 Higher Order Vagueness

Let us now return to vagueness. Sorites paradoxes occur because the nature of the transition in a sorites progression is problematic. The straight-forward picture:

$$\begin{array}{c} a_0 \ldots\ldots\ldots\ldots\ldots a_n \\ [-\ P\ -]\ [-\ \neg P\ -] \end{array}$$

jars because of the counter-intuitive nature of the cut-off point between the true and the false. The solution that we have been looking at removes this cut-off point. But though the machinery does so, it produces, instead, two others—one between the true only and the both true and false, and one between the false only and the both true and false:

$$\begin{array}{llll} a_0 \ldots & a_i \ldots & a_k & \ldots a_n \\ [-\ - & P\ - & -] & \\ & [-\ - & \neg P\ - & -] \end{array}$$

and these would seem to jar just as much. As with the extended liar paradox, the machinery of the proposed solution allows us to produce a phenomenon of the same acuity. What is one to say about this?

The natural thought is that these cut-offs should be handled in exactly the same way. Consider, first, the right-hand boundary. This is located between those a of which P is true and those of which it is not. Let us now use T, not for the truth predicate, but for the binary truth-of (satisfaction) relation. Specifically if α is a formula of one free variable, say y, let the S-schema be:

$$T\ \langle\alpha\rangle\ x \leftrightarrow \alpha_y(x)$$

where the right hand side is the result of replacing all free occurrence of y with x (clashes of bound variables being handled by suitable relabelling). In particular, for our vague predicate, P, we have $T\ \langle Py\rangle\ x \leftrightarrow Px$. When the variable is clear from the context, I will omit it to keep notation simple. Thus, I will write $T\ \langle Py\rangle$ simply as $T\ \langle P\rangle$. Then the S-scheme amounts to this:

$$(*)\ T\ \langle P\rangle\ x \leftrightarrow Px$$

Now, the predicate $T\ \langle P\rangle$ would seem to be just as vague as the predicate P. In particular, it would seem to be just as tolerant to small changes in its argument as the predicate P. Indeed, (*) would seem to tell us that the tolerances of P and $T\ \langle P\rangle$ march together. It follows that the predicate is just as soritical; and just as the original sorites was generated by a set of sentences:

$$Pa_0, \neg Pa_n$$
$$Pa_i \equiv Pa_{i+1} (0 \leq i < n)$$

So a sorites is generated by the sentences:

$$P\ \langle T\rangle\ a_0, \neg T\ \langle P\rangle\ a_n$$
$$T\ \langle P\rangle\ a_i \equiv T\langle P\rangle\ a_{i+1} (0 \leq i < n)$$

The predicate $T\langle P\rangle$ also gives rise to an inclosure. Let $\varphi(x)$ be $T\langle P\rangle x$, so $\Omega = \{x : T\langle P\rangle x\}$; $\theta(X)$ is the vacuous condition. Ω is a subset of $A = \{a_0, \ldots, a_n\}$—indeed, a proper subset, since a_n is not in it. If $X \subseteq \Omega$ then, since X is a proper subset of A, there must be a first member of A not in it. Let this be $\delta(X)$. By definition, $\delta(X) \notin X$, Transcendence. Now, either $\delta(X) = a_0$ (if $X = \phi$), and so $T\langle P\rangle\,\delta(X)$; or (if $X \neq \phi$) $\delta(X)$ comes immediately after something in $X \subseteq \Omega$, so $T\langle P\rangle\,\delta(X)$, by tolerance. In either case, $\delta(X) \in \Omega$, Closure. The contradiction is that $T\langle P\rangle\,\delta(\Omega) \wedge \neg T\langle P\rangle\,\delta(\Omega)$.

Similar considerations apply at the left-hand boundary. Let us write $F\langle P\rangle x$, 'P is false of x', for $T\langle\neg P\rangle x$. A predicate is vague (tolerant) iff its negation is. In particular, $\neg P$ is just as vague as P. And since, as the S-Schema tells us:

$$F\langle P\rangle x \leftrightarrow \neg Px$$

$F\langle P\rangle$ is a vague predicate, as, then is $\neg F\langle P\rangle$. We therefore have a sorites generated by the sentences:

$$\neg F\langle P\rangle a_0, F\langle P\rangle a_n$$
$$F\langle P\rangle a_i \equiv F\langle P\rangle a_{i+1} (0 \leq i < n)$$

Note that $\alpha \equiv \beta$ is logically equivalent to $\neg\alpha \equiv \neg\beta$.

And as is to be expected, this boundary is another inclosure. Let $\varphi(x)$ be $\neg F\langle P\rangle x$, so $\Omega = \{x : \neg F\langle P\rangle x\}$; $\theta(X)$ is the vacuous condition. Ω is a subset of $A = \{a_0, \ldots, a_n\}$—indeed, a proper subset, since a_n is not in it. If $X \subseteq \Omega$ then, since X is a proper subset of A, there must be a first member of A not in it. Let this be $\delta(X)$. By definition, $\delta(X) \notin X$, Transcendence. Now, either $\delta(X) = a_0$ (if $X = \phi$), and so $\neg F\langle P\rangle\,\delta(X)$; or (if $X \neq \phi$) $\delta(X)$ comes immediately after something in $X \subseteq \Omega$, so $\neg F\langle P\rangle\,\delta(X)$, by tolerance. In either case, $\delta(X) \in \Omega$, Closure. The contradiction is that $\neg F\langle P\rangle\,\delta(\Omega) \wedge \neg\neg F\langle P\rangle\,\delta(\Omega)$—or just $\neg F\langle P\rangle\,\delta(\Omega) \wedge F\langle P\rangle\,\delta(\Omega)$.

20.8 The General Case

Of course, the situation repeats, generating new boundaries. Thus, the next iteration gives us:

$[- \; - \; \neg F\langle P\rangle \; - \; - \; - \; -]$

$\qquad\qquad [- \; - \; F\langle P\rangle \; - \; - \; - \; - \; -]$

$\qquad\qquad [- \; - \; \neg P \; - \; - \; - \; - \; -]$

$\mathbf{a}_0 \ldots \ldots \ldots \mathbf{a}_i \ldots\ldots \ldots \mathbf{a}_k \ldots \ldots \ldots \mathbf{a}_n$

$[- \; - \; - \; P \; - \; - \; - \; - \; -]$

$[- \; - \; - \; T\langle P\rangle \; - \; - \; - \; - \; -]$

$\qquad\qquad\qquad [- \; - \; - \; \neg T\langle P\rangle \; - \; -]$

Look below the as. We have just considered the division between P being true and its not being true. We now have the divisions between $T\langle P\rangle$ being true, and its not being true, and the division between $\neg T\langle P\rangle$ being true and its not being true. The first of these is the same as that between P being true and its not being true, since P and $T\langle P\rangle$ are co-extensional. But the second is new. Above the as we have the symmetrical situation concerning F.

And so it goes on. We need to consider all predicates that can be obtained by iteration. Generally, given the vague predicates Q, $\neg Q$, at the next level we have $T\langle Q\rangle$, $\neg T\langle Q\rangle$, and $T\langle\neg Q\rangle$, $\neg T\langle\neg Q\rangle$ (i.e., $F\langle Q\rangle$, $\neg F\langle Q\rangle$). Thus, the hierarchy of predicates looks as follow. To keep notation simple, I will henceforth omit the angle brackets. (Thus, I will write $F\langle T\langle P\rangle\rangle$ as FTP, etc.)

$$
\begin{array}{cccccccc}
 & & & \bullet & & & & \\
 & & \swarrow & & \searrow & & & \\
 & P & & & & \neg P & & \\
 \swarrow & & \searrow & & \swarrow & & \searrow & \\
TP & & \neg TP & & FP & & \neg FP & \\
\swarrow\ \searrow & & \swarrow\ \searrow & & \swarrow\ \searrow & & \swarrow\ \searrow & \\
TTP\ \ \neg TTP & & FTP\ \ \neg FTP & & TFP\ \ \neg TFP & & FFP\ \ \neg FFP & \\
\vdots\ \ \ \vdots & & \vdots\ \ \ \vdots & & \vdots\ \ \ \vdots & & \vdots\ \ \ \vdots &
\end{array}
$$

By exactly analogous consideration, each pair in the family is vague, and each gives rise to an inclosure contradiction.

20.9 A "Soritically Closed" Language

How do we know that all these contradictions can be accommodated in a uniform way? With the self-referential paradoxes and their extended versions, we know this because we can construct a single semantically closed language, which accommodates all the contradictions in one hit. Exactly the same is true in this case. We can construct a "soritically closed" language. Specifically, we take a language that has the truth-of predicate T, and a naming device, $\langle . \rangle$. For definiteness, let us suppose that the language contains that for arithmetic, and that the naming is obtained by Gödel coding. We suppose, in addition, one vague predicate, P, and a sorites sequence $(a_0, \ldots, a_n)$. Let this be $0, \ldots, n$. Let σ be any string of 'T's and 'F's (including the empty string), and let $\#(\sigma)$ be the number of 'F's in σ. (The parity of this tells us, in effect, whether we are doing a left-to-right sorites, or a right-to left sorites. Even is the first; odd is the second.)

Our theory comprises the S-schema, plus the following:

$$
\begin{array}{ll}
\sigma P a_i \equiv \sigma P a_{i+1} & (0 \leq i < n) \\
\sigma P a_0, \neg\sigma P a_n & \text{when } \#(\sigma) \text{ is even} \\
\sigma P a_n, \neg\sigma P a_0 & \text{when } \#(\sigma) \text{ is odd}
\end{array}
$$

The theory is inconsistent. For every σ, the theory entails:

$$\bigvee_{0 \leq i \leq n} (\sigma Pa_i \wedge \neg \sigma Pa_{i+1})$$

(The proof when σ is the empty sequence was already given; in the general case, the argument is exactly the same.)

Moreover, the theory is non-trivial. We can construct an interpretation which shows this, as follows. Start with a language without T. Take an interpretation, $\mathfrak{I}$, which is standard with respect to the arithmetic machinery. Let $0 < m < n$. The extension of P is $\{0, .., m\}$, and the anti-extension is $\{m, .., n\}$.[8] So, in the model, $Pm \wedge \neg Pm$ holds, as does every biconditional $Ph \equiv P(h + 1)$, for $0 \leq h < n$. We now construct a model of the S-schema on top of $\mathfrak{I}$ as in Priest (2002, § 8.2). (The model constructed there is of the T-schema, but this generalises to one for the S-schema in an obvious fashion.) In this model, we have not just the S-schema, but its contraposed form. Hence, every $\sigma\alpha$ is logically equivalent to α or $\neg\alpha$, and so all the cases of the axioms where σ is non-empty collapse into the case where it is. Moreover, in the construction of the model, it is only sentences involving 'T' that change their value. So the truth values of all other sentences are as in $\mathfrak{I}$. (In particular, then, any purely arithmetic sentence false in the standard model is not provable in the theory.)

What we see, then, is that from a dialetheic perspective, higher order vagueness is essentially the same as the extended paradoxes of self-reference,[9] can be handled in exactly the same way, and is no more problematic.

20.10 Conclusion

Prima facie, sorites arguments and the paradoxes of self-reference are completely distinct. They are certainly distinct. But what I have tried to establish is that, at a fundamental level, they are the same. Both are inclosure paradoxes, where the underlying form is given by the Inclosure Schema. The two kinds of paradox must therefore have the same kind of solution. Given that the correct solution to the paradoxes of self-reference is a dialetheic one, then so must be a solution to the sorites paradoxes. I have discussed such a solution at length, and argued that, despite certain superficial differences, it is also essentially the same.

The fact that one has a single family of paradoxes, and a uniform solution, does not, of course, mean that various sub-families cannot have their own specificities. Even within the paradoxes of self-reference, the semantic and the set theoretic paradoxes have differences of vocabulary; more importantly, diagonalisation may

---

[8]Strictly speaking, $\{x; x \geq m\}$, since every natural number must be in either the extension or the anti-extension of P. But what happens for numbers greater than n in irrelevant for our example.

[9]This important observation is due to Colyvan (2007).

be achieved in various different ways (by employing literal diagonalisation, a least number operator, etc.) Thus, it is entirely possible for the sorites paradoxes to have their own specificities, which they do. For example, tolerance plays a distinctive role, and "higher order vagueness" must be accommodated. All this we have seen. But the specificities are superficial, just as the specificities of the set theoretic and semantic paradoxes are superficial when it comes understanding the paradoxes and framing an appropriate solution. Such, at least, has been the import of this paper. This paper is part of a longer one, which appeared as Priest (2010).

References

Colyvan, M. 2007. What is the principle of uniform solution and why should we buy it? Presented at the annual meeting of the Australasian Association of Philosophy, University of New England, Armidale NSW.

Hyde, D. 1997. From heaps and gaps to heaps of gluts. *Mind* 106: 641–660.

Priest, G. 1987. In *Contradiction: A study of the transconsistent*. Dordrecht: Martinus Nijhoff; second, extended, edition, Oxford: Oxford University Press, 2006. References are to the second edition.

Priest, G. 1995. *Beyond the limits of thought*. Cambridge: Cambridge University Press; second, extended, edition, Oxford: Oxford University Press, 2002. References are to the second edition.

Priest, G. 2002. Paraconsistent logic. In *Handbook of philosophical logic*, 2nd edn., vol. 6, ed. D. Gabbay and F. Guenthner, 287–393. Dordrect: Kluwer.

Priest, G. 2010. Inclosures, vagueness, and self-reference. *Notre Dame Journal of Formal Logic* 51: 69–84.

Index

K. Tanaka et al. (eds.), *Paraconsistency: Logic and Applications*, Logic, Epistemology, and the Unity of Science 26, DOI 10.1007/978-94-007-4438-7,

P

Q

R

S

T

MIX
Papier aus verantwortungsvollen Quellen
Paper from responsible sources
FSC® C105338

If you have any concerns about our products,
you can contact us on
ProductSafety@springernature.com

In case Publisher is established outside the EU,
the EU authorized representative is:
Springer Nature Customer Service Center GmbH
Europaplatz 3, 69115 Heidelberg, Germany

Printed by Libri Plureos GmbH
in Hamburg, Germany